T0181000

World Sustainability Series

Series editor

Walter Leal Filho, Hamburg, Germany

More information about this series at http://www.springer.com/series/13384

Walter Leal Filho
Editor

Implementing Sustainability in the Curriculum of Universities

Approaches, Methods and Projects

 Springer

Editor
Walter Leal Filho
Faculty of Life Science
HAW Hamburg
Hamburg
Germany

and

School of Sciences and the Environment
Manchester Metropolitan University
Manchester
UK

ISSN 2199-7373   ISSN 2199-7381   (electronic)
World Sustainability Series
ISBN 978-3-319-88915-3   ISBN 978-3-319-70281-0   (eBook)
https://doi.org/10.1007/978-3-319-70281-0

Printed on acid-free paper

This Springer imprint is published by Springer Nature
The registered company is Springer International Publishing AG
The registered company address is: Gewerbestrasse 11, 6330 Cham, Switzerland

Preface

One of the major barriers to the wide incorporation of matters related to sustainable development at higher education institutions is the fact that sustainability is seldom systematically embedded in the curriculum. Yet, proper provision for curricular integration of sustainability issues at the part of teaching programmes across universities is an important element towards curriculum greening. Despite the central relevance of this topic, not many events have specifically focused on identifying ways of how to better teach about sustainability issues in a university context.

It is against this background that this book "Implementing Sustainability in the Curriculum of Universities: Teaching Approaches, Methods, Examples and Case Studies" has been prepared. It contains a set of papers presented at a Symposium with the same title, held at Manchester Metropolitan University (UK) in March 2017. The event was attended by a number of institutions of higher education active in this field. It involved researchers in the field of sustainable development in the widest sense, from business and economics to arts and fashion, administration, environment, languages and media studies.

The Symposium focused on the means to implement sustainable development in teaching programmes, and this book is a contribution to the further development of this central topic. The aims of this publication are as follows:

i. to provide teaching staff at universities active and/or interested in teaching sustainable development themes with an opportunity to document and disseminate their works (i.e. curriculum innovation, empirical work, activities and case studies practical projects);
ii. to promote information, ideas and experiences acquired in the execution of teaching courses, especially successful initiatives and good practice;
iii. to introduce methodological approaches and projects which aim to offer a better understanding of how matters related to sustainable development can be tackled in university teaching.

Last but not least, a further aim of this book, prepared by the Inter-University Sustainable Development Research Programme (IUSDRP) and the World

Sustainable Development Research and Transfer Centre (WSD-RTC), is to catalyse a debate on the need to promote sustainable development teaching today.

I thank the authors for their willingness to share their knowledge, know-how and experiences, as well as the many peer reviewers, who have helped us to ensure the quality of the manuscripts.

It is hoped that this publication will help to outline the need for integrated approaches towards teaching on sustainable development, and hence contribute towards advancing this field of work even further.

Enjoy your reading!

Hamburg, Germany   Walter Leal Filho
Winter 2017/Spring 2018

Contents

Transforming Collaborative Practices for Curriculum and Teaching Innovations with the Sustainability Forum (University of Bedfordshire) 1
Diana J. Pritchard, Tamara Ashley, Helen Connolly
and Nicholas Worsfold

Enabling Faith-Inspired Education on the Sustainable Development Goals Through e-Learning 17
Judith Gottschalk and Nicolai Winther-Nielsen

Sustainable Architecture Theory in Education: How Architecture Students Engage and Process Knowledge of Sustainable Architecture . . . 31
Elizabeth Donovan

Education for Sustainability in Higher Education Housing Courses: Agents for Change or Technicians? Researching Outcomes for a Sustainability Curriculum 49
Alan Winter

A University Wide Approach to Embedding the Sustainable Development Goals in the Curriculum—A Case Study from the Nottingham Trent University's Green Academy 63
Jessica Willats, Lina Erlandsson, Petra Molthan-Hill,
Aldilla Dharmasasmita and Eunice Simmons

Sustainability Curriculum in UK University Sustainability Reports 79
Katerina Kosta

Discomfort, Challenge and Brave Spaces in Higher Education 99
Lewis Winks

The Teaching-Research-Practice Nexus as Framework for the Implementation of Sustainability in Curricula in Higher Education . . . 113
Petra Schneider, Lukas Folkens and Michelle Busch

Education for Sustainable Development: An Exploratory Survey of a
Sample of Latin American Higher Education Institutions   137
Paula Marcela Hernandez, Valeria Vargas and Alberto Paucar-Cáceres

Biorefinery Education as a Tool for Teaching Sustainable
Development   155
Ari Jääskeläinen and Elias Hakalehto

Reflections on Using Creativity in Teaching Sustainability and
Responsible Enterprise: A First and Second Person Inquiry   173
Helena Kettleborough, Marcin Wozniak and David Leathlean

A SDG Compliant Curriculum Framework for Social Work
Education: Issues and Challenges   193
Umesh Chandra Pandey and Chhabi Kumar

Research Informed Sustainable Development Through Art and Design
Pedagogic Practices   207
Fabrizio Cocchiarella, Valeria Vargas, Sally Titterington and David Haley

A Critical Evaluation of the Representation of the QAA and HEA
Guidance on ESD in Public Web Environments of UK Higher
Education Institutions   223
Evelien S. Fiselier and James W.S. Longhurst

Curriculum Review of ESD at CCCU: A Case Study in Health and
Wellbeing   247
Adriana Consorte-McCrea, Chloe Griggs and Nicola Kemp

A Unifying, Boundary-Crossing Approach to Developing Climate
Literacy   263
Ann Hindley and Tony Wall

Monitoring Progress Towards Implementing Sustainability and
Representing the UN Sustainable Development Goals (SDGs) in the
Curriculum at UWE Bristol   279
Georgina Gough and James Longhurst

Teaching Accounting Society and the Environment: Enlightenment as
a Route to Accountability and Sustainability   291
Jack Christian

Professional, Methodical and Didactical Facets of ESD in a Masters
Course Curriculum—A Case Study from Germany   307
Markus Will, Claudia Neumann and Jana Brauweiler

Incorporating Sustainable Development Issues in Teaching Practice . . .   323
Walter Leal Filho and Lena-Maria Dahms

Contributors

Tamara Ashley University of Bedfordshire, Bedford, UK

Jana Brauweiler University of Applied Sciences Zittau/Görlitz, Zittau, Germany

Michelle Busch University of Applied Sciences Magdeburg-Stendal, Magdeburg, Germany

Jack Christian Department of Accounting, Finance and Economics, Manchester Metropolitan University Business School, Manchester, UK

Fabrizio Cocchiarella Manchester Metropolitan University, Manchester, UK

Helen Connolly University of Bedfordshire, Bedford, UK

Adriana Consorte-McCrea Wildlife and People Research Group-ERG, Canterbury Christ Church University—CCCU, Canterbury, UK

Lena-Maria Dahms Faculty of Life Sciences, Research and Transfer Centre "Sustainable Development and Climate Change Management", Hamburg University of Applied Sciences, Hamburg, Germany

Aldilla Dharmasasmita Nottingham Trent University, Green Academy, Nottingham, UK

Elizabeth Donovan Research Lab: Emerging Architecture, Aarhus School of Architecture, Aarhus C, Denmark

Lina Erlandsson Nottingham Trent University, Green Academy, Nottingham, UK

Evelien S. Fiselier Faculty of Environment and Technology, University of the West of England, Bristol, UK

Lukas Folkens University of Applied Sciences Magdeburg-Stendal, Magdeburg, Germany

Judith Gottschalk Department of Communication and Psychology, Aalborg University, Aalborg, Denmark

Georgina Gough University of the West of England, Bristol, UK

Chloe Griggs Foundation Degree in Health and Social Care, Canterbury Christ Church University, Canterbury, UK

Elias Hakalehto Finnoflag Oy, Kuopio, Finland; Department of Agricultural Sciences, University of Helsinki, Helsinki, Finland; Faculty of Science and Forestry, University of Eastern Finland, Joensuu, Finland

David Haley Zhongyuan University of Technology, Zhengzhou Shi, China

Paula Marcela Hernandez Processes Engineering Department, Universidad EAFIT, Medellín, Colombia

Ann Hindley International Thriving at Work Research Group, University of Chester, Chester, UK

Ari Jääskeläinen Environmental Engineering Department, Savonia University of Applied Sciences, Kuopio, Finland

Nicola Kemp Early Childhood Directorate, Futures Initiative ESD Lead, CCCU, Canterbury, UK

Helena Kettleborough Manchester Metropolitan University Business School, All Saints Campus, Manchester, UK

Katerina Kosta Faculty of Humanities and Social Sciences, Oxford Brookes University, Oxford, UK

Chhabi Kumar Indira Gandhi National Open University, Jabalpur, Madhya Pradesh, India

Walter Leal Filho School of the Science and the Environment, Manchester Metropolitan University, Manchester, UK

David Leathlean Manchester Fashion Institute, Manchester Metropolitan University, Manchester, UK

James Longhurst University of the West of England, Bristol, UK

James W. S. Longhurst Faculty of Environment and Technology, University of the West of England, Bristol, UK

Petra Molthan-Hill Nottingham Business School, Nottingham Trent University, Green Academy, Nottingham, UK

Claudia Neumann International Institute IHI Zittau, Technical University Dresden, Zittau, Germany

Umesh Chandra Pandey Evaluation Centre, Indira Gandhi National Open University, Bhopal, Madhya Pradesh, India

Alberto Paucar-Cáceres Manchester Metropolitan University Business School, All Saints Campus, Manchester, UK

Diana J. Pritchard University of Bedfordshire, Luton, UK

Petra Schneider University of Applied Sciences Magdeburg-Stendal, Magdeburg, Germany

Eunice Simmons Nottingham Trent University, Green Academy, Nottingham, UK

Sally Titterington Manchester Metropolitan University, Manchester, UK

Valeria Vargas School of Science and Environment, Manchester Metropolitan University, Manchester, UK

Tony Wall International Thriving at Work Research Group, University of Chester, Chester, UK

Markus Will University of Applied Sciences Zittau/Görlitz, Zittau, Germany

Jessica Willats Nottingham Trent University, Green Academy, Nottingham, UK

Lewis Winks School of Geography, University of Exeter, Exeter, UK

Alan Winter Division of Urban Environment and Leisure Studies, London South Bank University, London, UK

Nicolai Winther-Nielsen Fjellhaug International University College Denmark, Copenhagen, Denmark

Nicholas Worsfold University of Bedfordshire, Luton, UK

Marcin Wozniak Manchester Metropolitan University Business School, All Saints Campus, Manchester, UK

Transforming Collaborative Practices for Curriculum and Teaching Innovations with the Sustainability Forum (University of Bedfordshire)

Diana J. Pritchard, Tamara Ashley, Helen Connolly
and Nicholas Worsfold

Abstract Evolving higher education policy, and the production of guidelines and frameworks by higher education authorities, aim to support universities embed education for sustainability and reflect recognition of the need to prepare graduates for the challenges and opportunities of the 21st Century. Yet, advances have been limited. This chapter examines developments underway at the University of Bedfordshire, offering insights for ways forward which are distinct from prevailing institutional management processes. It presents the work of a community of practice, created by a group of academics from a spectrum of disciplines. Here, core players from this 'Sustainability Forum' describe their community, activities and synergies with the wider University. The authors highlight the learning opportunities they generated by their collective actions resulting in curriculum developments and enhancements. These served their own undergraduate and postgraduate students, other groups within the university community and beyond. As such the chapter serves as a case study of what can be achieved by an informal group of highly motivated academics in a new university. The authors conclude by considering the value of this model to other institutional contexts, especially in the context of the constraints imposed by expanding external performative initiatives and quality processes.

Keywords Sustainability forum · New university · Model · Curriculum enhancement · Learning opportunities

D. J. Pritchard (✉) · N. Worsfold
University of Bedfordshire, Park Square, Luton LU1 3 JU, UK
e-mail: diana.pritchard@beds.ac.uk

N. Worsfold
e-mail: nicholas.worsfold@beds.ac.uk

T. Ashley · H. Connolly
University of Bedfordshire, Polhill Ave, Bedford MK41 9TD, UK
e-mail: tamara.ashly@beds.ac.uk

H. Connolly
e-mail: helen.connolly@beds.ac.uk

© Springer International Publishing AG 2018   1
W. Leal Filho (ed.), *Implementing Sustainability in the Curriculum of Universities*,
World Sustainability Series, https://doi.org/10.1007/978-3-319-70281-0_1

1 Introduction

In the aftermath of the 2005–2014 UNESCO Decade of Education for Sustainable Development, higher education authorities in the UK recognise that universities need to ensure that curriculum are 'fit for purpose', and enhance "graduates' capabilities to contribute to sustainable and just societies" (HEA 2005). Guidelines and frameworks, developed in consultation processes with education for sustainability experts, have sought to support and evolve higher education policy to advance its provision (HEA 2005; HEFCE 2013; Longhurst et al. 2014). But preparing graduates to understand, "cope with, manage and shape social, economic and ecological conditions characterised by change, uncertainty, risk and complexity" (Sterling 2012) has proved a challenge.

To the extent that universities have incorporated knowledge and cognitive elements regarding the environment, climate, economic and social change issues into courses, advances have been made. But this response merely delivers education *about* sustainability (Sterling 2014). By contrast, they have struggled to deliver on education *for* sustainability (Cotton and Winter 2010; Tilbury 2011). Beyond the cognitive elements, this includes the development of understandings, ethos, values and other aptitudes which are required "to equip students to make informed decisions in their home, community and working lives" (Fien and Tilbury 2002) and to be effective in the face of both predicted and unknown changes, as agents of change (Orr 2004). Amongst other issues, this includes the concepts and practice of global citizenship, environmental stewardship, social justice, responsibility and wellbeing, and long term and future's thinking (Sterling 2014; Longhurst et al. 2014). These are widely recognised to be important to deliver deeper and transformative types of education and are thus important aspects to support students' longer term development and employability rather than short term academic attainment and their employment in a changing world.

This chapter highlights the possibilities and scope for deeper transformative education opportunities that have been created not by formalised institutional processes, but rather by an informal community of academics coming together to participate in our Sustainability Forum (SF). Through the exploration of our work and experiences within this group, we seek to understand and explain a model through which sustainability can be embedded into university curricula and off co- and extra-curricular education opportunities created for applied, relevant and transformative learning. It is based on data derived from the minutes of meetings, reports from workshops, written feedback from participants of the events that were organised, and emerges from the interpretations and reflective inputs of the authors themselves who comprise the core of the SF.

2　University of Bedfordshire's Baseline Scenario in 2013

As with any large institution, the history, geography and ethos of the University of Bedfordshire have shaped current policies, practice and opportunities for change. It is located in the south of the UK, approximately 50 km north of London. Founded in 2006 through the merger of the University of Luton and the Bedford campus of De Montfort University,[1] it provides a range of vocational, business, science, social science, arts and humanities courses at undergraduate and postgraduate levels. Its student body comprises 22,000 full- and part-time students, including home and international students who are ethnically very diverse (only 50% are white). The University is an overtly widening participation institution and receives high proportion of students through the clearing process. Many students come predominantly from non-traditional backgrounds with limited higher education experience, or understandings of graduate career options or appreciation of the relevance of the skills they must acquire for employment or alternative career paths. The University's strategies and supporting activities are designed to ensure that there are opportunities for everyone to benefit from higher education.

The following section details the institutional scenario regarding education for sustainability before the SF formed, providing a baseline from which to appreciate subsequent developments.

2.1　Education for Sustainability Commitments; No Mechanisms for Implementation

The University was granted university in 1993, and as with 'new' universities (Huet 2014), its management assumes a central role in defining and directing research and teaching activities. In 2012, it formally established elements of an enabling—or at least not a benign—environment that proved favourable for the development of education for sustainability. Specifically, its Strategic Plan (2012–2017) identified 'sustainable development' as a central commitment to deliver on sustainability goals and established the importance of its "integration into the curricula and professional development". Yet, there was no definition of education for sustainability, no body or personnel assigned to develop it and no mechanism established by which this could be accomplished (Pritchard and Atlay 2015). Further, a teaching and learning culture and practice valuing the co-creation of knowledge amongst students and academics was defined by its Learner Experience Strategy (2013–2018). This affirmed the pedagogic need to "apply creative approaches" for curriculum enhancement and for teaching practices to support its "diverse learner populations".

---

[1]With campuses in Luton, Bedford, Milton Keynes and Aylesbury.

This was similarly reflected in its formal curriculum framework, called CRe8, which itself outlines institutional expectations for learner development, curriculum enhancement, teaching practices and assessment.

2.2 Limited Unit Provision in Sustainability

In 2013, offerings in any aspect of sustainability were limited and there were no courses dedicated directly to it. Only a handful of units existed with explicit reference to sustainability (including in business studies, life sciences, tourism and construction) which delivered the core knowledge needed to keep their subject offerings relevant before the mounting social and environmental changes. In other areas, including English, Art and Design, and Dance Performance, such attempts were mainly ad hoc and—as a consequence of being dependent on single individuals—actually or potentially transient at the institutional level.

2.3 Isolated Academics

Within a range of departments, a handful of academics engaged with the agenda in their curriculum, motivated by social, moral and professional interests and responsibilities. They were concerned to update their subject, respond to change the status quo, prepare students for the challenges ahead and respond creatively to emergent challenges.

Three academics in particular had developed theoretically grounded teaching practices. The senior lecturer in Dance, had developed eco-pedagogical frameworks and approaches. These integrated dance improvisation and Freirian pedagogical encounters to stimulate students to respond to nature both pragmatically and artistically and involved developing a pedagogy that is a site-sensitive, time-sensitive and situated. In Sociology, the senior lecturer provided offerings in human rights, international criminal law and contemporary forced migration where her approach to teaching incorporated critical pedagogy for democracy and social justice and transformative experiential learning. Her students have benefited from a range of direct learning experiences, including field visits to the Balkans and to the International Criminal Court in The Hague, providing them with opportunities to get at 'eye level' with the issues they are studying as a way of energizing ownership, personal commitment and civic and global responsibility.

The potential of academics to expand local curriculum offerings in relation to sustainability—much less collaborate with others—was not only curtailed by the solitary nature of their efforts but also by the prevailing reticence amongst senior management and other levels. Despite the establishment with national higher education policy, education about or for sustainability was not viewed as a priority. Proposals for new courses and units which highlighted sustainability in their title— even those in subject areas dealing with core knowledge aspects about the material, ecosystem or carbon reduction aspects of sustainability—were seen as high risk in not appealing sufficiently to deliver on recruitment targets, especially for clearing students. An apparent causality dilemma prevailed whereby the absence of demands from current students at the University for sustainability-related offerings was conflated to mean that there was a lack of *interest* from potential and current students in these issues.

3　Academics Form a Community of Practice

Given this situation, a first call to academics in June 2013 to meet to discuss the sustainability agenda was welcomed by academics. The informal meeting was convened by an academic who contacted academics based on a preliminary list of names provided by respective heads of departments. Twelve scholars from departments in three faculties attended, as did an active and multidisciplinary scholar in environmental sustainability who also works as principal curriculum developer at the Centre for Learning Excellence (CLE). They talked about how their work engaged with sustainability, recognising that it is pertinent to virtually all subject areas whilst also meaning different things. The breadth of their disciplinary backgrounds suggested that there was limited obvious scope for research collaborations which institutional norms favour for grant income or research outputs.

Expanding this emergent group through progressively wider announcements, a series of face-to-face events were organised. Typically, between twelve to fourteen academics participated, including a range of seniority (professors, and principal, senior and other lecturers), from departments across all four faculties. The initial two workshops, were coordinated and supported by the CLE.

A handful of undergraduate and post graduate students, encouraged by participating academics, also attended. So did the one representative of the BedsSU, a paid project coordinator of a National Union of Students (NUS) funded project, the Green Hub (2013–16). The limited interest of BedsSU to engage with the sustainability agenda may be more reflective of the situation across the sector. While the annual NUS reports conclude that 60% of student demand greater offerings and 87% consider their universities should take the agenda "more seriously" (NUS 2016), these findings are likely biased since respondents are less representative of the entire student population, since they are self-selecting and sufficiently motivated by the agenda to respond to this internet survey. In this context the role of academics as primary generators of education for sustainability initiatives is key.

At this stage, the group agreed that it would be called the SF, and that coordination would be assumed by the CLE-based academic, given her scholarly credibility as a researcher with teaching experience, and her primary role in professional services, not teaching, which gave her greater flexibility to support the group. The SF also decided to focus on teaching and curriculum, not on research, and as a result had less appeal for those with limited teaching responsibilities.

At these early workshops, key shared understandings and principles were explored. Participants agreed that no discipline or academic level held privileged knowledge, and that everyone had relevant insights to offer from the perspective of their backgrounds and experiences. In so doing they had effectively externalised the theories and practice of multi-disciplinarity, "horizontal learning" (Freire 2005), co-learning and collaboration. They recognised these as the most fertile basis from which to explore the skills, critical facilities and modes of awareness that can equip students to respond and adapt to our changing world scenarios and propose ways to take the agenda forward in the institution.

Academics exchanged ideas about appropriate teaching practices, agreeing on the importance of participatory and action oriented learning, and of getting students "out there" to be exposed to real-life situation. Ideas about the campus as a learning resource were shared and expanded, aided by participation of the Head of Sustainability from Estates and Facilities. Effective ways to engage students in understanding the role of values were also explored such as by activities asking them whether the consequences of something are good, bad, right, wrong and for whom and what. Ways to explore uncertainty, complexity and the limits of knowledge were identified given the particular challenge this presents for academics in empirically based subjects who are typically unaccustomed to acknowledge and communicate these.

Clearly, a self-selected group of individuals had come together on a voluntary basis with a common endeavour to learn, build their own capacity, transfer best practices and engage in a participatory and non-hierarchical dynamic. Such characteristics indicate that it had established itself as a 'community of practice' (Wenger and Snyder 2000; Barton and Tusting 2005).

4 Generating Curriculum Consequences and Demonstrating its Success in Informal Leadership (Bourhis et al. 2005) Learning Opportunities

In Table 1, include details of the key activities and processes that the SF organised and set in motion and the consequences regarding curricular, co and extra-curricular learning opportunities that these created. These were largely shaped by a core within the SF who are committed to the deeper transformative objectives of the education for sustainability agenda and who defined the design of the pedagogical initiatives.

4.1 Defining Sustainability and Its Incorporation into the Curriculum Framework

The initial work of the SF defined the dimensions of education for sustainability. By including knowledge, understandings and reflections of ethics and values,

Table 1 Changing curriculum scenarios resulting from work of Sustainability Forum; EfS = education for sustainability

Before 2013	Actions/events 2013–16	Curriculum change and learning opportunity outcomes
Favourable institutional strategies but no mechanisms for implementation: – Strategic commitment to EfS but no mechanism for implementation – No reference in curriculum framework to dimensions of EfS – No working definition of EfS nor links made to other curriculum priorities	SF established as community of practice, identifies shared principles for collaboration and a focus on teaching Student participation in SF enhances principles and practices of non-hierarchy and co-production SF defines meaning and dimensions of EfS tailored to institutional context and student demographics SF focusses on organisation of university wide pilot initiatives: (1) Model UN World Climate negotiations interactive event (July 2016) (2) Climate Change Collaboration event (Nov 2016)	– Curriculum framework (Cre8) revised to incorporate EfS principles – 'Action for Sustainability' credit bearing unit proposed by SF, and approved to 'enhance employability' – Model UN recognised as powerful experiential learning opportunity for students; University recommends further curricular development – University executive and senior management recognition that SF demonstrate capacity to consolidate University regional leadership
Academics work in isolation: – Isolated pedagogic approaches – Intellectual isolationism		– Opportunity to collaborate and exchange across disciplines and develop new practices – Able to design and implement innovative approaches with reduced risk – Adopt new teaching, curricular and assessment practices – Teaching collaborations lead to active research agenda amongst core group of SF
Limited course and unit provision and multi-disciplinary offerings – Few units relating to sustainability – No mechanisms to support		– New 'Action for Sustainability' credit bearing unit approved and running open to all students – Effective teaching models developed that will be replicated that facilitate multidisciplinary

(continued)

Table 1 (continued)

Before 2013	Actions/events 2013–16	Curriculum change and learning opportunity outcomes
inter-departmental offerings		curriculum offerings and for co- and extra-curriculum offerings – Plans in development for new cross-department course offering
Limited student engagement – Limited student involvement in curriculum development – No University wide offerings for student engagement with global issues or for portfolio experience		– SF open for student participation – University-wide curriculum and co and extra-curricular education opportunities created – Relevant portfolio opportunities established
Stakeholder engagement in curriculum design and delivery – Limited employer input to sustainability offerings – No employer inputs regarding global themes – No community involvement in curriculum, delivery or teaching activities		– Networks extended and practices established for ongoing involvement of organisations from public, private and civic sectors in education and learning activities – University regional leadership consolidated and asserted vis engagement in issue of global significance – University demonstrates innovative education leadership

interconnectivity and systems, geographical scales linking local and global, critical and creative thinking, agency for change making and participation, and long term future's thinking, the community had clearly defined its agenda as one of transformation. It had also unwittingly mirrored key components identified more universally as associated with education for sustainability (HEFCE 2013; Longhurst et al. 2014). The SF also clarified the importance of linking the agenda to 'employability' mindful of the primary concern of students for social mobility; effectively aligning it to the with strategic curriculum priorities of the University.

These elements were subsequently incorporated by CLE into the institution's curriculum framework. This was renamed 'Changing Worlds and Futures Thinking' to reflect the centrality of the agenda. It is used as a key reference for teams during course design and periodic review processes, requiring them to incorporate central elements. It is also used in induction for new academic staff, and for the professional teaching scheme, for which academics applying for Higher Education Authority (HEA) recognition are required to elaborate case material illustrating how their practices reflect the framework. In effect, the SF had delivered the Strategic Plan commitment to incorporate sustainability into professional development.

4.2 University Wide Curriculum Offerings

From January 2016, the SF focus took a practical turn with the decision to organise two pilot University-wide learning events. The design and planning for these were conducted in a series of more than ten two hour meetings with various configurations of the SF, using videoconferencing to link discussions across campuses. The events aimed to inspire interest from all groups within the University community in the sustainability agenda, highlight its relevance to real-world challenges, civic responsibility and work opportunities, showcase existing academic expertise in sustainability and pilot pedagogical approaches involving experiential, interactive and simulated and experiential learning. Climate change was selected as the theme, as one of the most urgent issues facing our local and global societies and environments.

4.2.1 Model UN World Climate Negotiation Event

In July 2016, a Model United Nations (MUN) World Climate Negotiations event was organised. Of course, the MUN is a well-established activity, involving a simulation of the UN where participants represent delegations from different countries and role-play meetings such as the General Assembly or Security Council. Across the globe it is adopted in a variety of HE contexts, most typically as an extra-curricular activity but—in a few institutions—within academic courses.

This event involved 35 people including UG and PG students and alumni, students from local schools, seven academics (including three who had not previously participated in SF activities), senior managers and support staff. Five members of local civic organisations (UN Association and the Bedfordshire Climate Change Forum) also participated and were invited because of their technical expertise regarding the UN and the theme. Participants were allocated to mixed groups to form 'delegations' which represented country blocs, reflective of their actual negotiation positions. Thus groups were composed of different ages and academic levels.[2] Participants conducted prior research, and at the event

---

[2]A (free) state-of-the-art computer software (C-Roads) was used. Developed by MIT, it is based on current climate science, and serves to analyse the results of the mock negotiations and give participants immediate feedback on the implications of their proposals. It added a real-time simulation of the impacts of political decisions.

summarised their positions, developed alliances and debated and negotiated to reach decisions regarding climate change targets.

The debriefing session and individual participant evaluations reveal the intensity, depth and breadth of the learning experience. Students commented that this "was the most I have ever had to think in my entire life", while academics noted that it had been "very challenging". Participants identified that the mixed delegations facilitated co-constructed learning. They claimed that it had developed their knowledge and understandings about the complexity of the climate change system and causes and solutions. They appreciated distinct national challenges and perspectives, requiring them to challenge own values. They also learnt about diplomacy in the practice and exercised public speaking, critical thinking, teamwork and leadership skills. Such dimensions evidence 'deep learning' (Biggs and Tan 2011), education for sustainability (Longhurst et al. 2014), and deliver transferable employability and significant life skills.

In response to a report subsequently submitted to the University's Quality Enhancement Committee–which evidenced the power of such experiential and simulated learning—University authorities requested that it be developed further. Plans are underway for its implementation both as a multidisciplinary curricular offering and as a curriculum enhancement offering as a discrete extra-curricular credit bearing unit.

4.2.2 Climate Change Collaborations

In November 2016, the SF delivered an event, Climate Change Collaborations (CCC), with support garnered from other units in the University, including financial and organisational support from CLE, and private sponsorship. Over 150 people participated.

The event reflected the disciplinary diversity of the SF. The venue was transformed into an exhibition space for the work of the art and design students, who had been prepared work addressing the theme of 'Fire', which depicted the environmental and human costs of climate change. Also, the event began with a dance performance by students, an eco-sensitive improvisation. Six parallel workshop sessions took place on: ecological and human consequences; alternative energies; mobility and transportation; communicating climate change; wellbeing, mindfulness and the arts; and sustainable materials and recycling. These had been proposed and were chaired by academics, some of whom used the event to deliver to their student cohorts curriculum content and even assessments.

The event was designed to facilitate discussion on real-world problems and responses and to maximise student participation. Each session included a mixed panel of practitioners and experts from civic, business, public sectors and an academic who during the first part of the event were allowed an initial five minutes to present their perspectives on the key challenges of sustainability and the activities they were involved into address these. For the second part, the sessions were tasked to propose cross-sector collaborations. These were subsequently discussed at a final expert and policy panel held at the event's end which was chaired by the Vice Chancellor of the University. In effect, the event provided a problem-based learning opportunity for cross-sections of the University and its extended community to engage in critical discussion and for students to identify the relevance of their own, and other disciplines, as a source of solutions, reflecting good sectoral practice (Longhurst et al. 2014: 16).

Academics used the event in distinct ways to develop their curriculum, teaching and learning with tangible outcomes resulting. From Biological Sciences, the senior lecturer designed new summative assessments for three distinct student cohorts for whom he made attendance at the event compulsory. They were set tasks based on discussions that arose from the panel presentations, resulting in their work being oriented to actual contexts of UK adaption to climate change. The provision of such an "authentic aspect" is likewise a best education for sustainability practice (Longhurst et al. 2014: 13).

Also within Life Sciences, and as direct result of the event, the head of department secured collaborations with council that will result in Masters students being involved in project work and research in public health priorities.

Within sociology it was used it as an opportunity to get final year students to engage critically with their own academic discipline: in a context where sociology has lacked an established ecological tradition and an engagement with the relationships between societies and their natural environments. The broad invitation to her students was that they therefore challenge the dominant paradigm, reflect on the environment as a social category, and as such begin to take ownership of the complex and pressing questions around scale, impacts, responsibility, justice and politics. Students suggested the introduction of a distinct unit for the course which is being explored.

The aim of the dance lecturer for students to be exposed—through their performance at and participation in the event—to new environments and debates about climate change effectively enabled them, in their own words, to "think about how to engage with a wider and different kind of audience" and of distinct "environments for making dance". Their involvement also comprised relevant portfolio experience, as it did also for student cohorts from art and design in exhibiting their work, the events management students who took part in stewarding, and media students who filmed proceedings. The event thus provided them with evidence of CV enhancing work related experience.

Further, the enhanced profile which the event gave across the University to the applied relevance of the sustainability agenda, resulted in the approval of a University-wide five-credit extra-curricular 'Action for Sustainability' unit. This forms part of the University's employability agenda further embedding the SF's approach into the University's institutional strategies.

Feedback from the event demonstrates that it effectively exposed students (academics and practitioners) to the deep structures of the world, and to complexity in experience and ideas (Warburton 2003). It is also suggestive of a practice which manifests the connected, socially and institutionally situated "Ecological University" (Barnett 2011, 2015). The event illustrates the value of such a cross-university and multidisciplinary event, where institutions elsewhere have demonstrated provides as an effective approach to engage with university communities in education and learning (Puntha et al. 2015). Plans are underway to support the further ten academics (who attended but are who not yet active in the SF) incorporate sustainability into their curriculum. A similar event is also being planned for November 2017.

5 Broader Strategic Benefits Delivered by the SF

Beyond the curriculum development and enhancement and learning opportunities created by the initiatives of the SF, other strategic benefits continue to become apparent. In comprising best practices in education for sustainability, as defined by the sector in its People and Planet University League, the SF has ensured the institution's rank within the top twenty for two consecutive years. More substantially, it has demonstrated regional leadership of the University to engage students, citizens and organisations with issues of global significance using dynamic and engaging learning approaches. This is validated by local communities, organisations and companies. By example, teachers of feeder schools whose students participated in the events, praised their effectiveness for the energised and informed interest students demonstrated in the topic. For their part, local residents felt "proud and excited to live in a town where the local university has taken a lead", others valued "the networking opportunities" and "opportunity to learn alongside students". Similarly, the format of the climate change collaborations event has since been adopted by a local organisation for their own conferences aiming engage local citizens in sustainability challenges.

6 Explaining the Sustainability Forum as an Ongoing Generator of Curriculum Change

The SF emerged less than four years ago from a group of academics with a shared interest in education and sustainability and has continued to generate impacts and changes in a range of domains. This could not have occurred without academics giving up their time (not recognised in their formal work-load planning model) for something that, at least initially, did not appear to offer tangible rewards and where the complex institutional reality of this University, as with others, where there are multiple other priorities which require long teaching hours and administrative burdens (Pearson 2015). We suggest that this is a consequence of some essential characteristics.

First, the SF's concern is inclusive to all subject areas, and motivates involvement from across subject areas where individuals have a deep emotional affinity if not political commitment to it. This deeper connection with the theme lowers the threshold for a requirement that it generate promotion linked outputs. Further, the concept of community of practice has deep resonance with the group as a source of identity (which the drafting of this paper has served to mature). The SF provides an informal autonomous space in which like-minded individuals can discuss interesting topics, develop professional practices and understandings of the ecology of the University and has developed capacity as an agent of change. This compares to processes, expectations and functions perceived to prevail amongst the formalised structures and practices of university management which typically fulfil 'top down' directives.

At no time has there been the intention to formalise "membership" of the SF. While there is a definite core of active participants, SF activities are open and able to attract the "peripheral participation" (Lave and Wenger 1991) of the occasional involvement of academics, research-only staff, professional services and those in senior management roles. In this regard, the SF's loose and open enough agenda means it resembles an inclusive "affinity space" (Gee 2005), drawing together people from across the University. In this sense, the group name of choice is instructive—a 'forum'; not a 'working-group' or 'task-force' (and certainly not an 'institute'). For academics in this new university the professional autonomy the SF affords them may be particularly pertinent as both a respite from the daily norms of teaching delivery and administration and (as research has been indicated elsewhere) because they perceive their values and practice "to be more at risk from contemporary policies than the academics at [older universities]" (Pearson 2015). Yet, their relative freedom clearly still enables these academics to determine how they dedicate some of their professional time.

Second, as a teaching-focused university lacking the academic departments most likely to have staff with research interests in sustainability (e.g. geography, environmental science, politics), academics here are scattered amongst departments and face the situation of being a 'minority of one' (Katz 1959). The concept of 'academic loneliness' in either small or interdisciplinary institutions has been long recognised although it has attracted little attention (but see Sapp 2006; Jauhiainen and Laiho 2009). It accounts for a further appeal of the SF as a space to engage in discussion about a shared intellectual, social and political passion with other academics from distinct and across subject areas.

Third, the emergence of pedagogy (rather than discipline-specific research) as the purpose of the group enabled teaching-active staff to find common-ground despite differences in ontology or epistemology—or even differences in their broader understanding of sustainability as a concept. Coupled to this, the coalescence of the SF around the University-wide activities enabled participants to receive shorter-term gains from participation; some viewed the organisation of these events through the lens of teaching delivery and development, allowing their involvement to be moved from one 'workload compartment' to another. This not only rationalised their devotion to its events—particularly where these could be harnessed to create interesting formal assessments—it also resulted in recognition from amongst senior members of academic departments and faculties that the SF has capacity to deliver high quality teaching opportunities. This offers the prospect of similar 'forums' emerging at other institutions where staff with similar or different interests and disciplines can be accommodated if the central theme (and output) delivers engaging teaching developments and delivery.

Fourth, as the SF began to develop into a mechanism through which academics can interact with other University agendas and structures, it generated an additional (if still nebulous) reward for members. By example, as a consequence of the Climate Change Collaborations event, the SF and its core participants increasingly engage with other stakeholders across the university and beyond where the access to them was formerly limited.

Finally, coordination of the SF by a non-teaching academic from within professional services enabled it to be maintained by regular contact, when other commitments may have led to its fracture. But the type of informal 'leadership' of the SF is not incidental. Rather it is held by someone with academic and personal integrity, and a style of consultation, support and steering which is reflective of good higher education leadership practices (Bryman 2007). This raises an important consideration for the management of such processes in other institutional contexts where respect for academic autonomy and professional identity has been progressively undermined over recent decades in UK universities (Henkel 2005; Slaughter and Leslie 1997). Further, the support provided by CLE is significant in defining its role as a sponsor and is a obvious consequence of the SF's capacity to deliver innovative teaching and learning approaches in line with the curriculum framework and Learner Strategy, where other earlier managerial initiatives had stalled (Atlay et al. 2008).

7 Conclusion

By creating a non-hierarchical space of "democratic … processes of change" the SF demonstrated characteristics similar to other higher education groups committed to education for sustainability (Tilbury 2011). This examination of its dynamic and impacts has revealed that its model is productive and fulfils the vision of HE authorities to influence the educational 'core business' and create a system-wide curriculum and other learning opportunity changes that address education for sustainability.

Although the precise conditions facilitating the emergence of the SF at the University of Bedfordshire are institution-specific, its model demonstrates the potential for quite distinct change management approaches and ways that other universities may more productively organise themselves to advance education for sustainability practices or other cross-disciplinary topics. Where an albeit small body of academics promote this agenda, centralised directives or management which respect, give room to, and support creativity and innovation, and succeed in making connections to other university priorities and communities, clearly yields professional and personal rewards for academics and delivers significant institutional benefits.

References

Atlay M, Gaitan A, Kumar A (2008) Stimulating learning—creating CRe8. In: Nygaard C, Holtham C (eds) Understanding learning-centred higher education (Chapter 13). Copenhagen Business School Press, Copenhagen

Barnett R (2011) The coming of the ecological university. Oxford Rev Educ 37(4):439–455

Barton T, Tusting K (2005) Beyond communities of practice: language power and social context. Cambridge University Press, Cambridge

Biggs J, Tang C (2011) Teaching for quality learning at university, 4th edn. Buckingham: The Society for Research into Higher Education & Open University Press

Bourhis A, Dubé L, Jacob R (2005) The success of virtual communities of practice: the leadership factor. Electron J Knowl Manage 3(1):23–34

Bryman A (2007) Effective leadership in higher education: summary of findings. Leadership Foundation for Higher Education, London

Cotton D, Winter J (2010) Its not just bits of paper and light bulbs: a review of sustainability pedagogies and their potential use in higher education. In: Jones P, Selby D, Sterling S (eds) Sustainable education: perspectives and practices across higher education. Earthscan, London

Fien J, Tilbury D (2002) The global challenge of sustainability. In: Tilbury D, Stevenson R, Fien R, Schreuder D (eds) Education and sustainability: responding to the global challenge. Commission on Education and Communication. IUCN, Gland, Switzerland

Freire P (2005) Pedagogy of the oppressed. Continuum, New York

Gee J (2005) Semiotic social spaces and affinity spaces: from the age of mythology to today's schools. In: Barton D, Tusting K (eds) Beyond communities of practice: Language, power and social context. Cambridge University Press, Cambridge, pp 214–232

HEA (2005) Sustainable development in higher education: current practice and future development. Higher Education Academy, York, UK

HEFCE (2013) Sustainable development in higher education: consultation on a framework for HEFCE. Higher Education Funding Council for England, England

Henkel M (2005) Academic identity and autonomy in a changing policy environment. High Educ 49(1):155–176

Huet I (2014) Research and teaching nexus in post 92 Universities: Tensions and challenges. In: Society for Research on Higher Education, Conference 2014, https://www.srhe.ac.uk/conference2014/abstracts/0193.pdf. Accessed 2 Feb 2017

Jauhiainen A, Laiho A (2009) The dilemmas of the 'efficiency university' policy and the everyday life of university teachers. Teach High Educ 14(4):417–428

Katz J (1959) The law and behavioural science program at yale: a psychiatrist's first impression. Law Sch Dev 12(1):99–106

Lave J, Wenger E (1991) Situated learning: legitimate peripheral participation. Cambridge University Press, Cambridge

Longhurst J, Bellingham L, Cotton D, Isaac V, Kemp S, Martin S, Peters C, Robertson A, Ryan A, Taylor C, Tilbury D, Quality Assurance Agency, Higher Education Academy (2014) Education for sustainable development: guidance for UK higher education providers. Technical Report. QAA, Gloucester. Available from: http://eprints.uwe.ac.uk/23353. Accessed 20 Dec 2016

NUS (2016) Sustainability skills. 2015–2016 Research into students' experiences of teaching and learning on sustainable development. National Union of Students

Orr D (2004) Earth in mind: on education, the environment and the human prospect. Island Press, Washington, DC

Pritchard D, Atlay M (2015) Sustainability and employability: alliances at the university of Bedfordshire. In: Leal Filho W, Azeiteiro UM, Caeiro S, Alves F (eds) Integrating sustainability thinking in science and engineering curricula. World Sustainability Series. Springer, pp 31–48

Pearson R (2015) Academic identity in a performative and marketised environment: a comparative case study. Ph.D. thesis, University of Nottingham, Nottingham

Puntha H, Molthan-Hill P, Dharmasasmita A, Simmons E (2015) Food for thought: a university-wide approach to stimulate curricular and extra-curricular ESD activity. In: Leal Filho W, Azeiteiro UM, Caeiro S, Alves F (eds) Integrating sustainability thinking in science and engineering curricula. World Sustainability Series. Springer, pp 31–48

Sapp D (2006) The lone ranger as technical writing program administrator. J Bus Tech Commun 20(2):200–219

Slaughter S, Leslie LL (1997) Academic capitalism: politics, policies, and the entrepreneurial university. John Hopkins University Press, Baltimore

Sterling S (2012) The future fit framework—an introductory guide to teaching and learning for sustainability in HE. The Higher Education Academy

Sterling S (2014) Higher education, sustainability and the role of systemic learning. In: Corcoran P, Wals A (eds) Higher education and the challenge of sustainability: contestation, critique, practice and promise. Kluwer Academic, Dordrecht

Tilbury D (2011) A global overview of commitment and progress. http://insight.glos.ac.uk/sustainability/Education/Documents/. Accessed 1 Apr 2014

UNESCO (2005) UN decade for education for sustainable development 2005–2014: http://unesdoc.unesco.org/images/0014/001403/140372e.pdf

University of Gloucester http://efsandquality.glos.ac.uk/connecting_efs_and_quality.htm

Warburton K (2003) Deep learning and education for sustainability. Int J Sustain High Educ 4(1):44–56

Wenger E, Snyder W (2000) Communities of practice: the organisational frontier. Harvard Bus Rev 78(1):139–145

Author Biographies

Dr. Diana J. Pritchard works as Principal Curriculum Developer. She draws on research and teaching experiences in a variety of developed and developing country contexts and on multi-stakeholder and participatory processes from a career in social development and environmental management involving community, national and international organisations.

Dr. Tamara Ashley is a senior lecturer in dance, leading the M.A. Dance Performance and Choreography programme. Since her doctoral research on dance practice in the context of environmental change, she has undertaken large-scale site-sensitive and ecological performance works and presents her work nationally and internationally.

Dr. Helen Connolly coordinates the B.A. Sociology degree. She is a sociology academic with research, policy and teaching interests in forced migration, international human rights and humanitarian law, international children's rights law and trafficking. She is a member of a number of European academic and policy networks.

Dr. Nicholas Worsfold is a Senior Lecturer in Environmental Sciences at the University of Bedfordshire's School of Life Sciences. He teaches ecology and environmental management and is interested in how both nature and people respond to environmental change.

Enabling Faith-Inspired Education on the Sustainable Development Goals Through e-Learning

Judith Gottschalk and Nicolai Winther-Nielsen

Abstract The Sustainable Development Goals (SDGs) are composed from a variety of universal goals. They come with a heavy load on ethical demands while they do not provide any ethical guidance. One possibility to fill this void is to teach the SDGs with a faith-based narrative. Grown out of a workshop by Bread for World we will present our approach of Ownership-inspired Behavior-driven development, which is a strategy for an e-learning governance to introduce an Android app to support e-learning on the SDGs in theological education in remote areas in the Global South. Our goal is to develop a governance strategy for higher theological education to fill the SDGs with a faith-based narrative by using e-learning technology. Based on the staircase curriculum for the education of future church leaders developed by the British FBO Relay Trust we introduce the theoretical framework for this goal, which makes the transfer from a hierarchical governance structure found in many African societies into a dynamic e-learning framework.

Keywords Sustainable development goals · Ownership · e-learning
Governance · Africa · Theology · Faith-based organizations

1 Introduction

The SDGs of the United Nations, are to be reached within the next 13 years. They are part of the 2030 Agenda for Sustainable Development. The aim of these goals is to motivate efforts across the world to prevent poverty, inequality, and climate

J. Gottschalk (✉)
Department of Communication and Psychology, Aalborg University,
Rendsburggade 14, 9000 Aalborg, Denmark
e-mail: gottschalk.judith@gmail.com

N. Winther-Nielsen
Fjellhaug International University College Denmark, Leifsgade 33,
2300 Copenhagen, Denmark
e-mail: nwn@dbi.edu

© Springer International Publishing AG 2018   17
W. Leal Filho (ed.), *Implementing Sustainability in the Curriculum of Universities*,
World Sustainability Series, https://doi.org/10.1007/978-3-319-70281-0_2

change. The SDGs cover universal topics: planet, people, prosperity, peace, and partnership. All of these topics are associated with many ethical demands, but the SDGs as do not provide any ethical guidance which is necessary in human society to do morally valid development work that does no harm. There is an ethical void within the SDGs leaving open how they can be reached, if they can be reached through more globalization through structural adjustment programs as proposed by, e.g. the World Bank, through grass-root approaches and empowerment as, e.g. proposed Bread for the World or the Relay Trust; generally the question is, through which ethical standards the SDGs are to be reached. One possible way to fill this void is to communicate and teach the SDGs with a faith-based narrative. And this is the project of the Relay Trust and the goal this paper achieves: To develop a governance strategy for higher theological education to fill the SDGs with a faith-based narrative by using e-learning technology and a staircase curriculum for the education of future church leaders. The theoretical framework for this goal is developed as a governance model called Ownership-inspired Behavior-driven development which makes the transfer from a hierarchical governance structure found in many African societies into a dynamic e-learning framework.

We are dealing with the following problems in the development of a curriculum for teaching the SDGs:

(1) What does an educational governance system that can implement an e-learning approach within a theological curriculum on the African continent look like?
(2) How can technology support the implementation of a curriculum concerning the SDGs within theological education?
(3) How can we enable theology to support SDG-based training in a theological curriculum to fill the ethical gap in the SDGs?

Faith-based organizations (FBOs) and religious leaders play a significant role in HIV and AIDS prevention, and they support the community through hospitals and doctors. More importantly do religious leaders prevent, subdue or possibly annihilate the stigmatization of HIV and AIDS victims. Stigmatization is the true killer of HIV/AIDS victims as it causes the disease to be taboo, leading to denial and no spreading of information.

Due to their understanding of the society in which they are located, FBOs and religious leaders can begin a dialog about issues that are connected to faith, such as family planning (cf. Karam 2010). Most people are faithful to a religion. This means that FBOs can have a huge impact on people and can inform them about the SDGs (cf. Karam 2015). For this reason, it is necessary not only to teach the SDGs, their philosophy, and their motivations to future church leaders but also to connect them to an ethical narrative based on the Bible, which influences all the actions pastors take and decisions they make as church leaders. That the future church leaders understand the SDGs will help empower African countries to develop. African countries need their own capacities to overcome their poverty, which may come from religious actors, to support its development.

Countries like Sierra Leone, The Gambia, Liberia and Cameroon, where the Relay Trust does its work within the Anglican Church are on the last ranks of the human development index. Being on these ranks has tremendous consequences for education in these countries: Many, are missing a well-established educational system. A high percentage of adults are illiterate. In case schools are found, these are situated close to bigger cities like Freetown in Sierra Leone or Kinshasa in the Democratic Republic of Congo. This urbanization increases problems of hunger, diseases and environmental problems within urban regions at least for the people who had to move from the countryside to these overcrowded cities where they can only with difficulty effort their livelihood. Therefore connecting remote theological seminaries to universities in urban areas and to enable education in the country-side is a reasonable and sustainable approach.

With this paper, we aim to tell a story about how we will use a robust persuasive e-learning governance system for higher education of church leaders in Africa.

We aim to achieve a robust e-learning governance for higher education on the African continent to make effective use of computer technology to support a remote training of future church leaders on the SDGs which can be realized with a smartphone app. The result is a theological curriculum involving the SDGs as well as design an interactive knowledge app for an SDG-based dogmatics and ethics course.

Through its design the app is usable not only for higher literate education but since it is designed as an app for managing oral knowledge likewise in a way that it can be used for lower education. Based on the Paris Declaration of 2005 and discussions surrounding it (cf. Müller 2009; cf. Mummert 2009), we will illustrate how modern persuasive computer systems support developmental cooperation when used in the context of e-learning governance at universities in countries in the Global South, especially in light of the debate on ownership, participation, and empowerment. This paper is organized as follows: in Sect. 2 we discuss the background of our research, in Sect. 3 we introduce our e-learning governance system, and then in Sect. 4 we present our approach to a curriculum concerning SDGs.

2 Background

The persuasive framework introduced by Winther-Nielsen (2013b, 2014) and Gottschalk and Winther-Nielsen (2013) emerged out of B.J. Fogg's work on computers as persuasive technology (cf. Fogg 2003). Winther-Nielsen (2013b) and Gottschalk and Winther-Nielsen (2013) implemented this system in a learning software called Bible Online Learner using a plug-in, Learning Journey, and Laurillard's (2012) model for learning practice capabilities from an external environment (cf. Winther-Nielsen accepted). Winther-Nielsen (2014) developed a persuasive model called RAMP, which he uses to explain how corpus-driven language learning can be used within a persuasive model. RAMP explains that corpus technology excels in four factors that must be considered. The primary goal for a learning technology must be to empower social relationships (R) among learners. For the technology to stimulate interest and offer a pleasant user experience, learners must be absorbed in learning tasks that increase autonomy (A) and mastery (M) simultaneously. This type of intrinsic motivation is necessary as no technology will succeed if the learner does not have an identifiable purpose (P).

Of relevance in this paper is the line between a persuasive software and a persuasive strategy for implementing a successful e-learning governance system. This system will be implemented at a university in a developing country in which teaching through technology is unusual given environmental constraints, like electricity blackouts and unstable Internet connection, and user-related constraints, which occur because most e-learning systems are designed for the needs of societies of the Global North.

An important concept for the e-learning governance is the Kairos. The Kairos is the best moment at which to present a message (cf. Fogg 2003), and the ideal moment to motivate a student can be determined through the use of technology and used to encourage students to make the most of their learning (cf. Winther-Nielsen 2014).

E-learning projects from the Global North usually face challenges in Africa. Teaching methods like the flipped classroom (cf. Winther-Nielsen 2014) are designed to suit students from the Global North, which is composed primarily of individualistic societies (cf. Danner 2012). African societies are based on elder systems, in which the elders in a village have a say in all decisions made. The concept of the individual is not as present as it is in, European societies. In Africa, people identify themselves through their community and traditions (cf. Danner 2012). A student would not stand up in a flipped classroom and ask a tricky question because he might question the authority of the professor and hence question his role as an Elder. Winther-Nielsen has however shown that this teaching method can apply within an African context when an appropriate learning context is created which is based on mutual trust. From a governance perspective, African society is structured as a classical hierarchy, while in e-learning systems students are treated as individual actors who can determine the pace of their own learning.

The potential of ICT in development projects was acknowledged by UNESCO (2005), which views "knowledge societies" as an accelerator of development in developing countries. An example is the $100 laptop project, applied to schools in developing countries (cf. Ciborra 2003). Many articles cover issues of ICT projects in Africa, like public infrastructure, governance, accountability, civil society, entrepreneurial and economic activity, and access to global markets and resources (Thompson and Walsham 2010). Bada (2002) discusses the adaptation of a local bank's Enterprise Resource Planning System, while Braa and Hedberg (2002) use ICT to empower health workers in South Africa. Ngwenyama et al. (2006), dealt with health education in West African countries and determined causality from statistical correlations between ICT investments in health and education. One criticism of these studies presented by Thompson and Walsham (2010) is that they do not focus on strategic development, but on the design and implementation of solutions for specific developing countries. In this paper, we aim to develop a governance strategy for e-learning in African countries that will provide future faith actors with in-depth theological knowledge and enable teaching of and improve sensitivity to the SDGs.

In 2012, during the Rio+20 summit of the United Nations, governments committed to creating a set of SDGs that would be integrated into the follow-up Millennium Development Goals (MDGs) after which were reached their deadline in 2015 (cf. Griggs 2013). In this way, the development paradigm, which included issues concerning the economy, society, and the Earth's life-support system, was reached (cf. Griggs 2013). The MDGs are a historic achievement because they helped mobilize countries across the globe to address social priorities. The SGDs go a step further, adding the concept of sustainability to all efforts to improve global development, including efforts related to education.

An engagement of FBOs in reaching the SDGs occurs because most people follow a religion. Thus, FBOs have a huge impact and can cause people to work to achieve the SDGs. 80% of the earth's population are religious, guided by religious leaders, who can be motivated to work for the implementation of the SDG. As a conclusion faith-based actors have a great impact on their communities and can be used, for example, to work towards peacemaking, family planning, gender equality, education, and prosperity (cf. Karam 2010, 2015; Hayward 2012).

Following the Paris Declaration of 2005, we hypothesize that, through ownership, it is possible to facilitate developmental cooperation in which both parties share the same rights and recognition (cf. Müller 2009) in order to achieve an SDG. Such a partnership would support the training of future church leaders and the development of partnerships between organizations from the Global North and Global South to successfully work towards the UN's 2030 Agenda.

In our paper we extend the concept of behavior-driven development (BDD) from the IT sector to develop a strategy that we call ownership-inspired BDD (OIBDD). BDD was introduced by North (2006) as an improvement of test-driven development, in which programmers first write a software test that fails and then make small changes until it works successful. BDD uses stories as starting point. In our context formulating a story means developing a causal chain which should be reached through the development work project. Which role the concept of ownership plays is described in the next section.

3   An e-Learning Governance Model for Countries in the Global South

In the context of development cooperation, ownership requires partners to assume responsibility for their own development because they have enough self-confidence to cooperate with an overseas partner organization. Self-confidence of the local partner is not always certain in a development work due to a dependence relation, which sometimes occurs in these partnerships. However, ownership creates partnerships, and partnerships between local partners and operating non-governmental organizations (NGO) are required to achieve the SDGs (cf. Müller 2009). We are talking about a mutual beneficial relationship within our project, where the African partners are empowered to be independent and equal to not support colonial concepts. Applying a sound user-centered design, we aim to empower staff and students at a university in the Global South to enable future church leaders to fill the ethical gap in the SDGs.

One of the main problems concerning development cooperation and the debate on ownership is the gap between the technical approach to implementing an ICT solution and management of the problem with achieving sustainability (cf. Müller 2009; cf. Thompson and Walsham 2010).

The goals of a persuasive technology in a development context are to achieve behavioral change and create intrinsic motivation to exhibit a specific behavior (cf. Fogg 2003; cf. Winther-Nielsen 2014). According to Müller, bridging the gap between ideas and practical solutions in the ownership debate requires the concept of ownership to be enabled within the context of developmental cooperation. However, it is important that ownership is an objective that needs to be reached; the overarching objective of developmental cooperation is to achieve ownership (cf. Beier 2009). An extensive study of various approaches to development cooperation, concluded that a mix of leadership and participation creates ownership (cf. Mummert 2009). This is also what the Paris Declaration on Aid Effectiveness stated in 2005: ownership will be achieved when partner organizations lead the development and implementation of strategies through broad consultative processes (cf. Mummert 2009). Following the World Bank to create ownership, there must be sufficient support through feedback by stakeholders in the overseas partner organizations to enable a country to achieve ownership (cf. Mummert 2009). Following these perspectives, the main objectives of creating ownership and responsibility for partner organizations in a context such as Africa are to provide them with feedback and monitor the actions they take to complete the developmental cooperation project.

The new persuasive development model, we suggest for creating ownership during developmental cooperation and from which we derived an e-learning governance system, is comprised of phases 1 and 2 of Oinas-Kukkonen and Harjumaa's (2009) model and Winther-Nielsen's (2013a, 2013b, 2014) model for persuasive learning technology. Oinas-Kukkonen and Harjumaa (2009) discuss how the design of every persuasive sys-tem begins: analysis of the persuasive context. This is indispensable when determining the Kairos in a persuasive system and identifying the point at which intrinsic factors for changing a behavior and learning in a persuasive con-text, like Winther-Nielsen's (2013a, 2013b, 2014) approach, can be developed. Methods to do this are e.g. structured interviews, observation or participatory methods like a well-being ranking or a cause-and-effect chain. It is important to determine who is being persuaded, and who the persuader is. This determination is what Oinas-Kukkonen and Harjumaa (2009) call the intent. They state that the context must be established before a persuasive system can be established. Without knowledge of the goal of the persuasive design, it is not possible to determine which persuasive functions will better persuade the users of the system (cf. Oinas-Kukkonen and Harjumaa 2009).

When developing a persuasion strategy, one must determine the message and how it will be transmitted. In other words, the following question must be answered: How will a user utilize the software to reach a specific state or change a specific habit (cf. Oinas-Kukkonen and Harjumaa 2009)? In Gottschalk (submitted) concepts like Participatory Well-being, Situational Analysis and Goal Establishment, Performance Appraisal Groups and Participatory Impact Analyses, which originally come from the Participatory Rural Appraisal are described in the context of modern software development. For our knowledge app, the message is that learners should improve their knowledge of the ethical consequences of their actions by filling the ethical void in the SDGs with moral concepts from Christian theology to achieve the SDGs within a morally valid framework. One way to transmit this message is through software, which must be easy to use and useful within a learning context. This is a narrow use case with a broad perspective. The goal of the app is to improve the educational governance of an African university, help the university develop a curriculum involving the SDGs, and use this persuasive concept as a strategy for development of e-learning and IT governance in developing countries.

There are two phases of persuasion that a persuasive software must handle: (1) persuasion at the macro level, which is external to the system and focuses on persuasive postulates and analysis before a persuasive system is established, and (2) persuasion on the micro level, which is occurs through various persuasive functions. The concept of Kairos is essential to answering this question as our hypothesis is that ownership is a Kairos itself. The goal of the persuasion is important. For example, people who would like to play more sports might prefer jogging on a warm, sunny day and reach their Kairos when a persuasive system suggests that they go jogging on such a day (cf. Øhrstrøm et al. 2009). A number of identifiers might help to estimate when a specific Kairos occurs. These identifiers could be keyboard or mouse activity or physical measures, like heart or breathe rate (cf. Øhrstrøm et al. 2009). Such a factor usually has a causal im-pact on the presence of a Kairos, while the Kairos has a causal impact on the identifier (cf. Øhrstrøm et al. 2009).

One Kairos in our model is ownership. As Fig. 1 shows, the following factors influence ownership and create a Kairos: (a) leadership, (b) participation, (c) feedback, and (d) monitoring. Within our model, these four factors, which were derived from Müller (2009) and Mummert (2009), are random variables that constitute the Kairos of ownership. This is what Fig. 1 describes: It is an ontology of the Kairos of Ownership, which is defined through the above mentioned four variables. (cf. Øhrstrøm et al. 2009).

Again, we must differentiate between the macro level of persuasion, which takes place before the persuasive system is developed and implemented, and the micro level of persuasion, which occurs within the persuasive system once it is programmed and implemented. However, when no micro level can be reached because the system has not been developed yet, the system is not capable of determining the

Fig. 1 Network of ownership

Kairos of ownership. In this case, it is necessary to rely on persuaders who want to encourage ownership and participation to empower their local partners within the persuasive context. Within our context of a persuasive e-learning system and its implementation within Africa, we can illustrate this reliance using a flow chart of an IT strategy for ICT during developmental cooperation, which is motivated by the analysis of persuasive context developed by Oinas-Kukkonen and Harjumaa (2009). It is important to keep in mind what we noted in Sect. 2 regarding African sociology: many Africans grow up within a hierarchical elder system, but e-learning systems from the Global North use a network governance model that is decentralized and not hierarchically structured. A method for gradually transferring a hierarchy to a network is shown below:

1. The Intent

    a. In a collaboration between the local partner and the NGO, assign the persuader roles to the responsible actors—random variables: leadership and participation
    b. Determine the change context and what the indicators for the change are—random variables: Leadership,participation, monitoring and feedback

2. The Event

    a. Use the input from the persuadees to learn more about the use case to implement a technical solution for it—random variables: leadership, feedback and participation
    b. Learn more about the persuadee to tailor your persuasive system accordingly. Act according to the needs ofthe user—random variables: feedback, participation, and monitoring
    c. Learn more about the context in which your system should operate—random variables: feedback and participation

3. The Strategy

    a. Determine the message the persuasde needs for the persuasive process and react to feedback from her/him—random variables: feedback, participation and leadership.

Each of the three aspects of the persuasive analysis uses the four variables—leadership, participation, feedback, and monitoring—from our ontology of ownership described in Fig. 1, as persuasive functions to create a Kairos of ownership within a development project. This occurs at the macro level of persuasion, before the technology is implemented. At the micro level of persuasion, we used the four random variables to create a Kairos. With the knowledge app, both teachers and students should be able to assume leadership by uploading content to the e-learning system. User-centered design is a strategy in which persuasive design and governance go hand in hand. In addition, we are able involve the partner organization in the process of developing an app to motivate the organization to create ownership, which serves as participation in developmental cooperation, the basis for empowerment of a community. Then, the second element of our governance approach, BDD, comes into play to create the case for our approach to e-learning governance: OIBDD. Through OIBDD, teachers will be able to assume leadership in an e-learning context and be part of the curriculum development process in their teaching. This is a way to make them responsible for their teaching and give them the ability to participate in processes going on at a university. Through ownership and the possibility of taking over responsibility for their teaching, we gradually turn the hierarchical model of an elder system at an African university into a network structure.

The idea for our knowledge app was developed during a workshop on knowledge transfer in oral cultures organized by the German NGO Bread for the World. The app can achieve those goals using the following developmental methods, which are specifically relevant within oral cultures:

- It is not specifically text-based; it visualizes data and knowledge, meaning that the app uses icons that are culturally appropriate and developed through user-centered design.
- It makes extensive upload and download of videos and speech messages possible. These messages are available in both the language of the country the app is used in and the local dialect of the community whose knowledge is being transferred. Also, an English transliteration of each message exists.
- It is collaborative; similar to a social media platform, everyone is involved in the development of content for the app.

We will use the knowledge app as an oral knowledge hub. This is where the persuasive OIBDD approach comes into play; through the free upload and generation of teaching content, the teachers using the knowledge app can generate stories based on the staircase curriculum, which we will describe in Sect. 4, and from these stories, the teacher can tailor and tunnel—in the sense of Winther-Nielsen (2014) and Fogg (2003)—to guide the students to the desired learning object. Then, the micro level of persuasion described in Winther-Nielsen's RAMP model comes into play, because the app itself is used now.

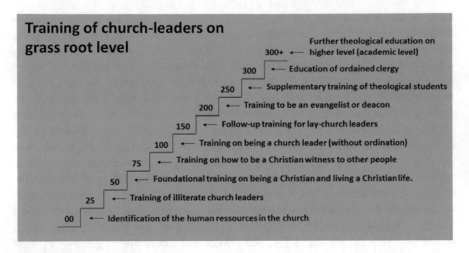

Fig. 2 Staircase curriculum developed by Relay Trust (We publish this figure by friendly permission from the Relay Trust)

4 The Governance Model at Work: Technology and Curriculum Design Through Ownership-Inspired Governance

The curriculum developed by Relay Trust is a staircase model and aims to work on Christian formation at all levels of church leadership. Within the program, practical and theoretical training are combined and supported by experiences from church life. Through this, the practical implications and constant contextual adaption of the lessons are ensured so that the training is relevant to students and the church. The Relay Trust makes training available in local and remote contexts instead of bringing students into foreign contexts. Especially at higher educational levels, the program enables learners to gain deep theological knowledge, which will support their efforts to obtain bachelor's, master's, or doctorate degrees in theology. In the staircase model students can continue to further their training and advance to higher levels provided they are motivated and have the necessary skills. The advantage of the staircase model is that students can follow their learning progress. Every step in the system is a closed unit and prepares the student for a certain level of leadership. The staircase model is shown in Fig. 2.

The SDGs do not have any ethical intent; thus, they have an ethical gap. This is where FBOs like Relay Trust and churches come into play. Churches and FBOs are very engaged in a number of issues, including HIV and AIDS awareness and prevention, peace and reconciliation after a war, building and running schools when the state is unable to do so, agricultural development, malaria prevention, skills acquisition among young people, and running hospitals etc. In addition, churches support refugees, engage in peace work, and organize campaigns to prevent terrorism. All of these issues are covered by the SDGs. Thus, there is an indirect link between the SDGs and theological education because of the responsibilities churches take on. In our project, we aim to directly link them using e-learning to fill the ethical void in the SDGs. The cornerstone of the staircase model is that students gain an understanding of the Christian faith as they continue on their academic paths; an essential part of theological education is learning the foundations of Christian ethics. This is how the SDGs are connected to our knowledge app. The app offers learners video and audio material concerning all the broad topics related to the SDGs. It relates this knowledge to Christian theology, filling the ethical void in the SDGs. For instances, when talking about the planet, the app makes reference to the Christian perspective on taking care of the plant, and when discussing the issue of prosperity, we can use the e-learning system to teach the church-leaders the basics of entrepreneurship and trade to make them tent-missionaries. This way, it frames the SDGs within the moral and ethical context of the Christian faith. Future church leaders are enabled to act in contexts in which churches build hospitals and do peace work. Importantly, all moral perspectives of the Christian faith are based on human rights.

Relay Trust defines itself as an FBO that aims to help struggling evangelical, mostly Anglican churches fight against organizational challenges due to poor education of their pastors. We focus specifically on francophone Anglican churches in remote rural areas because Francophone Anglican churches are often left behind. The reasons are that Anglican churches are considered as English churches and hence most teaching material is in English so that the Francophone churches are challenged by the language barrier. All the churches we partnered with are situated in countries listed in the last third of the Human Development Index (Gambia is 172nd, Sierra Leone is 183rd, and the Democratic Republic of Congo is 186th).

These countries' low rank on the Human Development Index means that they are lacking well established educational systems with primary schools. Our development projects seek to improve the SDG-inspired education of pastors in remote areas, as educated church leaders are often the only people in a village who receive higher education. However in most cases church leaders work as teachers in schools because they are literate and lead a church as a side-job. In these cases they earn the most of their income through teaching in schools.

During development projects, several models of higher education for pastors have been tested. Especially in francophone countries, students tend to be sent to France to receive higher education. However, these students must deal with inter-cultural challenges: they must keep up with students who grew up within a European educational system and are not expected to meet the demands of their family and village. As a consequence, many students do not finish their degrees or, if they do finish their degrees, do not return to their home countries. Generally for the students living abroad in a foreign contexts is a challenge because many students cannot afford the tuition fees themselves and so often not the best qualified student is selected to study abroad but the one with the best connections. Nevertheless the students receive degrees, even Ph.D.s. The problem resulting from this is that many universities do not consider an African Ph.D. holder as likewise

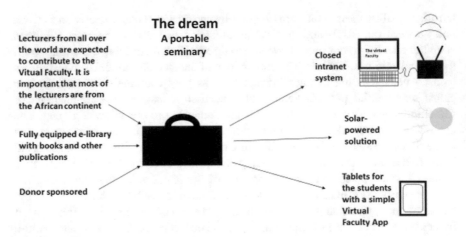

Fig. 3 Portable seminary for church leader training (We publish this figure by friendly permission from the Relay Trust)

qualified as European Ph.D. holders. Therefore our SDG-inspired curriculum can be applied in francophone countries in Western Africa through modern technology, especially mobile applications (Android smartphones are extremely widespread on the African continent).

We based our technology on smartphones and tablets which are found throughout the African continent. We developed the following app-based infrastructure in seminaries we support with the app (Fig. 3).

Relay Trust's concept is simple: given that many theology students in Sierra Leone and other African countries are situated in remote seminaries and that mobile phones are widely available on the African continent, the Trust set up a portable seminary based on the knowledge app presented in this paper. With this app, students can work at home or at a university and progress through the staircase training at a grassroots level. Thus, theological training can support SDG-based training through technology.

5 Conclusion

In this paper we have shown that concepts like Behavior-driven Development which are originally from the IT sector can be extended with one of the key concepts from the Paris 2005 declaration: Ownership. Based on this we have developed a governance strategy for e-learning to teach on the SDGs which is suitable for countries of the Global South. This strategy is specifically suitable as it takes into account the specific socio-structures found in African societies. By closely referencing church-based training, we are able to not only teach about the SDGs within the staircase curriculum but also fill the ethical void of the SDGs with moral faith-based narratives we are also able to show how technology can support e-learning in the Global South. For this we developed the concept of a portable seminary specifically for church-based training. Thus, we successfully situated the SDGs in theological education. This study has however also limitations and constraints: (1) Thus far our strategy has not been tested empirically though it was developed based on best practices we learned through our own work in development cooperation and thus used an inductive and empirically-based method to derive a governance strategy for e-learning from it. (2) Our specific governance for e-learning is constraint in the way that it is mainly usable for African contexts. The same e-learning governance could not necessarily be used in, e.g. in China as these societies are not community-based societies like in Africa and more individualistic. Hence the task for future work on OIBDD is to extent our research to other societies than just African societies and finally to empirically test our strategy, to implement our app and to documents the progresses on teaching the SDGs.

Acknowledgements We would like to thank Mette and Alex Bjergbæk Klausen from *Relay Trust* for helpful comments and support on earlier drafts of this paper. Judith Gottschalk would like to thank Marika Rothenberger from *Bread for the World—Protestant Development Service* for her helpful input regarding this paper.

References

Bada AO (2002) Local adaptation to global trends: a study of an IT-based organizational change program in a Nigerian bank. Inf Soc 18(2):77–86

Beier C (2009) Ownership: a guiding principle in development. In: Beyond Accra: Practical implications of ownership and accountability in national development strategies. 22–24 April 2009, London. Conference Papers. http://www.lse.ac.uk/internationalDevelopment/pdf/beyondAccraPapers.pdf. Last Accessed 1 Sept 2016, pp 22–27

Braa J, Hedberg C (2002) The struggle for district-based health information systems in South Africa. Inf Soc 18(2):113–127

Ciborra CU (2003) Unveiling e-government and development: Governing at a distance in the new war. Working Paper No. 126, Information Systems and Innovation Group, LSE, London

Danner H (2012) Das Ende der Arroganz: Afrika und der Westen - ihre Unterschied verstehen. Brandes & Apsel, Frankfurt am Main

Fogg BJ (2003) Persuasive technology. using computers to change what we think and do. Morgan Kaufmann, San Francisco

Gottschalk (Submitted) Sustainable corpus-driven computer-assisted language learning in Madagascar

Gottschalk J, Winther-Nielsen N (2013) Persuasive skill development: on computational surveillance strategies for modeling learning statistics with PLOTLearner. In: Behringer R, Sinclair G (eds) IWEPLET 2013 Proceedings http://www.lulu.com/shop/reinhold-behringerand-georgina-sinclair/iweplet-2013-proceedings/paperback/product-21189131.html. Last Accessed 14 Jan 2014, pp 109–116

Griggs D (2013) Sustainable development goals for people and planet. Nature 495:305–307

Hayward S (2012) Religion and peace building reflections on current challenges and future prospects. In: United States Institute of Peace Special Report

Karam A (2010) Concluding thoughts on religion and the United Nations: redesigning the culture of development. CrossCurrents 60(3):462–474

Karam A (2015) Realizing the faith dividend: religion, gender, peace and security in Agenda 2030. Technical Report UNFPA. UNFPA, New York

Laurillard D (2012) Teaching as a design science: building pedagogical patterns for learning and technology. Routledge, London

Müller U (2009) Mind the gap! ownership in practice. In: Beyond Accra: Practical implications of ownership and accountability in national development strategies. 22–24 April 2009, London. Conference Papers. http://www.lse.ac.uk/internationalDevelopment/pdf/beyondAccraPapers.pdf. Accessed 1 Sept 2016, pp 5–15

Mummert U (2009) Factors for successful development strategies: A pragmatic analysis. In: Beyond Accra: Practical implications of ownership and accountability in national development strategies. 22–24 April 2009, London. Conference Papers. http://www.lse.ac.uk/internationalDevelopment/pdf/beyondAccraPapers.pdf. Accessed 1 Sept 2016, pp 91–133

Ngwenyama O, Andoh-Baidoo FK, Bollou F, Morawczynski O (2006) Is there a relationship between ICT, health, education and development? An empirical analysis of five West African countries from 1997–2003. Electron J Inf Sys Developing Countries 23(5):1–11

North D (2006) There's more to add to BDD than evolving TDD. https://dannorth.net/2006/06/04/theres-more-to-bdd-than-evolving-tdd/. Accessed 14 Jan 2017

Øhrstrøm P, Aagaard M, Moltsen L (2009) It might be Kairos. In: Oinas-Kukkonen H, Hasle P, Harjumaa M, Segerståhl K (eds) Persuasive 2008: The Third International Conference on Persuasive Technology, pp 94–97. University of Oulu, Oulu, Finland

Oinas-Kukkonen H, Harjumaa M (2009) Persuasive systems design: Key issues, process model, and system features. Commun Assoc Inf Sys 24(28):485–500

Thompson M, Walsham G (2010) ICT research in Africa: need for a strategic developmental focus. Inf Tech Dev 16(2):112–127

UNESCO (2005) Towards knowledge societies. UNESCO World Report. UNESCO Publishing, Paris

Winther-Nielsen N (2013a) PLOTLearner as persuasive technology: Tool, simulation and virtual world for language learning. In: Behringer R, Sinclair G (eds) IWEPLET 2013 Proceedings. http://www.lulu.com/shop/reinhold-behringerand-georgina-sinclair/iweplet-2013-proceedings/paperback/product-21189131.html. Last Accessed 14 Jan 2014, pp 21–28

Winther-Nielsen N (2013b) PLOTLearner for a corpus of the Hebrew Bible: The case for repurposing in language learning. In: Behringer R, Sinclair G (eds) IWEPLET 2013 Proceedings. http://www.lulu.com/shop/reinhold-behringerand-georgina-sinclair/iweplet-2013-proceedings/paperback/product-21189131.html. Last Accessed 14 Jan 2014), pp 53–60

Winther-Nielsen N (2014) PLOTLearner's persuasive achievement: force, flow and context in technology for language learning from the Hebrew Bible. HIPHIL Novum X(2):78–94

Winther-Nielsen N (Accepted) The corpus as tutor: data-driven persuasive language learning. Digital Humanities Quarterly

Sustainable Architecture Theory in Education: How Architecture Students Engage and Process Knowledge of Sustainable Architecture

Elizabeth Donovan

Abstract This paper aims to outline and suggest an alternative pedagogy approach for teaching sustainable architecture theory within a studio-based environment, utilising the pedagogical methods for solving complex problems and applying 'designerly thinking.' Emphasising the qualities of a hybrid approach of electives and integration in an architectural education framework. It will reflect on under standings of architectural pedagogy and the integration of sustainability into architectural theory education. Beginning with a discussion of contemporary literature and policy concerning education for sustainability in architecture, it is accompanied by critical understandings of how architects synthesise different types of knowledge while designing, raising questions about the 'match' between theoretical educational experiences and subsequent behaviours in practice. Taking an example from Denmark, we outline the approach of Aarhus School of Architecture, where the majority of teaching occurs within design studio courses with short supplementary theory courses each year. During this pilot theory course, a direct observational study was made of 3rd-year architecture student during the two-week theory course on sustainable architecture theory to understand how they engage with the sustainable architecture discourse. The pedagogical approach involved multiple modes of learning including; participating in lectures, readings and doc umentaries followed by group work producing mapping, essay writing, and a final visual representation. The findings emphasise the importance of not only students' personal critical reflection and active engagement but in order for students to find sustainable architecture theory engaging and relatable to their design practice. Additionally, students need to be supported in gaining different types of knowledge through different modes of learning.

Keywords Sustainability · Sustainable architecture · Pedagogy · Education Designerly thinking

E. Donovan (✉)
Research Lab: Emerging Architecture, Aarhus School of Architecture, 20 Norreport, 8000 Aarhus C, Denmark
e-mail: ed@aarch.dk

W. Leal Filho (ed.), *Implementing Sustainability in the Curriculum of Universities*, World Sustainability Series, https://doi.org/10.1007/978-3-319-70281-0_3

1   Introduction

The built environment is critical to sustainable development. Haines (2010), provides the example that a 22 storey concrete office building contains the same
embodied energy as six atomic bombs the size of those which were dropped on
Hiroshima. This farfetched comparison illustrates the magnitude of resources
concealed within our buildings.

   In the age of the Anthropocene, the role of the architect is even more crucial in
the design of our built environment, not only concerning energy and resources.
A multitude of disciplines and values (social, technological, ecological, economic,
political) overlap with architecture, and this offers a unique opportunity for the
profession. Van der Ryn and Cowan (1996) explain "[…] the environmental crisis
is a design crisis. It is a consequence of how things are made, buildings are constructed, and landscapes are used." Hartman (2011) acknowledges that sustainability is no longer "at the margin of the profession," however, real expertise in the
field has yet to be developed in practice. Continuing she believes that to meet the
future challenges which face the profession "[…] more creative cross-disciplinary
thinking to stimulate innovation" is needed in education.

   The cross-disciplinary role of the architect extends beyond the design and
construction of buildings; it also includes wider urban, mobility, social and cultural
implications. A crucial skill which architects learn in their education is the ability to
handle and resolve complex problems. Customarily, architects approach
problem-solving in a holistic manner by integration rather than dissection. This is
advantageous in addressing the complex issues found in sustainability.
Traditionally, architectural education occurs in studio-based environments where
teaching methods have evolved from practice based (students design work critiqued
by teachers) to a strategy based on students' cognitive processes. The evolution has
been fostered by ideas about 'designerly ways to think' or 'designerly thinking'
(Turkienicz and Westphal 2012). The notion of 'designerly ways of knowing' was
developed by Nigel Cross in the late 1970s emerging out of new approaches in
design education (Cross 1982). This concept is now widely used in architecture
education and is an important consideration for pedagogy methods.

   Sustainability is unfortunately usually taught as either a parallel unit or as an
elective course. There is still some controversy around how sustainability should be
taught in architectural education. The Architectural Journal (Unknown 2013) presented five contrasting views in response to the question "should architecture students be taught sustainable design?' The responses ranged from a deep green
strategy informing everything an architect does, collaborative approaches, sustainable design to be prioritised by educators, landscape architecture must play a
key role in architecture students' understanding of what sustainability is and lastly,
the 's' word should be completely ignored. This illustrates the challenge educators
committed to teaching sustainable architecture face within the larger architectural
education framework.

This paper, therefore, aims to outline and suggest an alternative pedagogical approach for teaching sustainable architecture theory within a studio-based environment utilising the pedagogical methods for solving complex problems and applying 'designerly thinking.' Emphasising the qualities of a hybrid approach of electives and integration in an architectural education framework. This will be supported by first, explaining two different approaches to framing sustainability for architectural students and then outlining current literature and policy in regards to sustainable architecture education. Secondly, exploring the role of the architect and theory, accompanied by a discussion on how designers learn, both in practice and in education. Thirdly, a case study from the Bachelors' studies at Aarhus School of Architecture is supplied and critically examined as an example of this pedagogy approach. Lastly, reflections on the students' learnings are presented and discussed.

2 Changing Perspectives: Problem or Opportunity

The reality students now face upon leaving university is grim. Climate change, rising sea levels, droughts, floods, rising populations, hunger, destruction of human and nonhuman habitat and the depletion of resources in a finite world offer a realistic but somewhat depressing welcome to the workforce. However, by changing the perspective in which these problems are framed can result in a variety of different outcomes. While overwhelming and difficult to comprehend, these complex issues offer many opportunities to the architecture profession. It is important to address these issues within architectural education. To change how problems are framed, from a pessimistic perspective to an optimistic challenge where students can see an opportunity for sustainable designs solutions.

There are two common approaches, dissection, and integration. The first approach involves dissecting the problem into tangible-bite-sized issues which allow students to tackle a problem without being overwhelmed. While this approach makes the problems relatable, it also reduces it to a singular rather than a complex issue. This is counterintuitive to the normal 'designerly' way of thinking which employs a more holistic attitude towards problem-solving. An example of this is waste.

The building and demolition industry alone is responsible for over 900 million tonnes of waste each year in Europe (Butera et al. 2015). Osmani et al. (2008) estimate that one-third of this waste originates from design decisions, therefore, it is clear that architects have the ability to make simple changes to reduce waste and consequently contribute to a more sustainable future.

The second approach to framing the problem uses integration to interpret the situation more holistically and in harmony with a normal 'designerly' way of thinking. While this may lead to students becoming overwhelmed, it offers a richer and more comprehensive understanding of the problem. For example, in 2015, the UN set seventeen new and ambitious sustainable development goals for the 2030 Agenda for Sustainable Development (United Nations 2015). Impressively, architecture can influence every one of the seventeen goals from 'no poverty' to 'sustainable cities and communities.' This emphasises how unique and crucial the field of architecture is in its ability to make a difference in multiple scenarios. This approach understands architectures role in a more complex system.

It is imperative that sustainable architecture education tackles not only design, form, and aesthetics but also addresses the relating complex issues and interdependent systems which make up the built environment. While architecture is just one part of this complex problem, we offer the above integrated approach as a starting point and a way of illustrating the challenges and opportunities current students will face when entering professional practice.

3   Literature and Policy

There is limited literature and policy dedicated to addressing sustainability within architecture education, especially regarding theory. Likewise, there is little education about how designed-based practices, such as architecture, may contribute to sustainable development (Farmer 2013). Policy is often only guidelines and left up to interpretation. RIBA (The Royal Institute of British Architects) in the UK does not dictate a fixed curriculum and only states that "[…] students must have a 'knowledge' of sustainability" (Mark 2013). However, Mark (2013) expands on this explaining that each school is responsible for deciding what approach, type and amount of knowledge is included. Mark (2013) continues to articulate that leading architecture institutes support an integrated approach to sustainability. Concerns which arise from this lack of policy is that the capability of graduating students' in this discipline may vary and employers subsequently have no clear expectations (Mark 2013).

Within architecture, in some respects, the profession and education are closely linked. Mark (2013) conveys that it is clear that all schools need to produce students who can design for the challenges of the future. Elaborating that the future of architecture is developing and consequently the skills future architecture graduates, must also change and adapt.

Contemporary research explores architectural practice to understand what students need to know when entering the workforce, especially regarding sustainability. The EU-funded project EDUCATE (2012) was undertaken by a partnership of seven European institutes researching into policy and pedagogy of architectural education. As previously mention, it highlights the imperative for academic programs to acknowledge and accept the extensive array of values and knowledge contributing to sustainable development. Including, the inherently complex endeavour of embracing sustainability as a multi-, inter- and trans-disciplinary approach to education. Through doing this, it allows students to learn to grapple with complexity and fosters critical engagement between students and the concept of sustainable development (EDUCATE 2012).

4 The Role of the Architect

As the social context and challenges change, so does the need for architecture to address these complex and cross-disciplinary problems in a creative manner (Hartman 2011). Architecture is developing and while there is an emerging need for specialists, architects as a whole remain, generalist. While dealing with expanding complex problems the role of the architect is also growing in complexity. Haines (2010) describes the role of the architect as needing to understand their context and site, while not being geologist or landscapers. They need to understand the social context in which they are designing, however, they are not sociologists. They must develop a structure which can support their design, but the are not engineers. They must make material choices, however, they are not chemists or physicists. Their design must meet codes and requirements, while they are not lawyers. In summary, they must respond to an expansive range of criteria and synthesise information from a panoply of academic disciplines and sustainability is a new addition to this demand.

As mention previously, there is a causational link between practice and education. While architects (in general) remain generalist so does the conventional nature of the architectural curriculum. Altomonte (2009) outlines that the curriculum.

> [...] covers an extremely broad range of technical and non-technical areas, seeking to equip students with the awareness, knowledge, understanding and ability needed to perfect their design skills whilst making informed decisions in response to project briefs.

As project briefs, both in practice and in education, grow in complexity and become even more multi-disciplinary a plurality of creative approaches and problem-solving is needed.

In the five years it takes most students to complete their architecture education, it is not feasible to thoroughly address and provide education concerning all of these disciplines. However, it is possible to address how we approach a problem and seek design solutions. Architects find design solutions in unpacking a problem, examining relationships, changing the scale to get a broader view and finding new ways to *see* the problem.

5 The Role of Theory in Design Education

Theory offers students a different way in which to 'see' a problem or the context in which it is situated. Hillier (1993) expresses in order to move beyond direct experiences of the world, abstract or propositional forms of knowing, theories, must be engaged with to be able to *see* problems differently, both theoretically and in conceptualising design solution. Cole (1980) makes the argument that sustainable design is in its infancy and consequently experiential knowledge is not fully established. Accordingly, theory needs to be incorporated into building design, monitored and then disseminated into the design profession to develop more experiential knowledge.

Typically within architecture education, an immense amount of tacit knowledge is gained and expressed through existing built examples or drawings. However, if built examples and drawings are not sufficiently developed, new methods of design and building need to be employed to move beyond current unsustainable methods of designing, constructing and living. To create more sustainable buildings theories or philosophies can be utilised to underpin or frame strategies (Farmer 2013). Theory is essential to transition beyond our current methods of practice to question or critically examine what we already experience or what exists in the world which is available for reproduction.

Hillier (1998) explains:

> Any theory about how we should act to produce a certain outcome in the world must logically depend on some prior conception of how the world is and how it will respond to our manipulations. Careful examination will show that this is always the case with architectural theories. We invariably find that the precepts about what designers should do are set in a prior framework which describes how the world is.

If students can 'see' the world and its processes differently, then there is the ability to be able to address the complexities and the multitude of values generated by sustainable development. If we are not able to understand these complexities, then the framework for design will only represent a dissection approach to understanding this reality.

Theory can, however, be a controversial topic within architectural education as not all educators agree with its importance at the same level. This poses challenges when educational approaches are mainly studio-based within the architectural education. Therefore, it is even more essential to make sure the pedagogical approach is advantageous to 'designerly' ways of thinking and synthesising knowledge.

6 How Designers Learn

How designers learn will be discussed in relation to research in practice and also within the framework of education and studio-based learning. As mention previous, the knowledge required and how it is subsequently synthesised is comparable between practitioners and students.

How architectural practitioners develop and syntheses knowledge within the design process of practice offers an interesting insight into how to best provide architectural students' knowledge of sustainability. The above discussion highlighted the need for theory or propositional knowledge to change how students see the world and the design process, to get new insights into changing existing unsustainable practices. In contention, the design profession has been shown to draw on experiential and epistemic knowledge in pursuit of innovative design solutions.

> […] few designers claim to create design from theory, and many would go out of their way to deny it. But this does not mean that they do not design under the influence of theory. Much use of theoretical ideas in architecture is tacit rather than explicit. (Hillier 1998).

This disconnect between propositional and experiential knowledge is an interesting interaction regarding education. While theoretical knowledge is essential to change students' world view, experiential knowledge is important to understand it. Therefore, it is important to provide pedagogical approaches which take this into consideration; this is supported by Lawson's' following discussion on how designers' synthesis knowledge.

Elaborating on the notion of experiential and epistemic knowledge in practice Lawson (2004) discusses how ideas are developed within the design process as a result of how designers handle different information. Within design conversations 'schmata' (spatial concepts) and 'gambits' (approaches within the design process) are explored to address ill-defined design problems and refer to experiential knowledge for guidance. Lawson (2004) also articulates the relationship between the design problem and design solution as 'messy mapping' highlighting that there is not one clear answer to a clearly articulated problem. Rather that design problems can be redefined and refined throughout the design process.

Architecture projects or designs evolve as the design problems are redefined and receive input from actors from multiple disciplines such as engineering, construction, product manufacturers, and financiers. With the growing focus on sustainability this now includes a range of expertise from disciplines not traditionally associated with the architecture profession. The design team draws on knowledge from this vast range of fields to be included in design conversations. Practices and architectural practitioners also operate in learning communities which pass knowledge between projects and from different perspectives with the understanding that values can change over time (Wenger 2000).

Similar to practice, schools of architecture still construct a division between theoretical and applied teaching in the basic curricular (Farmer and Guy 2009). This is usually in the form of design studios and theoretical elective courses. Within the studio setting, which can be relatable to architectural practice environments, students learn to 'think architecturally' as Attoe and Mugerauer (1991) refers to it (or as outlined in this paper 'designerly thinking') and to secondly visualise and graphical represent phenomena. Attoe and Mugerauer (1991) also highlights that students should spend their time learning how to bring things together, synthesise, as it is a required central skill. This is especially relevant within the framework of sustainable architecture as a trans-disciplinary approach which was outlined earlier.

A collection of research has been completed to understand pedagogical approaches to best address architecture and design students' way of learning within a studio environment. A selection outlined by Altomonte et al. (2014) demonstrates how architecture students learn. Rutherford and Wilson (2006) reveal that for students to be engaged in lectures and design tasks they must be active knowledge seekers in an environment of cooperation and activity. Relatedly, Cole (1980) explains that if lecture content is excessive, inaccessible or not understood as relevant, it will not be efficiently engaged with. Leroy et al. (2001) propose problem-based learning as a successful way of introducing a new way of thinking and providing the subsequent problem-solving skills. Students can also become engaged when problem-based learning is applied to a practical task, joining experience and observation to contextualisation and experimentation (Kolb and Kolb 2005). Problem-based learning can be a precursor to the previous discussion of ill-defined problems in practice. Brandt et al. (2013) also suggest that understanding how to conceptualise these ill-structured problems should be addressed with the same creativity and responsiveness that is applied to the content of a design problem. Levy (1980) understands engagement as one of the key processes to foster motivation and increased efficiency of learning as a result of students' interest in subject matter. Lastly, Child (2004) discusses supporting student motivation in terms of group support, peer influences on behaviour and a form of extrinsic motivation.

How designers or architects learn is comparable in both practice and education. There is a need to foster 'designerly thinking,' address ill-defined problems, and draw on and synthesise both propositional and experiential knowledge from a variety of disciplines. Reflecting on the challenges of educating architecture students while addressing the complexities and multitude of values within sustainability, Altomonte et al. (2014) aspires for an architecture curriculum that

> seek to bridge [the] divide and develop pedagogies that combine both technical and holistic issues of sustainability with a design approach that is inventive, creative and responsive to pressing environmental needs.

The above discussion has influenced the applied pedagogical approach used in the following example course providing an engaging and creative method for understanding sustainable architecture theory.

7 Sustainability Education at Aarhus School of Architecture

Aarhus School of Architecture (AAA) is one of two independent arts-based architecture schools in Denmark. The schools' practices are firmly based in the Beaux Arts traditions with a focus on habitation, sustainability, and transformation (Reinmuth et al. 2011). The Beaux Arts influence is important as it was within the Ecole des Beaux Arts that the notion of the architectural studio was developed (Green and Bonollo 2003). Correspondingly, AAA offers a majority of studio-based learning environments from Bachelors to Masters level, exclusively for architecture students. The sustainable approach for the school can be summarised by the description of 'Teaching Programme Three' which is a Masters level studio programme that includes studying environmental challenges, sustainable agendas, solving challenges across scales, sustainable solutions for specific contexts—environment, social and physical and lastly addressing professional challenges of emerging practices (Arkitektskolen 2015). This description covers an extensive range of topics, and this is in agreement with the universities approach to sustainability. Similarly to the previously mentioned RIBA policy, there is no explicit focus, and each student is encouraged to explore sustainability within their design work in a way which is most interesting to them. This is supported by Unterrainer (2015), professor in sustainability; he indicates that the projects from design studios across levels provide a rich cross-section of different perspectives on various environmental and social agendas. Highlighting that "there is not one singular green reality."

As the majority of teaching occurs within a studio-based environment and sustainability is only partially integrated into some programmes in the curriculum, the consequence is sustainability only becomes a focus if a student or studio teacher has a specific interest in the topic. A slightly more reductionist approach is applied to the Bachelor's program where second-year students have the opportunity of taking an elective two-week theory course in sustainability which will be outlined and discussed in the remainder of this paper.

8 Designing a Pedagogical Approach for Sustainable Architecture Theory

The previous discussions have framed the aim of this pedagogical approach. The pedagogical aim was to develop a new method for teaching a sustainable architecture theory course within a studio-based learning environment. The discussions above have highlighted the importance of integrating sustainability into the curriculum, and while it is not possible to achieve a complete integration within the university's framework, this pedagogical methodology intended to bridge this gap and provide theoretical content in a 'designerly' manner. The course used the method of visual representation familiar to a studio context, encouraged students to be active knowledge seekers, engaged them by providing flexibility in selecting topics and supported motivation with group work and discussions.

A Socratic method of teaching was practiced as outlined as a form of 'excellent studio teaching' by Attoe and Mugerauer (1991). It employed questioning to reveal and not to judge, rather than exposition and lecturing. There was a focus on the student's decision making and observational processes in preference to the product. This method taught students how to be critically reflective of their work. The following courses description was influenced by these previous pedagogical concepts.

'Sustainable Architecture Theory' was offered as a two-week intensive elective course as part of the second year bachelor education in January 2016. This course was run with 30 students as a pilot course for including sustainability as one of the three theory courses offered each year. With the above discussion in mind, in regards to the importance of addressing sustainability holistically, this course tested the notion of delivering a broad and holistic approach to the content but specifically framed within architecture. The course aimed to help students understand the complexity and systems within sustainable architecture theoretically, improve problem-solving skills relating to complex and ill-defined problems and to engage with theory with a more 'designerly' method. Figures 1 and 2 illustrate that even

Fig. 1 Student studio environment

Fig. 2 Students working in groups

though this was a theoretical course, it was still taught within a physical studio environment, allowing for lots of informal discussions and interactions across groups. The studio environment which is depicted here shows no hierarchy of space, discussions among teachers and student peers occurred in circles around tables or different student work.

The pedagogical approach was both descriptive and exploratory in nature. A linear structure delivered a combination of lectures, theoretical readings, videos and recorded lectures, group work (including drawing, mapping, and writing), discussions and presentations. To elaborate, the initial portion was teacher centred. Students were introduced to basic understandings of sustainability. From this groups of around four students were formed and chose a specific theory for the duration of the short course.

They were supplied with key readings, recorded lectures, videos and literature. They were required to complete three tasks, all of these tasks were supported but tutoring, group discussions, and group presentations throughout the course. Each theory had two groups assigned to it, to try and foster conversations across groups.

Their first task required the students to complete a 'Giga map' of the information they had gathered. This task was inspired by the 'System Oriented Design' mapping method—Giga mapping—developed by Birger Sevaldson (2012). The aim of this task was to engage the students to understand the multiple actors, disciplines, and values within each theory and also to recognise the complexity their theory concerned. As you can see in Figs. 3, 4 and 5 the visual outcomes varied for each group. They vary from mapping ideas with no hierarchy in Fig. 3, to starting to visually represent concepts in Fig. 4, and lastly endeavouring to understand connections between concepts, strategies, and actors in Fig. 5.

The second task was to critically reflect on the provided information and their Giga map, then to produce in their group a 2000-word essay about their architectural position. This task aimed foster critical thinking, reduce their findings and understanding the key issues they had explored, formed a position.

The final task was to visually represent their 2000-word text. There was no prescribed method or outcome for this task, and it was aimed to bring the theoretical

Fig. 3 Group one—giga map

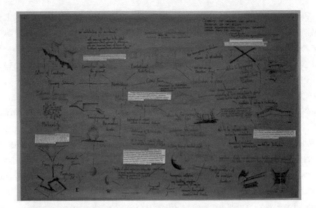

Fig. 4 Group two—giga map

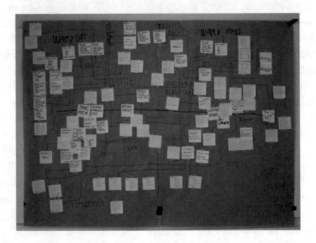

Fig. 5 Group three—giga map

task of writing back into their familiar design methods to again try and critically engage with the key concepts in a 'designerly' way. The majority of visual productions were video or animations.

This second part was learner-centred and based on the pedagogical belief that theoretical knowledge in architecture is best understood with personally engaging or translating it into action. This can be seen in Figs. 6, 7 and 8. Key values of

Fig. 6 Group one—visual representation (film stills)

Fig. 7 Group two—visual representation (drawing)

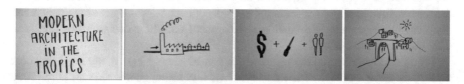

Fig. 8 Group three—visual representation (animation stills)

sustainability have been explored visually in a manner of different ways. Figure 6 assessed their street, waste, Cradle to Cradle principles and their role as students and designers; Fig. 6 shows some screenshots from their seven-minute video which had the focus of being more conscious designers. Figure 7 was a drawing which explored architecture and regionalism. Comparing the local resources and architecture with unsustainable extremes. Figure 8 explores architecture in the 'tropics' through a five-minute animation explaining the social implication of not only industrialisation but how this affected society and architecture and what they as students could learn from this different approach to construction and context.

9 Methodology for Observation and Analysis

A qualitative and informal direct observational method was employed to track and understand the students' different stages of awareness and how they responded to course objectives. It also followed how they solved problems, different group work interactions and participation, how accessible the content was and when to add more input in the form of discussions and additional content. 'Field notes' were taken throughout the course to document key turning points in their comprehension and produced work. How the students interacted with theory in a 'designerly' manner was also observed and towards the end of the course there was a discussion around if the students thought this method of teaching theory and sustainability was beneficial as a way of learning but also as a way of integrating the learnt concepts into their design and normal studio work.

10   Reflections on the Students Learning

Tracing the students' changing understandings across the workshop illustrated a transition from grappling with easy-to-understand notions related to architecture to critically reflecting on the role of architecture within sustainability. Both personal and critical awareness and reflection could be seen to develop through engagement with the different methods. The most substantial transition could be seen between the first task of Giga mapping where they focused on the architectural elements of the theory, then having to articulate these concepts and form a position. This made them engage more with sustainable development and their design ethos and how they live their lives.

Group one who created the video which can be seen in Fig. 6, reflected the most when they went out to film the city for their last visual representation task. They started to see the urban fabric unfold through the 'lens' of their mapping. When they returned, they were surprised how much unsustainable living and building was happening in their city which they had never noticed before. They especially noticed the amount of waste and over-packaging that was produced and unnecessary.

Group two who created the drawing in Fig. 7 spent a considerable amount of time reflecting of regionalism and looking at resources which are local to Denmark. They reflected critically on materiality and became very interested in traditional building techniques. They explored the embodied energy of materials and especially regarding transportation and resource extraction. After a study trip into the city to the new harbour development, they reflected on how unsustainable these were regarding designing for the Danish climate, the life span of the buildings, materials, transportation and also cultural aspects. They were critical as to why local materials were not utilised.

Group three which created the animation in Fig. 8, were interested in the critical of the role of the architect concerning social and cultural aspects. They explored the implications of buildings and design in a social context and started to reflect on the importance of designing for the social, cultural and climatic context you are in.

Many students produced descriptive work in regards to the sustainable architecture theory, but in discussions, they engaged critically about how these theories relate to sustainable development and how sustainability, in general, affected their personal lives. They started to unfold and grasp the complexity of the issues and understand the links between many of them and the implications of not viewing sustainability holistically. By engaging with it in a 'designerly' way, they were able to relate and explore the concepts in a method which was familiar and relatable to the rest of their studio education.

11 Conclusion

This paper has reflected on architectural pedagogy in relation to sustainability. It framed the discussion within an integrated approach to problem-solving. A reflection of contemporary research and policy signified the importance understanding sustainability as an inter-, trans- and multi-disciplinary endeavour which should be integrated into architectural curriculums. This was support by the changing role or architects becoming more complex and highlighted how theory and propositional knowledge could alter how students 'see' in new ways in order to address these issues. An exploration of how designers learn emphasised the need to synthesise both propositional and experiential knowledge from an array of different fields. An example of an elective course run in the Bachelor's program at Aarhus School of Architecture described a pedagogical approach which aimed to connect the topic of sustainable architecture theory to a design studio education framework.

Literature and research outlined the need for full integration of sustainability into architecture education. However, policy shows there is a hesitation to do so. The presented university framework only allows for partial integration, and the described elective course was offered as a pilot to understand if a hybrid approach was possible. Instead of introducing sustainable architecture theory into the design studio (the integrated approach), the design studio was introduced into sustainable architecture theory for the duration of the short course. A limitation to this is that as the course is presented as a separate elective, the students struggled to overcome this imperceptible barrier and were initially hesitant in connecting these two paradigms.

A second common issue which is present in the field of sustainable architecture education is the need to address multiple disciplines and actors. This was especially challenging as the university in which the course was taught is an independent architecture school. Students are not often confronted with other disciplines, so the importance of them understanding sustainable architecture as a plurality of disciplines was crucial. Giga mapping was utilised as a method because of its ability to visually showcase not only the different actors and ideas but also their relationships. It was essential that students understood these in a visual way to link with 'designerly thinking.'

Critical engagement was also addressed as a relevant issue from the field. To foster the students' critical engagement a varied mode of learning was included in the method. This was to encourage the students to change perspectives, change scales, change techniques of representation and also move between problem-solving, 'designerly thinking,' writing and visual representation. By moving through tasks, the students needed to critically assess which information to take with them to the next task. Corresponding group discussions also offered another opportunity to give and received critical feedback and ideas.

As discussed earlier the limitation of the integration of sustainability into architecture education is important as it will take policy and a considerable amount of engagement from teaching staff to achieve a total integration, especially in an independent school. In this transition period, new and innovative courses such as the one outlined in the paper can offer a hybrid course which in uncommon in many current architectural education environments.

This course, however, is only one small pilot course that has many limitations. These limitations include the unique circumstances of the independent educational framework of AAA, studio-based teaching as the main platform and the majority of teaching usually occurring in the studio context. In addition, the shortness of the course only offered a small glimpse into observational findings and in order to truly understand the implications and benefits of introducing the studio-based context and 'designerly thinking' into theory content a much longer study would need to take place. The isolated example of only 30 second-year students also limits the applicability of the observations to wider audiences.

This paper set out to suggest an alternative pedagogical approach for teaching sustainable architecture theory within a studio based environment. While this was successfully achieved it also acknowledges the ongoing dilemma for sustainable design education, the need to address and cover the breadth and complexity of sustainability in a holistic manner while maintaining a full architecture design curriculum. These gaps in the architecture education need to be bridged and explored with a pedagogical approach which is appropriate for the learning styles of architecture students. Without this, the sustainable agenda will not be fully integrated into architecture education and practice.

References

Altomonte S (2009) Environmental education for sustainable architecture. Rev Eur Stud 1:12

Altomonte S, Rutherford P, Wilson R (2014) Mapping the way forward: education for sustainability in architecture and urban design. Corp Soc Responsib Environ Mange 21:143–154

Arkitektskolen Aarhus (2015) Focus areas [WWW Document]. Arkitektskolen Aarhus. http://aarch.dk/info/research/about-research/focus-areas. Accessed 15 Nov 16

Attoe W, Mugerauer R (1991) Excellent studio teaching in architecture. Stud High Educ 16:41–50

Brandt CB, Cennamo K, Douglas S, Vernon M, McGrath M, Reimer Y (2013) A theoretical framework for the studio as a learning environment. Int J Technol Des Educ 23:329–348

Butera S, Astrup TF, Christensen (2015) Environmental impacts assessment of recycling of construction and demolition waste. Technical University of Denmark, Lyngby

Child D (2004) Psychology and the teacher. Continuum, London

Cole RJ (1980) Teaching experiments integrating theory and design. J Architect Educ 34:10–14

Cross N (1982) Designerly ways of knowing. Des Stud 3:221–227

EDUCATE (2012) Sustainable architectural education. University of Nottingham, White Paper

Farmer G (2013) Re-contextualising design: three ways of practicing sustainable architecture. arq: Architect Res Q 17:106–119

Farmer G, Guy S (2009) Pragmatism and the ethics of sustainable architecture. Presented at the ethics and the built environment. University of Nottingham, Nottingham, pp 90–106

Green LN, Bonollo E (2003) Studio-based teaching: history and advantages in the teaching of design. World Trans Eng Technol Educ 2:269–272

Haines C (2010) The role of the architect in sustainability education. J Sustain Educ

Hartman H (2011) The green building agenda is gathering pace but the task ahead is enormous. Architects' J 234:20

Hillier B (1993) Specifically architectural theory: a partial account of the ascent from building as cultural transmission to architecture as theoretical concretion. Harvard Architect Rev 9:8–27

Hillier B (1998) Space is the machine: a configurational theory of architecture. Cambridge University Press, Cambridge

Kolb AY, Kolb DA (2005) Learning styles and learning spaces: enhancing experiential learning in higher education. Acad Manage Learn Educ 4:193–212

Lawson B (2004) Schemata, gambits and precedent: some factors in design expertise. Des Stud, Expertise in Des 25:443–457

Leroy P, Ligthart S, van den Bosch H, Ligthart S (2001) The role of project-based learning in the "Political and social sciences of the environment" curriculum at Nijmegen University. Int J of Sus in Higher Ed 2:8–20

Levy A (1980) Total studio. J Architect Educ 34:29–32

Mark L (2013) The green curriculum. Architects' J 238:78–82

Osmani M, Glass J, Price ADF (2008) Architects' perspectives on construction waste reduction by design. Waste Manag 28:1147–1158

Reinmuth G, Nielsen AG, Toft AE (2011) A beaux arts education for the 21st Century. Architectural Publisher B

Rutherford P, Wilson R (2006) Educating environmental awareness: creativity in integrated environmental design teaching. 40th Annual Conference of the Architectural Science Association ANZAScA. Presented at the Challenges for architectural science in changing climates. Adelaide School of Architecture, Adelaide, pp 261–269

Sevaldson B (2012) GIGA-mapping [WWW Document]. Systems oriented design. http://www.systemsorienteddesign.net/index.php/giga-mapping. Accessed 12 Oct 2016

Turkienicz B, Westphal E (2012) The cognitive studio: exercises in design learning. In: Shaping design teaching: explorations into the teaching of form. Aarborg University Press, pp 185–202

United Nations (2015) Transforming our World: The 2030 agenda for sustainable development (No. A/RES/70/1)

Unknown (2013) How to teach sustainable design. Architects' J 238:83–85

Unterrainer W (2015) Sustainability: an imperative for plurality and context. In: Sustainability: an imperative for plurality and context. Arkitektskolens Forlag, Aarhus, p 7

Van der Ryn S, Cowan S (1996) Ecological design. Island Press

Wenger E (2000) Communities of practice and social learning systems. Organization 7:225–246

Education for Sustainability in Higher Education Housing Courses: Agents for Change or Technicians? Researching Outcomes for a Sustainability Curriculum

Alan Winter

Abstract This paper summarises the findings of research that explored the nature of the curriculum for sustainability within the field of Housing Studies at British universities. The aims of the research were to understand the attitudes of Housing academics to sustainability, to identify the factors that have influenced those attitudes, and to highlight ways in which the inclusion of environmental sustainability in the Housing curriculum can be encouraged. This paper outlines the key results of the research, which was based on an online survey and interviews with Housing academics and professionals across the UK. The paper argues that the nature and content of a sustainability curriculum within Housing courses is contingent upon a range of factors such as the individual research, teaching and professional backgrounds of academics, the existence or otherwise of support mechanisms or strategies for sustainability within universities, the position of professional bodies and employers and, indirectly, of national and devolved government policies. Housing Studies as a discipline is focused principally on the social and economic pillars of sustainability: Housing academics generally do not have a well-developed understanding of the environmental pillar, which few consider to be core to the Housing profession.

Keywords Housing · Education for sustainability · Curriculum · Agents of change

1 Introduction: Housing Studies and Its Place in Sustainability Education

The Report of the World Commission on Environment and Development: Our Common Future (UN 1987) identified three areas of concern for the sustainability of the planet: social, economic and environmental. This paper argues that profes-

A. Winter (✉)
Division of Urban Environment and Leisure Studies, London South Bank University, 103 Borough Road, London SE1 0AA, UK
e-mail: alan.winter@lsbu.ac.uk

© Springer International Publishing AG 2018
W. Leal Filho (ed.), *Implementing Sustainability in the Curriculum of Universities*, World Sustainability Series, https://doi.org/10.1007/978-3-319-70281-0_4

sional Housing education focuses on the first two but largely omits reference to the environmental strand. The social housing profession is potentially a key player in helping achieve the UK's carbon emission reduction targets but there seems to be little pressure from within the profession for housing officers to have a good understanding of environmental issues. The emphasis is increasingly on "frontline issues" (Richardson et al. 2014) in which a concern for climate change is not central.

The importance of sustainability and climate change to the Housing profession is clear. Residential buildings contribute about a quarter of the UK's carbon emissions annually (Palmer and Cooper 2013) and, in 2014, local authorities and housing associations owned 17% of homes in the UK, many of which are old and of low quality in terms of energy efficiency (Shelter England 2016). Housing associations build around 22,000 homes every year (Shelter England 2016) and their activities (together with private sector housing developers) are significant contributors to the UK's carbon emissions. The demanding targets on the reduction of UK carbon emissions by 2050 set by the Climate Change Act 2008 will require a 3% annual reduction in domestic emissions. Given the low take up of the Green Deal it is clear that more needs to be done within the housing sector; some argue that meeting the 2050 target will be impossible without more action from the sector (Palmer and Cooper 2013). The National Energy Efficiency Action Plan (NEEAP), the 2010 Energy Performance of Buildings Directive (EPBD) and the Government's Fuel Poverty Strategy, which sets out long term targets for fuel poor homes to be EPC band C or above by 2030 (UK Green Building Council 2016), are all examples of legislative and policy initiatives that will require the social housing sector (and housing professionals) to have a much higher level of knowledge and understanding of sustainability for the future.

2   The Research Study

The research was undertaken as part of the author's thesis for a Professional Doctorate in Education; it developed from his concern that Housing education was not delivering the learning outcomes necessary to ensure that environmental sustainability was seen as mainstream.

The research study consisted of an academic survey, conducted online between April and June 2015, a professional survey conducted online in November 2015, and interviews with three representatives from the Chartered Institute of Housing (CIH), fifteen Housing and other academics from universities, two academic managers and two housing association sustainability managers. The interviews were carried out between May and December 2015. The main part of the research was developed from a pilot study at one English university, with the analysis of data undertaken using a grounded theory approach.

Fifty one academics who teach on Housing courses at twenty two universities and colleges across the UK were invited to participate in the survey; twenty eight responded. The respondents represented thirteen of the twenty two institutions (60%), and their courses around 85% of Housing students. The professional survey was aimed at senior staff in housing associations, with the purpose of providing evidence about the attitudes of housing professionals to sustainability. Twelve academics were interviewed, representing an even balance from universities across the UK. Respondents to the survey were asked to indicate their own level of knowledge and understanding of sustainability; eleven interviewees (50%) indicated that they had a "good" level of knowledge and understanding, and this proportion was repeated within the Housing academics group (six out of twelve interviewees).

Although the research draws on data from universities in all four countries of the UK, the sample used has some limitations. Most of the major centres that deliver Housing courses have been included but data from two larger courses, and from a number of smaller institutions that have recently started offering Housing courses, were unavailable due to lack of responses. The extent of the professional data was much narrower than on the academic side but nevertheless included key representatives of the professional body; both sustainability managers are recognised leaders in their field.

3 What Is Being Taught as "Sustainability" on Housing Courses?

There was a broad range of topics covered across the courses in the survey but the level of detail and the time devoted to some of them is limited These can be divided into seven categories:

- Housing technical
- 'Green' technical
- Policy background
- Policy implementation
- Social and communities
- Economic
- Climate change principles.

Most of the topics included reflect the professional nature of the courses, job relevance for the students and the needs of employers. There is an emphasis on Housing technical issues, social and community issues and policy concerns across the Built Environment. There is less coverage of the principles and underlying global politics of climate change and 'green' technical issues, such as renewable energy, biodiversity and 'green' infrastructure.

Of the Housing technical topics covered, the most common were housing development, retrofitting, housing design and energy efficiency; all of which reflect the priorities of social landlords. Although there is some focus on the mechanisms for delivery of sustainable housing (such as the Code for Sustainable Homes and the Building Regulations), there was little evidence that students are exposed to more innovative approaches (such as low carbon buildings and environmental impact assessments). In terms of policy, the emphasis appeared to be on implementation rather than discussion of the background and drivers of policy; a knowledge of regeneration, Town Planning and the Green Deal might be deemed to be more important than an understanding of the policy background leading to them. Only four respondents made reference, for example, to key policy statements such as the Stern Review and the Egan Wheel.

4   How Important Is Education for Sustainability to Housing Academics?

Academics were asked to indicate the importance, on a scale of 1–6 (with 6 being the highest), they gave to education for sustainability. The majority considered it to be very important. Nineteen respondents rated its importance at scales 5 and 6; seven rated its importance at scales 3 and 4. (Table 1). The more engaged respondents, with a higher level of personal knowledge and understanding, rated its importance higher than the others.

Location in itself does not seem to make a difference to attitudes or practice on sustainability. There are centres and individuals who are more engaged on sustainability than others but this derives more from the teaching, research and professional interests of academics rather than where they are based. What does impact in this respect is the extent of devolution and difference that comes from a different policy approach from government. It is this, in particular, that drives the relatively higher levels of importance given to sustainability by respondents in Scotland and Wales (Table 1).

The ways in which Housing academics considered education for sustainability to be important could be categorised into three areas: future thinking, environmental awareness and professional responsibility and citizenship. There were a range of views about what education for sustainability should be seeking to achieve, suggesting that Housing academics recognise the importance of sustainability—all

Table 1 Importance of Education for Sustainability (EfS)—does location make a difference?

Importance of EfS	UK (28)	England (10)	Northern Ireland (2)	Scotland (5)	Wales (7)	London (4)
Scale 6	12	6	0	2	4	0
5	7	2	1	2	1	1
4	5	1	1	0	1	2
3	3	1	0	1	1	0
2	0	0	0	0	0	0
1	1	0	0	0	0	1

three pillars—and some, though not yet a majority, have the knowledge, skills and interests to include a substantial element of the environmental pillar within their teaching programmes. The research findings suggest that Housing academics are open to the proposition that environmental sustainability should have a higher profile within the curriculum but that various factors operate as barriers to its inclusion.

5 What Factors Have Led to the Inclusion of Sustainability Within the Housing Curriculum?

The inclusion of sustainability in the Housing curriculum is influenced by a number of factors. It appears that the individual interests (including research) of academic staff are a highly significant factor; where academics have a high level of personal interest, sustainability is a significant aspect of a course curriculum. Where this is not the case, sustainability is not absent from the curriculum but it is dealt with in less depth.

By far the most significant factor was professional relevance (13 survey responses) with individual staff interests the next most important (five responses). Only one respondent thought that the requirements of the CIH were a significant factor: they do not use their accreditation of courses to promote environmental sustainability. Two respondents mentioned the lax approach of the CIH towards accreditation, in that they *"demand very little these days"* (Academic respondent 23, 2015) and that any focus on sustainability is simply *"rhetoric and/or lip service"* (Academic respondent 8, 2015), as a factor working against the inclusion of sustainability. Government policy (and the need for employers to implement it) was cited by three respondents. Five respondents identified sustainability as, for example, an *"overarching theme"*, an *"organising framework"* and a *"key issue"*. One respondent commented that it represented "an essential underpinning of professional education in housing" (Academic respondent 23, 2015).

No one from England and Wales mentioned employer demand as an important factor; both respondents from Northern Ireland considered this factor to be the most important. In broad terms the research study indicates that professional relevance and staff interests are the most significant factors influencing the inclusion of sustainability in the Housing curriculum.

6 The Research Findings

The research study showed that the sustainability curriculum within Housing courses is contingent upon a range of factors such as the individual research, teaching and professional backgrounds of academics; the existence or otherwise of support mechanisms or strategies for sustainability within universities; the position of professional bodies and employers and, indirectly, of national and devolved government policies. The presence of support mechanisms or strategies does not directly lead to the adoption of sustainability as a topic within a broader curriculum; although there were several examples of research centres, practical projects and organisational strategies identified by respondents, there were none which had changed the nature or level of their engagement with sustainability as a direct result.

Housing Studies as a discipline is focused principally on the social and economic pillars of sustainability: Housing academics generally do not have a well-developed understanding of the environmental pillar, and few consider it to be core to the Housing profession. Where it is included it is often within the context of issues such as fuel poverty rather than the impacts of climate change. Few Housing academics have an interest in or knowledge of environmental sustainability and the issues around climate change. Even among academics and professionals who claimed to have an understanding of the topic it was accepted that the issues are complex and difficult. This acts as a barrier to the adoption of sustainability within the curriculum. Factors cited in the research as linked to this were lack of time (too many other commitments) and the need to specialise in order to be able to make successful bids for research funding or for inclusion in the Research Excellence Framework.

Where sustainability is included, the main factor for this is the academic or professional interests of individual academics. This can come from a variety of sources, such as a research or professional background, but that is not necessarily environmental. Interest can stem from political activism, professional experience or a sense of social justice as well as an expertise in "green" issues. The professional backgrounds of Housing academics vary: many have a professional Housing background, having worked in different roles in local authorities or housing associations. It is not uncommon for those academics to maintain links with the profession, through a voluntary role, such as board member or activity within the professional body. Those who see themselves principally as researchers tend to come from a social science background. It was rare to find Housing academics with an environmentalist background; those who teach the elements of environmental sustainability within the curriculum are usually those members of staff with a different professional background, such as urban design or town planning.

Those academics who do teach sustainability—regardless of their motivation—see its inclusion as good practice: that a good education ought to include sustainability and climate change issues. There is little evidence of a radical approach but Housing as a discipline and profession is about delivering change and improving social conditions, so Housing students might be expected to be open to a radical agenda.

The approach of Housing academics reflects the emphasis within the Housing profession, which is on the social and economic strands of sustainability. There is no lead being given on developing a curriculum for sustainability; this might be expected to come from the CIH but they see themselves as a membership-led organisation and, therefore, as representing the Housing profession. The needs of the social housing sector are geared around maintaining the ability to build and manage housing stock, in order to meet housing need within a political environment that is subject to a high level of uncertainty and change (Lau and Grainger 2010). The financial viability of housing associations is crucial to their ability to respond to this agenda; strategies to maintain this position (such as mergers) are strategic priorities whilst front-line services are focused on helping tenants and residents cope with challenging social and welfare changes (Richardson et al. 2014). Where there is no legislative requirement to meet certain levels of environmental performance within domestic dwellings, housing associations and housing developers are likely to regard the additional costs of environmental improvements as barriers. Social landlords are also more likely to see retrofitting as a means of improving residents' well-being (in terms of fuel poverty, for example) rather than being driven by the need to reduce carbon emissions.

Two points that recurred within the research concerned the attitudes of academics and the political context of sustainability. One university senior manager referred, in a public meeting, to what he called "massive leadership challenges", suggesting that universities had a "moral obligation" to ensure that students finished their studies having achieved a sufficient understanding of sustainability. This was to be achieved through changed relationships with local communities and the management of the university estate as well as through the curriculum and learning and teaching. It was recognised that implementing those changes would not be easy, as universities are not "normal" organisations who are able to direct change in a top down way. Although there may be champions for sustainability within universities, it is likely that not all academics share the same passion for sustainability. Identifying the right approach to organisational and curriculum change is therefore important if the cultural shift required is to be achieved. There are differences between "maximalist" and "minimalist" approaches, with a maximalist approach likely to include initiatives such as forcing change through structural requirements, for example the introduction of common first year modules across all courses or using validation requirements to make sustainability a required topic within courses. A minimalist approach is, perhaps, more likely to adopt smaller, more incremental steps such as raising awareness through events and more visible recycling facilities.

A common attitude (and one found during the research study) seems to be an acceptance, particularly on the part of managers and non-academics, that academics are reluctant to change, and that academics generally prefer to work within their own areas of interest (Porritt 2015; Lampkin 2015). Responses from the academic interviews suggested that there is an element of truth in this but that it stems, not from any innate conservatism among academics, but from very practical concerns about moving away from areas of expertise, particularly when this is likely to involve teaching a subject they know little about. A lack of knowledge and expertise about sustainability is seen as a key factor in preventing academics stepping outside of their current area of activity; due to the wide-ranging nature of the topic, defining the complex issues of environmental sustainability and climate change can be difficult.

One of the factors that would make a difference in this respect is a broader recognition that environmental sustainability is important. If the Housing profession does not see it as a core issue, then it is arguable that the government does not see it as important either. From proclaiming the credentials of the "greenest government ever" in 2010, the government has shifted significantly away from that position. At a public talk at London South Bank University (LSBU), held to discuss the role of higher education in protecting the environment, Jonathan Porritt was "*pessimistic*" about the direction of travel of UK government policy on climate change, arguing that "*a dashboard of indicators are heading in the wrong direction*" and that the government saw economic growth as more important than environmental sustainability (Porritt 2015). The research suggests that, whilst Housing academics are open to the inclusion of sustainability within the curriculum and their own teaching, environmental sustainability is seen as a specialist issue. Where it is taught on Housing programmes it is often provided by "others", such as town planners or urban designers and others from built environment backgrounds where sustainability is viewed differently. Responses from the academic interviews suggested that Housing academics would, given a strong enough lead and sufficient guidance, be happy to include sustainability within their own teaching programmes where appropriate.

7   What Do the Curricula of Housing Courses at UK Universities Include Under the Heading of "Sustainability"?

On the majority of courses, and for the academics teaching on them, "sustainability" has a social rather than environmental meaning. This reflects the focus of the housing profession and the CIH. Some topics within the curriculum, such as development or maintenance, deal with an environmental element but generally only within the constraints of technical issues, such as the requirements of the Building Regulations or, before it was abandoned, the Code for Sustainable Homes. The focus is on the application of adopted policy or government legislation, such as Energy Performance Certificates or the Green Deal.

The level of knowledge and understanding is mostly at a basic level, reflecting the professional view of what is sufficient for the recognition of issues and problems that can be then referred on to specialist colleagues. The strongest views from the research saw the role of the Housing sector (and the professionals within it) as an informed client, able to specify clearly desired contract or project outcomes, leaving detailed content to specialists. In other areas, this approach is already a recognised role for housing professionals, who often (certainly in housing management terms) are required to have a broad general knowledge rather than a narrow specialist one. Nevertheless, this role requires a higher level of knowledge than students would be able to acquire from the current curriculum.

The research suggests that Housing courses do include a broad range of topics under the label of "sustainability" but that the content that is covered in this context will largely follow changes in government or devolved policy, and be concerned with implementation within organisational practice. There is very little included about the principles of sustainability and climate change or about sources of renewable energy. The level of knowledge and understanding of sustainability that Housing students will gain from their course varies; the majority of courses provide only a broad, superficial knowledge of sustainability. This reflects the level of knowledge that is seen as necessary for front-line staff (Richardson et al. 2014).

A recurring point from the research was made about the benefits or otherwise of treating environmental sustainability as a specialist topic. Opinions are divided about which approach is best (Shriberg 2002; Lampkin 2015). Five respondents highlighted the importance of integration of sustainability into the curriculum, so that it becomes mainstream. There was an element that saw environmentalism as getting too much attention and that regarded a "stand alone" approach as problematic:

> courses [should] promote the political, economic and social aspects of sustainability as requisites. Political, as a path to environmental, sustainability is often overlooked in my opinion (Academic Respondent 21, 2015).

Among all the respondents there was an acceptance that environmental sustainability should be included in the curriculum although there are clearly differences about its relative importance among the other elements of sustainability. It was very rare to see negative comments and there were certainly no indication of any denial of climate change that is visible in the political arena.

8   The Role of Professional Bodies and Subject Benchmark Statements

The specification and learning outcomes for the CIH Professional Qualification (PQ) are fairly detailed regarding the core Housing content but rather brief regarding environmental sustainability. The learning outcomes for the PQ have changed periodically, most recently in February 2012 and November 2014. The 2007 outcomes were updated and consulted on in November 2011; following comments from the Professional Housing Education Group (i.e. housing academics) it was agreed to include "green issues" in the specification (CIH 2011). In the event, the PQ outcomes were changed to include two references to the environment in indicative content: under the Core Content of Housing and Society it refers to "sustainability and regeneration", and under the Theme (i.e. optional) Content of Asset Management it refers to "environmental challenges". There was no mention of environmental issues under the Theme Content for Planning and Development nor in any of the eight Learning Expectations underpinning course design. The QAA Subject Benchmark Statement for Housing Studies identified the "impact of housing on the environment" as part of the core knowledge within Housing Studies, together with the ability to "review and appraise" environmental sustainability in the context of housing development and design (QAA 2014). Following this lead, the amended CIH PQ Guide of November 2014 did include as one of the ten key skills for CIH Chartered Members (CIHCM) "to be able to promote and engage with policies and processes which conserve resources and support sustainable development" (CIH 2014), but there was no mention of how this might be delivered within the curriculum indicative content.

The author believes that the subject benchmark statement and the CIHCM key skill above represent strong statements in support of environmental sustainability that places it at the core of housing education and the housing profession. On the basis of this research, it appears that not many people seem to take any notice of it, least of all the CIH.

9   What Does "Education for Sustainability" Mean to Those Who Set the Curriculum? Is There a Radical Agenda?

One of the aims of the research was to explore if there was a "radical" agenda on sustainability being delivered through the Housing curriculum. If Housing students and housing professionals can be thought of as "agents for change" in a broad sense of working towards delivering social justice, that label does not seem to apply in terms of the kinds of knowledge of sustainability they might acquire through the CIH accredited courses. Where students learn about environmental sustainability it is largely within the context of the implementation of government policy and professional practice; what the author has termed as a "technical" approach.

There is a strong recognition that sustainability education can be a force for change (Down 2006; Wade 2015; Xypaki 2015); the author argues that social change is an important aspect of Housing education—that Housing education and the profession are rooted in a radical tradition that understands that change and improvement of society and the conditions in which people live are a key part of its purpose. This argument showed itself in several of the academic and professional interviews, although it was not shared to the same extent by all survey respondents or interviewees.

Those tendencies towards change are informed by a broad (but not always detailed) awareness of environmental issues. These are, however, tempered by the realities of the profession itself and the political world in which it has to operate. That is not to say that Housing academics or the wider profession have a passion for and commitment to environmentalism; it is clearly the case that for the majority of people within the Housing profession "sustainability" is viewed through the prism of the communities in which their homes and tenants are situated. This is not to argue that such a view is wrong but, in the author's view, the environmental pillar of sustainability deserves and needs a far greater level of engagement from the

Housing profession. The absence of environmental sustainability from the curriculum is not often noticed; as one respondent put it, "*I do put a sustainability slant on things wherever possible but when I don't, it doesn't jar like it might with diversity issues*" (Academic Respondent 3, 2015).

We can identify four key influences on the curriculum: academic course teams, the CIH, the Housing profession and government policy. The research identified a clear, if indirect, connection between the emphasis given to environmental sustainability at the policy level and what is being taught. The existence of different approaches to sustainability by the devolved governments in Wales and Scotland were a major factor in the way that academics in those countries addressed sustainability. The need for the Housing profession to follow the lead of government is clear, as it is both a direct and indirect vehicle for delivery of a key government policy. That relationship emphasises the importance of the role of government in determining the nature and direction of professional education, particularly in fields such as Housing and Town Planning, that are frequently at the centre of political debate and change. It can argued that it is difficult, if not impossible, to teach Housing Studies without engaging with its political and professional context. Although it highlights the difficulty of pursuing an agenda that may be at odds with the dominant political discourse, it also emphasises the need for Housing students to understand the nature of the arguments behind that political debate in order to be able to contribute as reflective professionals.

10   The Next Steps

This research started from the premise that the Housing curriculum included relatively little on the environmental aspects of sustainability, and that what was included was approached from a technical perspective. What Ball describes, in the context of education research, as an "uncritical acceptance of moral and political consensus" (Ball 1995, p. 259) could, the author argues, describe the position of environmental sustainability within Housing Studies. The level of knowledge and understanding of, and the degree of interest in, climate change and sustainability, among Housing academics is generally low. It is certainly insufficient to overcome the "drag" of a number of factors, such as the lack of a lead from the profession, government and universities, the tendency for specialisation among academics and the fact that environmental sustainability is not seen as one of the most urgent challenges facing the social housing sector.

The challenges of integrating sustainability into the mainstream have been well documented (Down 2006; Rusinko 2009; Sterling and Witham 2008; Xypaki 2015; Wade 2015). We can identify two key factors (both of which are largely missing from Housing Studies) that may make a difference: finding ways to encourage and support academics in adopting sustainability into their teaching and research, and raising the levels of awareness, skills and knowledge around sustainability within higher education.

Numerous factors have been outlined that make a difference to what gets taught as sustainability and whether it is taught at all, but not all of those factors are within the control of academics, course teams and universities. Nevertheless, it is possible to take action in a variety of ways:

- raise awareness across the institution and provide opportunities for academic staff to become more knowledgeable about sustainability, particularly among academics for whom sustainability is not an obvious element within their teaching or research; the role of in-house research centres and other areas of expertise can play an important part here;
- provide support that will encourage adoption of what one respondent called a "scary" topic (Academic respondent 15, 2015); this can be either directly through teaching and learning materials that will help academics learn more about the topic, its impact and relevance to their own discipline, or through help with curriculum design so that sustainability is genuinely embedded within courses, not simply hide to find; Housing Studies is a discipline where the relevance of sustainability is clear—some other disciplines do not have that advantage—so the challenge for Housing is to make the environmental elements more explicit;
- central support is essential; identifying champions within Schools and Departments has advantages in bringing the change process closer to academics but too much reliance on a relatively small number of individuals can allow sustainability to be seen as "specialist", whereas the purpose of a programme of change should be to put sustainability firmly in the mainstream;
- be clear about what learning outcomes on sustainability are desired but do not get bogged down in definitions; welcome the broadness and richness of sustainability as a topic and embrace it; encourage and require course teams to rethink their curricula to provide space and opportunity for sustainability;
- identify clear threads or pathways through courses that allow students to see the relevance and importance of sustainability to their lives and careers; this could include those that lead to particular jobs or roles, such as housing development or maintenance, or just provide an in-depth understanding of climate change issues; it is important that "sustainability" modules have a relevance to all students, regardless of their routes through a course—it is a delicate balance to avoid being seen as too "specialist" and therefore as only something that "others" will focus on;
- similarly, the value of a common "sustainability" module across all courses is not universally accepted (Shriberg 2002; Lampkin 2015); it seems better to design a curriculum that has sustainability running through it and which requires students to engage with the topic as an integral part of their discipline. In Housing Studies, whilst not yet approaching best practice everywhere, we have an opportunity to emphasis the links between all three pillars of sustainability but that requires a higher profile for the environment. If environmental sustainability and climate change are not yet seen as core within Housing Studies, there are many opportunities to reemphasise the relevance of those issues to the

matters that really concern housing professionals. Giving sustainability a higher profile within Housing education could help change perceptions and awareness within the profession.

All these ideas and approaches represent a long-term project; too much change in a short period will only discourage academics from adopting a new approach.

References

Ball S (1995) Intellectuals or technicians? The urgent role of theory in educational studies. Br J Educ Stud 43(3):255–271

Chartered Institute of Housing (2011) CIH MCIH review. CIH, Coventry

Chartered Institute of Housing (2014) CIH professional qualification guide. CIH, Coventry

Down L (2006) Addressing the challenges of mainstreaming education for sustainable development in higher education. Int J Sustain High Educ 7(4):390–399

Lampkin S (2015) The challenges of introducing sustainable development in the curriculum at a UK university. Local Econ 30(3):352–360

Lau A, Grainger P (2010) What are the main factors influencing RSLs decisions to build beyond level 3 of the Code for Sustainable Homes? Paper to Housing Studies Association conference, April 2010

Palmer J, Cooper I (2013) United Kingdom housing energy fact file. Department of Energy and Climate Change, London

Porritt J (2015) Students organising for sustainability: old and new. Accessed at http://www.jonathonporritt.com/blog/students-organising-sustainability-old-and-new. 18 Oct 2015

Quality Assurance Agency for Higher Education (2014) Subject benchmark statement: housing studies. QAA, London

Richardson J, Barker L, Furness J, Simpson M (2014) Frontline futures—new era, changing role for housing officers. CIH, Coventry

Rusinko C (2009) Integrating sustainability in higher education: a generic matrix. Int J Sustain High Educ 11(3):250–259

Shelter England (2016) Shelter housing databank. Accessed online at http://england.shelter.org.uk/professional_resources/housing_databank. 17 Nov 2016

Shriberg M (2002) Institutional assessment tools for sustainability in higher education: strengths, weaknesses, and implications for practice and theory. Int J Sustain High Educ 3(3):254–270

Sterling S, Witham H (2008) Pushing the boundaries: the work of the higher education academy's ESD project. Environ Educ Res 14(4):399–412

UK Green Building Council (2016) Briefing paper: potential impacts of the EU referendum on green building policies. Accessed online http://www.ukgbc.org/news/uk-gbc-comment-eu-referendum. 17 Nov 2016

United Nations (1987) Our common future—Brundtland report. Oxford University Press, Oxford

Wade R (2015) Learning, pedagogy and sustainability: the challenges for education policy and practice. In: Atkinson H, Wade R (eds) The challenge of sustainability: linking politics, education and learning. The Policy Press, Bristol, pp 63–83

Xypaki M (2015) A practical example of integrating sustainable development into higher education: green dragons, City University London students' Union'. Local Econ 30(3):316–329

Author Biography

Alan Winter is an Associate Professor who has taught on the Housing Studies courses at London South Bank University since 1992. Prior to that he worked for 13 years in local authority housing departments in London, Kent and Berkshire. He was a County Councillor (for five years) and a housing association Board Member (for 18 years), including five years as vice Chair of Toynbee Housing Association. Alan has an MA in Industrial Relations from Brunel University and is a Fellow of the Chartered Institute of Housing (CIH). Alan's main areas of teaching are housing association governance, social welfare, social and housing policy, sustainability, and European housing. He has been Chair of the Faculty Learning and Teaching Committee and is working to integrate sustainability education into the Housing and Human Geography courses at LSBU. He has recently completed his doctoral thesis for the Professional Doctorate in Education (Ed.D.) and this paper is based on that research.

A University Wide Approach to Embedding the Sustainable Development Goals in the Curriculum—A Case Study from the Nottingham Trent University's Green Academy

Jessica Willats, Lina Erlandsson, Petra Molthan-Hill,
Aldilla Dharmasasmita and Eunice Simmons

Abstract Since 2015, governments, businesses and civil society, together with the
United Nations have been encouraged to work towards seventeen Sustainable
Development Goals (SDGs) as part of the 2030 Agenda for Sustainable
Development (United Nations, 2017). In this context, Higher Education Institutions
(HEIs) have a special responsibility to embrace the SDGs as they educate the
decision-maker of tomorrow. The aim of this paper is to give practical examples of
how to embed the SDGs in the curriculum of an HEI by outlining the process of
integrating the SDGs into the core curriculum at Nottingham Trent University
(NTU). This includes, among others; a Future Thinking Learning Room: an
innovative online resource library, discipline specific approaches, the use of the
estate as a 'Living Lab', community case studies and investing in staff development. As a result, ESD no longer needs to be an afterthought when it comes to

Green Academy: Nottingham Trent University, 50 Shakespeare St., Nottingham NG1 4FQ,
email: GreenAcademy@ntu.ac.uk.

J. Willats (✉) · L. Erlandsson · A. Dharmasasmita · E. Simmons
Nottingham Trent University, Green Academy, 50 Shakespeare St.,
Nottingham NG1 4FQ, UK
e-mail: jessica.willats@ntu.ac.uk; GreenAcademy@ntu.ac.uk

L. Erlandsson
e-mail: lina.erlandsson@ntu.ac.uk

A. Dharmasasmita
e-mail: aldilla.dharmasasmita@ntu.ac.uk

E. Simmons
e-mail: eunice.simmons@ntu.ac.uk

P. Molthan-Hill
Nottingham Business School, Nottingham Trent University, Green Academy,
50 Shakespeare St., Nottingham NG1 4FQ, UK
e-mail: petra.molthan-hill@ntu.ac.uk

© Springer International Publishing AG 2018
W. Leal Filho (ed.), *Implementing Sustainability in the Curriculum of Universities*,
World Sustainability Series, https://doi.org/10.1007/978-3-319-70281-0_5

curriculum content, allowing it to be an easily achievable priority across all academic departments. This chapter may be of interest to those looking for inspiration and ways to embed the SDGs within Higher Education and beyond.

Keywords Sustainability · ESD · Sustainable development goals
Higher education · Curriculum development

1   Introduction

The concept of Education for Sustainable Development (ESD) is beginning to become mainstreamed within the Higher Education sector (Wals 2014). By embedding this sustainable development theme in Higher Education curricula, Higher Education Institutions (HEIs) have great opportunities to make a difference and contribute to a more sustainable future (Leal Filho et al. 2015). Through a slight shift in priorities, Higher Education providers can contribute to change and offer solutions to many of the global issues we are facing today.

The main obstacles to overcome in order to fully integrate sustainable development themes in teaching and learning are the limited time and resources available to academics, as well as the often narrow interpretation of the sustainability concept. Academics are already under enormous time constraints and to add another thing to their already demanding workloads is often not feasible. With limited time to get familiar with the topics and few opportunities to fit it into the existing curriculum sometimes combined with a lack of support from senior management, it can seem like an impossible task (Molthan-Hill et al. 2015).

Another challenge is that HEIs tends to focus solely on their environmental performance and how to create a sustainable campus. This is not enough to foster sustainability literate graduates. Studies have shown that universities often promote behavioural change but a significant knowledge gap still exists when it comes to the reasoning behind such changes (Tilbury 2015). Themes such as social equity and economic prosperity in addition to environmental protection need to be dealt with in an adequate manner in order to equip students with the skills and knowledge to tackle sustainability issues, both today and in the future.

This paper will discuss and reflect upon NTU's approach to overcoming some of these issues and hence draw some conclusions, which will inform future practice. By focusing on the United Nation's Sustainable Development Goals (SDGs), the University has managed to start moving away from the narrow, environmentally focused definition of sustainability and raise awareness of the full breadth of the concept. The level of detail covered by the SDGs encompasses all 3 aspects of the triple bottom line (Sachs 2012) and hence the goals provide a well-rounded angle from which to approach sustainability within learning and teaching.

Every discipline seems to have their own definition of the sustainability concept (Molthan-Hill et al. 2015), and by adopting the SDGs a consensus on the matter can be reached. By creating a platform for ESD, which catalogues information relating to this concept, school-specific resources and ideas of teaching activities, academics can save time and will be able to embed sustainability in their teaching in a simple, yet effective, way. This paper will outline the process of creating this platform and provide recommendations to those wishing to take a similar approach. Other initiatives such as estate and community case studies and methods to map ongoing ESD practice are also introduced as examples of time-saving, innovative solutions to embedding sustainability in the curriculum.

2 Importance of ESD in HEIs

Education for Sustainable Development (ESD) is defined by UNESCO as follows:

> Education for Sustainable Development (ESD) is about enabling us to constructively and creatively address present and future global challenges and create more sustainable and resilient societies (UNESCO 2017).

HEIs can contribute to sustainable development through education, operations, and research (Lipscombe et al. 2008). The progression towards fully embedding these themes in the curriculum is emerging, something that can be seen in the growing number of peer-refereed journals, publications and focus of research in Higher Education (Leal Filho et al. 2015). However, there is still a long way to go before sustainable development is fully integrated in the curriculum across all universities and other HEIs.

University graduates will be leaders within all sectors of society, from government to the private sector. This is illustrated by the fact that 1 in 10 world leaders received their education from a UK university (British Council 2014). If graduates can bring values such as global citizenship, systems thinking, and lifelong learning from university into their professional lives, change is more likely to happen. Furthermore, it has been well documented that Higher Education often sets the example within fields such as sustainability, from which other institutions then follow (Puntha et al. 2015). An assertive strategy on ESD within Higher Education could therefore be the key to the solutions to many of the global issues we are faced with today, from fighting climate change to reducing poverty and social inequalities, hence contributing to reaching targets set out by the SDGs.

3 The Sustainable Development Goals

The Sustainable Development Goals (SDGs), also known as the Global Goals, came into action in January 2016 as part of the 2030 Agenda for Sustainable Development and took over from the Millennium Development Goals as the UN working framework for sustainable development. The 17 goals were developed after the final outcome of the Rio+20 Conference, in 2012 and adopted by more than 150 world leaders at the UN Sustainable Development Summit in 2015. The goals and targets aim to create change in 5 critical areas: People, Planet, Prosperity, Peace and Partnership (United Nations Sustainable Development Knowledge Platform 2017).

The multifaceted nature of the goals brings attention to the variety of themes within the concept of sustainable development, from poverty reduction and responsible consumption to gender equality and climate action. The focus on partnership for implementation of the goals is also a significant factor in achieving the goals and over the next 15 years, the SDGs will bring together stakeholders from all sectors in society to create change and transform the world.

For these reasons, the SDGs are suitable to adapt for HEIs to achieve a long-term plan for sustainable development both within their own institutions and in the wider society. Their broadness and ability to appeal to all disciplines and subjects creates opportunities for extensive strategies on sustainable development and a joint approach to the concept. By adopting the SDGs in the formal curriculum in Higher Education, less emphasis can be put on defining the sustainability concept and what it implies. Instead, focus can shift onto finding ways to achieve the goals and create a more sustainable future (Leal Filho et al. 2015).

For these reasons, NTU has chosen to adopt the SDGs as part of the formal curriculum for all courses at the University. The advantages of using the goals as a framework for sustainability, rather than just focusing on the concept in itself, has proven to be an effective way to engage the unengaged and bring the whole University together under a common theme.

4   Nottingham Trent University and ESD

ESD is often seen as a way to contribute to the fulfillment of other agendas (employability, internationalisation etc.) rather than a concept with intrinsic value. However, at NTU, ESD is starting to become a main focal point in the University-wide curriculum development, as well as in its general teaching and learning strategies. The current university Strategic Plan was launched in 2015, with the key objective focusing on *Creating the University of the Future* (Nottingham Trent University 2015). This ambitious commitment to ensure the continuous growth and improvement of the University shows an obligation to invest in the future and guarantee current and prospective students that they will gain the right skills, knowledge and attributes to succeed, no matter what challenges they might face in the future.

NTU is located in the East Midlands region of the United Kingdom. Approximately 28,500 students are currently enrolled at the university, 2600 of whom are international students and 4250 are postgraduate students. NTU is one of the most sustainable universities in the world as evidenced by a a range of awards for successfully integrating sustainability in both the estate and the curriculum. In

2016, NTU was awarded first place in the People & Planet University League[1]—the third time in recent year. It became the first university in the UK to achieve Gold in the Learning in Future Environment (LiFE)[2] accreditation in 2015. NTU have also won a number of Green Gown Awards, presented by the Environmental Association of Universities and Colleges (EAUC)[3] and recently received the Responsible Futures accreditation from the National Union of Students (NUS).[4]

Through the university-wide course review process, *Curriculum Refresh* (Simmons et al. 2016), sustainability has received a key role in the development of the formal curriculum. The *Curriculum Refresh* framework is aiming to reinterpret and refresh the curriculum and 7 elements of the framework have been developed to assess the sustainability content of the more than 640 courses at the university. This will ensure all taught courses at NTU will be in line with the objectives set out in NTU's Strategic Plan.

The *Curriculum Refresh* framework also demonstrates the university's commitment to the SDGs. All courses at the university are encouraged to explore how their disciplines can contribute to the realisation of the SDGs and how to work in collaboration across subject areas to make the goals a reality. The SDGs have proven to be a useful starting tool when talking about how multifaceted the concept of sustainable development is. By using the different goals and targets in lectures and staff development sessions, it is possible to show the different aspects of sustainable development and move away from the idea that sustainability is strictly an environmental issue. Due to this focus, the SDGs have played a central role in staff development sessions in which the goals are used as a kicking off point for further discussion on how to embed sustainable development within learning and teaching.

ESD takes a central role at NTU through the work of the *NTU Green Academy*, a team dedicated to embedding sustainable development in the formal and informal curriculum at the university. With its roots in the Higher Education Academy (HEA) initiative with the same name, the Green Academy has developed from a short-term project to a permanent team (Puntha et al. 2015). Today, the team consists of 2 full-time ESD Coordinators as well as an Academic Associate, a Team Leader and a Deputy Team Leader, all of which are part-time roles. The NTU

---

[1]The People and Planet University League is an independent league table of UK universities. It is compiled by People and Planet, a UK-based student campaigning network that ranks universities based on a selection of criteria focusing on environmental and ethical performance.

[2]The Learning in Future Environments (LiFE) index is a self-assessment tool developed for Higher Education Institutions by the Environmental Association for Universities and Colleges (EAUC) that gives HEIs an opportunity to evaluate their environmental performance and social responsibility.

[3]The Environmental Association for Universities and Colleges (EAUC) supports HEIs in the UK with environmental management and sustainability work through a range of initiatives, including the Green Gowns Award and the LiFE index.

[4]The Responsible Futures accreditation was created by the National Union of Students (NUS), to assess how well HEIs are embedding sustainability and social responsibility in the formal and informal curriculum.

Green Academy works to reach out with the ESD agenda to all schools and departments at the university through staff development, individual consultancy for academics and lecturers, and through developing innovative teaching materials. The team also leads on the Sustainability in Practice (SiP) Certificate, an optional online module exploring sustainability issues that is available for all students and staff at the institution (Molthan-Hill et al. 2015).

5 The Future Thinking Learning Room

The learning room, which was developed as a support for staff going through the *Curriculum Refresh* process, aims to increase the sustainability content in the curriculum. The sustainability related resources in the learning room are categorised by the 8 academic schools at NTU, with subcategories relating to the different departments in each of the schools. There are currently around 1000 handpicked resources, as can be seen in Fig. 1 with new resources added daily. The resources include such items as journal articles, websites, suggested books and films, ideas for tutorials, YouTube playlists, and information on how to access physical learning resources such as games and other activity based resources.

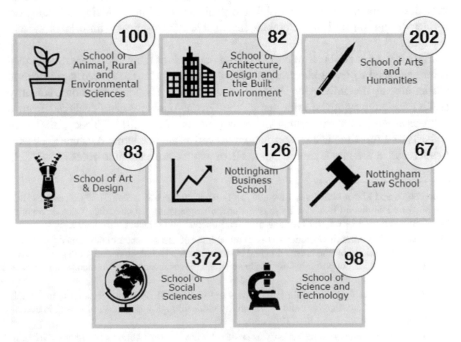

Fig. 1 Number of resources for each of the 8 schools at Nottingham Trent University. 1130 collated by 30.01.2017

Fig. 2 Number of resources for each of the sustainable development goals. 519 collated by 30.01.2017

A unique feature of the learning room is that the resources have also been mapped against the SDGs. There is a whole section of the learning rom dedicated to the SDGs focusing on how to link all 17 goals to the curriculum, as can be seen Fig. 2. This area of the site consists of some general background information about the Sustainable Development Agenda and a subsection for each SDG with resources relating to the specified goal.

In order to maximise engagement with the resources the learning room was carefully planned and laid out with the end user in mind. Moving away from the historical learning room list format it utilises embedded webpages allowing staff to move around the site intuitively and access a wide variety of resources with ease. The categorisation techniques used within this online library allow staff to hone in on particular aspects of sustainability that may be of particular relevance to their discipline, whilst also encouraging exploration into new areas of interest. The learning room also exhibits newly created case studies relating to estate and community based projects (see next section) that have been designed for teaching purposes.

The learning room helps to create a platform for dialogue with academic staff about how to best embed sustainability in the curriculum. This is not a one-way communication channel and the idea is that staff can affect the content of the learning room. Each of the schools uses an assigned Dropbox where members of staff can submit resources they think might be suitable for the learning room and that they want to share with other academics. The learning room also provides an additional mechanism for internal knowledge exchange encouraging staff to share their good practice.

The Green Academy team showcase the learning room during their monthly workshops for Course/Programme Leaders and others who are looking to integrate ESD in line with the Curriculum Refresh framework. The workshops make use of the learning room to help provide ideas for transforming learning through inclusion of material relating to the SDGs and hence facilitate sustainability literacy and other transferable skills. During the sessions, academics are given an opportunity to consider how to integrate ESD into the curriculum context in line with strategic aspirations and in ways most likely to benefit student learning. The use of the SDGs within the sessions is a useful tool for communicating the many dimensions of sustainability to academics, particularly those who are struggling with linking ESD to their own discipline.

6 Community and Estate Projects in Relation to SDGs

The Green Academy coordinated a project initiated by the TILT (The Trent Institute for Learning and Teaching) Education for Sustainable Futures group to create case studies focusing on both community volunteering projects and innovative environmental work happening on the NTU estate. Each of the case studies shows how the project relates to the various SDGs (as seen in Fig. 3). The resources are multi-disciplinary and are intended for use across a wide range of courses, throughout all 8 schools. There are many benefits to derive from these case studies, for example, they will showcase the innovative projects happening at NTU and the role of the university in the wider society. They will also create a new way of embedding sustainability into the curriculum and connect the 'subliminal' curriculum (Tierney et al. 2015) to the formal one.

Moving forward, challenges and 'wicked' problems (Newman-Storen 2014) in relation to the estate, for example packaging in relation to catering, will continue to be identified and turned into student projects as part of the core curriculum.

7 Challenges Faced

The success of this project relied somewhat on the collaboration of academics across all the schools at NTU to help provide credible and relevant resources to be featured in the learning room. As can be seen in Fig. 1, the number of resources available for each academic school ranges from school to school. This is mostly due to the varying levels of engagement and resources provided by the academics from

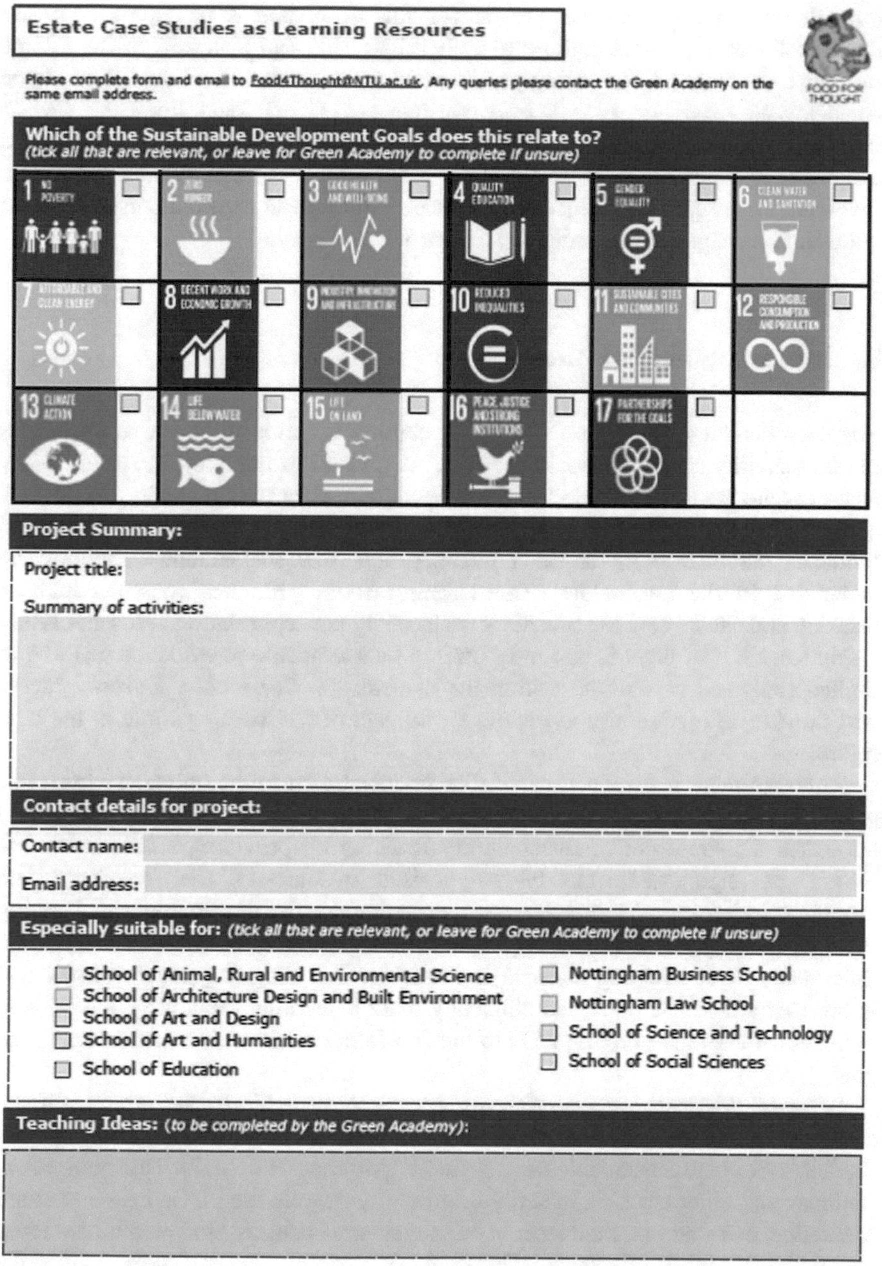

Estate Case Studies as Learning Resources

Please complete form and email to Food4Thought@NTU.ac.uk. Any queries please contact the Green Academy on the same email address.

FOOD FOR THOUGHT

Which of the Sustainable Development Goals does this relate to?
(tick all that are relevant, or leave for Green Academy to complete if unsure)

1 NO POVERTY ☐	2 ZERO HUNGER ☐	3 GOOD HEALTH AND WELL-BEING ☐	4 QUALITY EDUCATION ☐	5 GENDER EQUALITY ☐	6 CLEAN WATER AND SANITATION ☐
7 AFFORDABLE AND CLEAN ENERGY ☐	8 DECENT WORK AND ECONOMIC GROWTH ☐	9 INDUSTRY INNOVATION AND INFRASTRUCTURE ☐	10 REDUCED INEQUALITIES ☐	11 SUSTAINABLE CITIES AND COMMUNITIES ☐	12 RESPONSIBLE CONSUMPTION AND PRODUCTION ☐
13 CLIMATE ACTION ☐	14 LIFE BELOW WATER ☐	15 LIFE ON LAND ☐	16 PEACE JUSTICE AND STRONG INSTITUTIONS ☐	17 PARTNERSHIPS FOR THE GOALS ☐	

Project Summary:

Project title:

Summary of activities:

Contact details for project:

Contact name:

Email address:

Especially suitable for: *(tick all that are relevant, or leave for Green Academy to complete if unsure)*

☐ School of Animal, Rural and Environmental Science
☐ School of Architecture Design and Built Environment
☐ School of Art and Design
☐ School of Art and Humanities
☐ School of Education

☐ Nottingham Business School
☐ Nottingham Law School
☐ School of Science and Technology
☐ School of Social Sciences

Teaching Ideas: *(to be completed by the Green Academy):*

Fig. 3 Form designed by the Green Academy 2016 to capture case studies of estate projects to be used as teaching materials

the different schools. Another reason for this difference is perhaps the greater natural affinity of some disciplines to embedding ESD topics as well as the volume of course content and competing priorities in some academic schools. The Green Academy also sought feedback from various key stakeholders along the way in order to provide a resource that fitted the needs of the end user. This process posed a number of challenges due to the size of the university and number of people involved. The project was hence very time consuming and required persistence and good time management in order to achieve the end goal.

8 Impacts and Outcomes

The main benefit of the Future Thinking Learning Room is an increased awareness of sustainability across the board at NTU. As a result of this project, sustainability resources are readily available to all members of staff at the university via an easy and efficient mechanism. It also empowers members of staff in being able to influence the ESD work at the university and how sustainability can best be embedded in the curriculum. The learning room promotes cross-disciplinary thinking and shows how the broader sustainability concept relates to all schools and disciplines. Whilst there is currently limited data available to evidence this due to limited analytical capabilities within the system, the *Curriculum Refresh* process will provide an opportunity to review the impact of the learning room in the near future.

Although there is no direct calculable financial gain to the university from this project, the learning room offers a more efficient use of staff time. This easy to use hub takes the stress out of sustainability teaching by providing a comprehensive library of signposted, clearly categorised resources. The resources are non-prescriptive and aim to act as an inspiration for the development of learning ideas, in keeping with a holistic approach to ESD. For those with even more limited time, ready-made teaching material is available, including suggested seminar and tutorial activities. The improved efficiency that the learning room provides helps to break down barriers, bringing ESD to the forefront of achievable curriculum based goals.

Some of this teaching material will be made available to the public. Fifteen academics and both ESD Coordinators from NTU have published 'The Business Student's Guide to Sustainable Management' (Molthan-Hill 2017). This book offers seminars and other teaching material on how to integrate the SDGs in Accounting, Marketing, HR and other subjects in Management/Business Studies but also ideas on how to teach Systems Thinking, Corporate Peace-making and the Crowdsourcing of sustainable solutions, which would be of interest to lecturers and students from other disciplines.

The response to the Future Thinking Learning Room has been overwhelmingly positive. The innovative use of the mechanisms within the university's virtual learning environment (VLE) has resulted in this learning room being used as an example of good practice throughout NTU. It has been showcased at several university wide events, such as the annual Course Leader Conference, the eLearning Showcase 2016 and the Annual Learning and Teaching Conference. The two Education for Sustainable Development Coordinators who designed the learning room were the first people in the institution to receive a 'Creating Engaging Course Materials' badge as recognition for their work on this project.

The staff development workshops have also received very positive feedback from participating members of staff, who appreciated the clarification of the sustainability concept and the Future Thinking framework as well as the opportunity to share good practice and discuss ESD with colleagues from other schools and disciplines. A number of comments have also been received from individuals who gave positive feedback on the style and accessibility of the learning room as well as the breadth of content. For this reason the learning room has led the way for similar projects, for example, the parallel elements of Curriculum Refresh such as *Employability* and *Internationalisation*.

The Future Thinking Learning Room has increased the awareness in both the staff and student body of sustainability related projects across the university. For example volunteering opportunities, green estate initiatives and sustainability research. The learning room has also utilised the institutional *Curriculum Refresh* as an opportunity to embed sustainability into 'business as usual'—the related workshops are run in SCALE-UP (Student Centred Active Learning with Upside Down Pedagogies) active learning style (Beichner et al. 2007) and cross-reference ESD with other areas of the *Curriculum Refresh* such as student research and internationalisation to ensure a more holistic approach going forward.

9 Further Engaging Students and Staff with the SDGs

In addition to *Curriculum Refresh* and the supporting learning room and workshops the Green Academy have utilised the SDGs to further the ESD agenda at NTU in a number of other ways. The Sustainability in Practice (SiP) certificate is an optional, online module available to all staff and students across NTU (Molthan-Hill et al. 2015). This popular certificate, developed by the Green Academy has been running since 2013 and was shortlisted for the Green Gown Awards in 2014. In 2016/17 two new themes were added: Energy and Clothing, in addition to the original theme of Food. The SDGs are introduced in the third session of the certificate that focuses on giving students an interdisciplinary understanding of the sustainability concept and realising the importance of collaboration to overcome global issues. The goals are used as an example of how various disciplines and subject areas can work together to address sustainability issues. Students are also asked to reflect on their own role in realising the SDGs and how to work with other disciplines and sectors in order to achieve this. Due to its popularity and positive feedback from student and staff participants, the SiP certificate is now part of NTU's Higher Education Achievement Record (HEAR) scheme for its undergraduate completers.

More recently, the SDGs were used in order to capture examples of ESD in action at Nottingham Business School. The form, as shown in Fig. 4, was disseminated during departmental meetings and proved to be an efficient way of

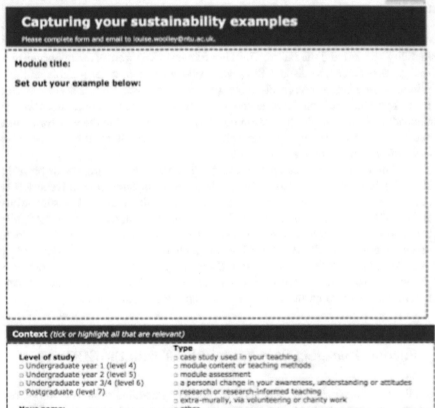

Fig. 4 Form designed by the Green Academy in cooperation with Adrian Castell and Louise Woolley from Nottingham Business School, to capture ESD examples

mapping where sustainability issues are addressed throughout the school, for example within teaching and assessment, via research or even through volunteering or charity work.

10 Conclusion

The Future Thinking Learning Room provides academics with a platform to share pockets of good practice and experiences within the field of ESD, as well as a quick and easy way to find sustainability and more specifically SDGs related resources to embed in their teaching. The learning room encourages cross-collaboration between disciplines, life-long learning, and global citizenship—all vital parts of the NTU Strategic Plan. It also encourages university staff to embed sustainability across the entire course curriculum, rather than just sporadically in one or two modules.

By focusing on the SDGs, a wider understanding of the sustainability concept is provided. Academics get the opportunity to embed themes such as poverty reduction, gender equality and quality education in addition to environmental themes and give students a more adequate picture of the sustainability concept.

Evaluation will be undertaken as part of the *Curriculum Refresh* impact review. In the future, the learning room will continue to provide a wide range of resources for academics to use as inspiration for teaching sustainability. The aim is to create more opportunities to include teaching materials created by members of staff, to curate the good practice that exist at the university, and to develop the connections to the SDGs section further. The latest project is a blog, *NTU & the Sustainable Development Goals*, with the aim to share knowledge of and celebrate the many achievements of students and staff at the university who are working towards achieving the SDGs. The blog, launched on 1st February 2017, is publicly available here: https://ntu-sdgs.blog/.

11 Recommendations for a Similar Approach

This project could be replicated in other HEIs if the technical infrastructure is in place through a fully functioning VLE or similar approach. By creating a platform like this, information on the SDGs can be shared to a very wide audience across the university. It is advisable to seek feedback from colleagues across the institution to make the resource library as user-friendly as possible. This will also give legitimacy to the resources, since they will have been approved by experts in each area.

At NTU, the Future Thinking Learning Room and the introduction of the SDGs on a university-wide level was helped by the Curriculum Refresh project, which drew attention to the learning room as a support for academics going through this process. It is recommended that others hoping to adopt a similar approach should try to coincide the introduction of the SDGs alongside or within a change that is already happening, for example a course review, a quality assessment, staff development or a larger scale sustainability project. It is recommended to work with academic quality departments to ensure that the themes become a key area in quality assurance processes and course or module reviews. This process might require different approaches in different institutions.

Further information can be obtained from the Green Academy Team at NTU: GreenAcademy@ntu.ac.uk.

References

Beichner RJ, Saul M, Abbott D, Morse JJ, Deardorff DL, Allain RJ, Bonham SW, Dancy MH, Risley JS (2007) The student-centered activities for large enrollment undergraduate programs (SCALEUP) project. Research-Based Reform of University Physics. [Online]. Available at http://www.compadre.org/per/items/detail.cfm?ID=4517. Accessed 14 Feb 2017

British Council (2014) UK Alumni Leading the World. Available https://www.britishcouncil.org/organisation/press/uk-alumni-leading-world. Accessed 17 Jan 2017

Leal Filho W, Manolas E, Pace P (2015) The future we want: key issues on sustainable development in higher education after Rio and the UN decade of education for sustainable development. Int J Sustain. High Educ 16(1):112–129

Lipscombe B, Burek C, Potter J, Ribchester C, Degg M (2008) An overview of extra-curricular education for sustainable development (ESD) interventions in UK universities. Int J Sustain High Educ 9(3):222–234

Molthan-Hill P (ed) (2017) The business student's guide to sustainable management: principles and practice, 2nd edn. Greenleaf/PRME book series. Greenleaf, Sheffield

Molthan-Hill P, Dharmasasmita A, Winfield F (2015) Academic freedom, bureaucracy and procedures: the challenge of curriculum development for sustainability. In Leal Filho W, Davim JP (eds) Challenges in higher education for sustainability. Springer, Cham, Switzerland, pp 199–215. ISBN:978-3-319-23705-3

Newman-Storen R (2014) Leadership in sustainability: creating an interface between creativity and leadership theory in dealing with "Wicked Problems". Sustainability 6:5955–5967

Nottingham Trent University (2015) Creating the university of the future. Available https://www4.ntu.ac.uk/strategy/. Accessed 17 Jan 2017

Puntha H, Molthan-Hill P, Dharmasasmita A, Simmons E (2015) Food for thought: a university-wide approach to stimulate curricular and extracurricular ESD activity. In: Integrating sustainability thinking in science and engineering curricula. Springer, Berlin, pp 31–47

Sachs JD (2012) From millennium development goals to sustainable development goals. Lancet 2012(379):2206–2211. doi:10.1016/S0140-6736(12)60685-0

Simmons EA, McNeil J, Lamb S (2016) Curriculum Refresh: a whole-institution approach to reviewing the curriculum. Trent Institute for Learning & Teaching. Nottingham Trent University, Nottingham

Tierney A, Tweddell H, Willmore C (2015) Measuring education for sustainable development: experiences from the University of Bristol. Int J Sustain High Educ 16(4):507–522

Tilbury D (2015) Higher education for sustainability. Routledge Handbook of Higher Education for Sustainable Development

UNESCO (2017) Education for sustainable development. http://en.unesco.org/themes/education-sustainable-development. Accessed 27 Jan 2017

United Nations Sustainable Development Knowledge Platform (2017) Transforming our world: the 2030 Agenda for Sustainable Development. https://sustainabledevelopment.un.org/post2015/transformingourworld. Accessed 27 Jan 2017

Wals AE (2014) Sustainability in higher education in the context of the UN DESD: a review of learning and institutionalization processes. J Clean Prod 62:8–15

Author Biographies

Jessica Willats B.Sc. (Hons) is one of two Education for Sustainable Development (ESD) Coordinators working in the NTU Green Academy, working to embed sustainable development across all academic schools at the university. Graduating from the University of Bristol with a B.Sc. in Environmental Geoscience and a concern about climate change, she is studying for an M.Sc. in Atmospheric Science at the University of Gothenburg realizing the impact education can have on behaviour change and the ability of young people to take action. Jessica co-designed and runs the NTU Green Academy's 'ESD-Future Thinking Learning Room', which is an innovative online resource library for staff. Jessica also designed and runs the new Green Academy blog, 'NTU & the Sustainable Development Goals', available at https://ntu-sdgs.blog/.

Lina Erlandsson MA, B.Sc. (Hons) is an Education for Sustainable Development Coordinator at Nottingham Trent University. In her role, Lina works to embed sustainable development themes across all schools and departments at NTU by providing expertise and consulting academics on ESD. Lina moved to the UK from Sweden in 2014 after graduating from the University of Gothenburg with a B.Sc. in Political Science. She went on to complete an MA in Politics at NTU, where she focused on how policies on both the national and international level can support sustainability and how public attitude towards the concept is shaped. Lina also co-designed and runs the NTU Green Academy's 'ESD-Future Thinking Learning Room' and recently co-led NTU and NTSU's submission for the Responsible Futures accreditation, which the partnership received at the end of last year.

Dr. Petra Molthan-Hill PFHEA, Ph.D., MBA, MDiv leads the Green Academy at Nottingham Trent University, developing curricular and extra-curricular activities in Education for Sustainable Development. In 2016, Petra won the Sustainability Professional Award in the Green Gown Awards. Petra is also the Academic Lead for PRME in Nottingham Business School. Since 2010 Petra has led the Greenhouse Gas Management Project at NBS, which won the Guardian University Award for 2015 in Business Partnership together with NetPositive Ltd. Petra is also the Co-Chair of the United Nation's PRME Working Group on Climate Change and Environment, the Chair of the TILT group 'Education for Sustainable Futures' at NTU and the Lead for the Pedagogical Research of the Responsible and Sustainable Business Lab at NBS.

Aldilla Dharmasasmita FHEA, M.Sc., BA (Hons) is the Academic Associate for the Green Academy at Nottingham Trent University. She co-developed and now manages the Sustainability in Practice Certificate, a project shortlisted for The Green Gown Award 2014. She is also an Associate Lecturer with Nottingham Business School (NBS), specialising in Business Ethics and Sustainability Strategy in Business. She completed her Masters in Corporate Social Responsibility after being awarded a full-tuition scholarship at The University of Nottingham. She is currently a Ph.D. candidate with NBS researching Education for Sustainable Development (ESD) competencies.

Prof. Eunice Simmons B.Sc. (Hons), M.Sc., PGCE, Ph.D., FRSB, FRSA, PFHEA is Deputy Vice Chancellor at Nottingham Trent University where she is the strategic lead for Education for Sustainable Development. She has degrees in biological sciences, soil conservation and forest ecology and has managed large land-based and environmental faculties. Her leadership of the *Curriculum Refresh* process at the University has increased the opportunities for all NTU students to engage with the wider sustainability agenda.

Sustainability Curriculum in UK University Sustainability Reports

Katerina Kosta

Abstract One of the major barriers to incorporating sustainability in the HE curriculum is its absence from the university sustainability strategy, the annual reflection of which is the annual sustainability report. While strategies specify targets, reports record what has already been achieved. In that respect, reports function as internally created reviews of universities' sustainability activity. Various reviews of sustainability teaching activity have taken place in the UK HE sector. The current study attempts to explore formal sustainability teaching provision exclusively through HEIs' annual sustainability reports. The sample consists of the most recent, whole-institution sustainability reports issued by UK HEIs from 2016 to 2018. An exploratory content analysis identifies sustainability curriculum coverage patterns, using a coding frame based on the STARS framework. Findings suggest that of the 167 UK HEIs 4% report on their sustainability curriculum provision comprehensively. The findings might be of interest to sustainability professionals in the reporting or the curriculum provision end. The study hopes to encourage wider coverage of sustainability curriculum provision in HE sustainability reports.

Keywords Sustainability reporting · Sustainability curriculum · Sustainability in higher education (SHE)

1 Introduction

While many universities have made impressive progress with regard to greening their campuses, embedding sustainability in the curriculum is often seen as a harder challenge (Ryan and Cotton 2013; Sterling 2012; Tilbury and Ryan 2011). One of the most important barriers for the implementation of sustainability in the

K. Kosta (✉)
Faculty of Humanities and Social Sciences, Oxford Brookes University, Headington Campus, OX3 0BP Oxford, UK
e-mail: katerina.kosta-2016@brookes.ac.uk

© Springer International Publishing AG 2018
W. Leal Filho (ed.), *Implementing Sustainability in the Curriculum of Universities*,
World Sustainability Series, https://doi.org/10.1007/978-3-319-70281-0_6

curriculum is its absence from the institutional strategy. Strategies set forth the agenda of each institution and if the curriculum is included in the institutional strategy, it is likely that it will be prioritised and promoted. Being mostly future oriented, strategies contain projected aims and goals, while the annual reflection of an institution's sustainability strategy is the sustainability report, which describes not what might happen in the future but has already been achieved. In that respect, sustainability reports function as internally created reviews of institutions' sustainability activity.

Reviewing institutions' sustainability activity and particularly their sustainability curriculum provision seems to be in high demand by various UK higher education stakeholders. First of all students, the majority of whom believe that 'sustainable development is something which university courses should actively incorporate and promote' (Drayson and Taylor 2015: 634). Recording and promoting institutions' sustainability curriculum provision may thus improve the experience of students, who have in a way become shareholders of universities through the tripling of their tuition fees. Moreover, the People and Planet University League, an influential environmental performance ranking of UK universities includes heavily weighted indicators for reporting on sustainability curriculum provision. In Wales, sustainability curriculum reporting is mandated by the government, which encourages the institutionalisation of sustainability in higher education (Glover et al. 2011). In the rest of the UK, the Quality Assurance Agency (QAA) for higher education has complemented its quality audits with an ESD (Education for Sustainable Development) component (QAA 2014). Thus, recording sustainability curriculum provision in an institution's sustainability report might satisfy the needs of the above internal and external higher education stakeholders.

Sustainability reports are an emerging trend in higher education (Ceulemans et al. 2015b; White 2014) and there is limited guidance on their format and content. Of the existing guidelines, the Sustainability Tracking Assessment and Rating System (STARS) by AASHE is one of the most systematic and complete. STARS recommends reporting all 'sustainability courses' and 'courses that include sustainability' giving numbers, titles and content descriptors (STARS 2016: 31). It is stated that conducting an inventory of academic offerings provides an important foundation for advancing sustainability in the curriculum (STARS 2016). While STARS originated from and mainly provides for the American and Canadian HE sector, the Learning in Future Environments (LiFE) index by the EAUC is a UK self-assessment and reporting system. LiFE recommends that ESD curriculum provision is embedded within the institutions' sustainability strategy while being routinely monitored and evaluated (LiFE 2017). This routine monitoring can take place through the university's annual sustainability report. Guidelines on Sustainability Reporting have also been issued by the Green Gown Awards, which recognise excellent sustainability performance in the tertiary education sector. In 2016, the new category of Sustainability Reporting was introduced and monitoring sustainability performance in *core* functions of an organisation is recommended (Green Gown Awards 2017). The recommendation of comprehensive reporting of an institution's sustainability curriculum by leading HE sustainability consultancies (AASHE, EAUC) is coupled with the importance of the principle of materiality for sustainability reporting (PwC 2016; KPMG 2012; GRI 2015). The most material function of universities is education and sustainability education is 'the greatest contribution HE can make to sustainable development' (HEFCE 2014). As a consequence, sustainability education and the curriculum are expected to be a basic component of university sustainability reporting (Ceulemans et al. 2015b; Fischer et al. 2015). In response, the current study explores how universities include sustainability curriculum in their sustainability reports. It does so by answering three research questions:

1. What percentage of UK universities publish a whole-institution sustainability report?
2. What is the up-to-date status of the reports?
3. To what extent is sustainability curriculum provision covered in the reports?

Answering the first and second research questions will result in a descriptive analysis of university sustainability reports as an emerging practice among UK HEIs. The third research question will explore the coverage of the material aspect of sustainability curriculum provision. The paper is structured as follows: First, a literature review will describe previous research on sustainability curriculum baselining and sustainability reporting. Next, the study's design—sample, data collection and data analysis—will be explained, followed by the findings, a discussion of the findings and the conclusion.

2 Literature Review

2.1 Sustainability Curriculum Audits

To identify and record sustainability curriculum provision, certain universities in the UK have organised institutional curriculum reviews. These are a relative new development and three recent reviews conducted at an institutional level are presented here. In 2015, Wyness and Sterling attempted to record sustainability curriculum provision at Plymouth University by distributing a review tool to academic leads, asking them to self-evaluate the degree of sustainability in their courses. For the construction of the tool and in order to define what constitutes sustainability teaching they followed Tilbury and Wortman's conceptualisation which recommends thematic 'entry points' around the triple bottom line concept of sustainability (Tilbury and Wortman 2008). The review included courses that address the environment, economy and equity aspects of sustainability in combination *or separately*.

At the University of Bristol Tierney et al. (2015) perform a sustainability curriculum review, based on the catalogue of taught courses and Key Information Sets (KIS) on student experience provided by HEFCE. They use the UNESCO definition of ESD, which covers four areas: socioeconomic justice, cultural diversity, human rights of future generations and restoration of the Earth's eco-systems (Tierney et al. 2015: 508). Finally, Bloemen (2013) at the University of Edinburgh conducted a baselining of undergraduate courses[1] related to Sustainability and Social Responsibility (SRS), a concept that encompasses environmental and social themes. Looking at the three reviews, it can be noticed that each university adopted a different definition of sustainability and structured its audit accordingly. All three audits recognised that their basic aim was to inform and help shape the institution's sustainability strategy, the annual reflection of which is the sustainability report. The current study explores how sustainability curriculum provision is covered by universities' annual sustainability reports.

While the sustainability curriculum reviews mentioned above took place at an institutional level, sustainability curriculum baselines have also been attempted at a national level. In Wales, the Higher Education Funding Council for Wales (HEFCW) provided funding for universities to conduct audits of sustainability curriculum provision. The Sustainability Tool for Auditing University Curricula in Higher Education (STAUNCH) was selected by the Welsh authorities as the auditing method for that purpose (Glover et al. 2011). This tool scores sustainability course content across three categories: economic, environmental and social. By adopting this triple bottom line model of sustainability, STAUNCH recognizes courses that address any of the three aspects in combination *or separately*. For instance, a module that focuses exclusively on Productivity or the Gross National Product would be recognised as a sustainability course. In contrast to this compartmentalized treatment of sustainability, the authors of a national sustainability curriculum review commissioned by HEFCE identify as sustainability courses the ones that combine at least two of the three aspects of the triple bottom line model (HEFCE 2008: 88). The project uses the following definition of sustainability teaching: 'teaching that contains a significant element of work related to *either or both* the natural environment and natural resources, PLUS a significant element of work related to either or both of economic or social issues' (HEFCE 2008: 89). This integrative '*plus*' requirement echoes the '*at the same time*' rule for designing sustainability curricula by the Forum for the Future, where it is expected that the environmental, economic and social aspects are *simultaneously* present in the design of sustainability courses (Parkin et al. 2004). The Welsh and English national audits described above were conducted using externally defined criteria, which may not provide for the heterogeneous treatment of sustainability within each institution. Annual sustainability reports on the other hand are issued internally

---

[1]The scoping exercise was carried out at the Colleges of Humanities and Social Sciences (CHSS) and Science and Engineering (CSE).

by each university and could provide a self-defined baseline of each institution's sustainability curriculum provision. An exploration of how UK universities' annual sustainability reports present institutions' sustainability curriculum provision has not yet taken place and the current study aims to address this gap.

2.2 Sustainability Reporting in Higher Education

Due to the relatively recent appearance of sustainability reporting in HE education, research literature on the topic is rather limited (Beveridge et al. 2015; Ceulemans et al. 2015a, b). Even more limited is literature on universities reporting on their sustainability curriculum provision. Of this limited research, five studies are highlighted below as they adopt a similar approach. In 2011 Lozano examined university sustainability reports that followed the GRI guidelines and found that only 6% of them contained references to educational dimensions. In the same year, Fonseca et al. analysed the sustainability reports of the 25 biggest Canadian universities to discover that curriculum references exist in a quarter of the documents. In 2014 White explored sustainability plans in the US seeing 81% of them containing references to the curriculum. A year later, Lidstone et al. (2015) analysed sustainability plans in Canadian universities to find that 86% contain goals within the domain of education. Finally, in 2016 Vaughter et al. looked at sustainability plans in Canadian FHE institutions to see that 12% contain references to the curriculum. In all five studies, sustainability curriculum is under-reported, since it does not appear in 100% of the documents explored (Table 1).

It can be noticed that each study analyses different types of documents. Lidstone et al. (2015) and White (2014) focus on sustainability *plans* with an institution wide coverage while Vaughter et al. (2016) explore both *plans* and *policies* (Table 1). Fonseca et al. (2011) and Lozano (2011) on the other hand analyse sustainability *reports* as stand-alone documents (Table 1). While plans and policies reflect potential sustainability activity, reports record what has already been achieved. The present study will focus on sustainability *reports* which adopt a whole institution approach and which have been published in the past four academic years (2012–2016). In doing so it will also fill a geographic gap in the literature as this type of study has taken place in the US and Canada but not in the UK.

Table 1 Previous studies on sustainability reporting in higher education

	Vaughter et al. (2016)	Lidstone et al. (2015)	White (2014)	Fonseca et al. (2011)	Lozano (2011)
Location	Canada	Canada	US	Canada	Worldwide
Disclosure analysed	Plans and policies	Plans	Plans	Reports	Reports
Curriculum references (%)	12	86	81	25	6

3 Research Design

3.1 Sample

The present study focuses on UK universities receiving public authority funding and being legally registered as a Higher Education Institution (HEI) (n = 167), as listed on the Higher Education Statistical Agency (HESA) website in January 2017. The selection criteria for the sample are highlighted below:

1. The most recent, whole-institution sustainability reports issued by UK universities in the past four years (2012–2016) are included in the sample.
2. Disclosure documents with a future orientation, projected aims and targets like *Sustainability Strategies, Policies, Missions, Visions* and *Plans* are not part of this study, which focuses on reports of what has already been achieved. This choice is influenced by a fairly consistent finding in studies of plan implementation of low levels of correlations between the specifics of plans and ensuing development (White 2014; Vaughter et al. 2016).
3. Reporting documents that focus on a specific area like *Environmental Management Systems* (EMSs), *Carbon Management Plans* or *Learning and Teaching Strategies* are not included unless they adopt a whole institution approach.
4. It is acknowledged that multiple, alternative sustainability reporting venues exist like web-pages, newsletters, YouTube videos, twitter and Facebook updates. While the contribution of these alternative forms of reporting is valuable, they are beyond the scope of this study, which attempts an elementary census of whole-institution sustainability reports as an emerging practice among UK HEIs.

3.2 Data Collection

To identify the sustainability report of each university, the following steps were followed. The name of the university alongside the term 'sustainability' was entered in the Google search engine to locate the university's sustainability website, where links to the reports or to a 'policies' section is usually provided. In case this search did not yield any results, the institutional website was searched using the keywords 'sustainability/environmental' and 'report/statement'.

3.3 Data Analysis

The analysis of the reports proceeded in three phases, in order to address the three research questions. To highlight the fact that specific examples are used solely for research purposes, university names have been substituted by a list number. Yet, since the information analysed is publicly available on institutional websites, the HESA list is provided in Appendix A.

The first phase of the analysis identified the existence (yes/no) of a sustainability report fulfilling the data collection criteria. The second phase explored the academic year the report was covering (2012/13, 2013/14, 2014/15, 2015/16) under the assumption that providing timely and up-to-date information constitutes an optimal sustainability reporting practice. The third phase recorded the extent of formal curriculum provision coverage. This last variable was operationalised using the following coding frame:

Code 1 = No mention of sustainability curriculum in the report.
Code 2 = Single reference to sustainability curriculum not elaborated further.
Code 3 = More than single reference, giving examples of some courses on offer but in total these courses constitute less than 50% of the total provision.
Code 4 = Comprehensive coverage of the sustainability curriculum, making reference to more than 50% of the total provision.

The coding frame was based on the STARS 'AC1: Academic Courses' guideline, which recommends recording all an institution's 'sustainability courses' or 'courses that include sustainability' in order to achieve the highest reporting score (STARS 2016: 32).

An exploratory content analysis was performed. At first, each document was searched using the following keywords: *curriculum, course, degree, module, education* and *teaching*. Since keyword searches do not provide for alternative terminology, a subsequent close reading of each document was performed to more efficiently determine the coverage of sustainability curriculum according to the coding criteria. While the absence of reference (Code 1) and the existence of a single reference (Code 2) were easy to identify, determining whether the reports covered more (Code 4) or less (Code 3) of the sustainability courses on offer was more difficult to establish. To optimize the accuracy of the results under Codes 3 and 4, an inventory of all sustainability courses was created for each of the universities concerned based on catalogues of undergraduate and postgraduate taught sustainability courses. If the report made references to less than 50% of these courses, it was coded as 3 while if it covered more than 50% of the courses it was placed under Code 4. To make the comparison between Codes 3 and 4 possible, an inventory was created for each of the universities concerned, based on catalogues of undergraduate and postgraduate taught courses as found in the UCAS (2017) website. UCAS stands for University College Admissions Service and is an online database that contains all undergraduate and postgraduate taught courses in the UK HE sector. The database was searched using the keywords sustainable, sustainability and environment which also turned results for environmental. It is acknowledged that these lists are not exhaustive inventories of all, existing sustainability curriculum, as sustainability courses exist that do not feature these keywords in their titles. Yet, this method captures in a replicable manner the majority of the available sustainability courses offered by UK HEIs, using an official, publicly available dataset.

4 Findings and Discussion

This section will present the findings from the three phases of the analysis and relate them to previous literature. The first phase of the analysis explored how many universities issued an annual sustainability report in the period defined. It appears that out of the 167 universities in the sample, 46 have published a whole institution sustainability report in the past four academic years (if multiple sustainability reports had been issued, the most recent one was selected). The remaining universities either published one-dimensional reports (usually focusing on estates and operations) or recorded sustainability performance information exclusively on their websites. A similar trend has been identified by Fonseca et al. (2011), who mention that less than a third of the 25 largest Canadian universities publish a sustainability report. In a similar vein, a study of all Canadian further and higher education institutions revealed that 50% of them published a formal sustainability reporting document (Beveridge et al. 2015). Thus, the current study's first finding is aligned with findings of previous research, which suggest that sustainability reporting in higher education is an emerging practice (Lidstone et al. 2015; Lozano 2011; White 2014).

The fact that sustainability reporting is an emerging area in higher education is further reflected in the diversity of the report labelling. In the current sample, the standard term 'Sustainability Report' was the title of five documents, while the rest adopted different terms, common examples of which can be found in Table 2. This diversity in titling has also been identified by Lidstone et al. (2015) in Canada and White (2014) in the US, who discover various titles for the documents in their samples. Hahn and Kühnen (2013) claim that it is the voluntary nature of

Table 2 Example of diverse sustainability report labelling	Title heterogeneity in the sample
	Sustainability Report
	Sustainability Annual Report
	Sustainability Review
	Sustainability Highlights Report
	Sustainability and Environmental Policy Annual Report
	Sustainable Thinking: Report on Sustainability and the Environment
	Environmental Policy Action Plan Annual Review
	Environmental and Sustainable Development Strategy Update
	Environment and Sustainability Annual Report
	Annual Report on Sustainability Performance
	Energy and Sustainability Report

Table 3 The most recent academic year covered by the reports

Academic year	Percentage of reports covering each academic year (%)
2012–13	4
2013–14	13
2014–15	39
2015–16	44

sustainability reporting that has led organisations experimenting with how they disclose information and subsequently how they label the reports.

Due to this heterogeneity in titling, deciding which documents would be included in the sample did not depend solely on their title but also on whether the content fulfilled the data collection criteria. For instance, university No 4 issues an extensive document titled *Environmental Management System Annual Report*, which covers a variety of issues including *Legislation* and *Annual Performance 2013/14*. The *Annual Performance 2013/14* part acts as whole-institution sustainability report and this document was included in the sample, despite the Environmental Management System title. Institution No 100 split their *Environmental and Sustainable Development Strategy Update* in two parts, the first part *Review of recent achievements* describes what has been achieved so far while the second part presents *Objectives and targets for further improvement*. This document was included in the sample as the first part serves as a sustainability report and fulfils the data collection criteria. Institution No 104 include their sustainability report in their *Annual Report* while also providing a separate, stand-alone sustainability report. In such cases the more comprehensive of the two documents was included in the sample. University No 164 devotes a big section of their *Sustainability Strategy* to their past sustainability achievements across the institution. This document was also included as it fulfilled the data collection criteria, despite the fact that it was labelled 'strategy'.

In the second phase of the analysis the up-to-date status of the reports was explored. It appears that 44% of the reports were updated to the most recent

Fig. 1 Sustainability curriculum provision coverage in university sustainability reports

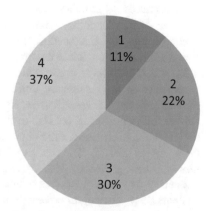

academic year 2015–2016, while 39% covered the previous year (2014–2015) and 4% dated back to 2012–13 (Table 3).

These findings echo the findings of Lock and Seele (2016) on CSR reporting in Europe, where similar patterns in the up to date status of the reports were identified. This diversity can be attributed to companies having different publication routines, with some publishing a report every year while others bi- or triennially (Lock and Seele 2016: 189). Timeliness of the reported content is seen as contributing to the quality of the reports.

The third phase of the analysis identified the sustainability curriculum provision coverage achieved by the reports (Fig. 1). It was found that 37% of the reports in the sample contain comprehensive coverage of the university's sustainability curriculum provision (Code 4). Extensive reference to the curriculum, yet without covering the majority of the sustainability courses on offer is found in 30% of the reports (Code 3). Reports with single reference to sustainability curriculum make up 22% of the documents (Code 2). Phrases typical of this highlight approach include 'embed sustainability across the curriculum', not followed by any elaboration or explanation. The remaining 11% of the reports do not refer to the curriculum (Code 1), even though they adopt a whole institution approach addressing other aspects of sustainability like research.

Having 33% (Codes 1 and 2) of the sustainability reports containing no or little reference to the curriculum, compares favourably with results from previous studies, where it is seen that sustainability curriculum is inadequately reported. Fonseca et al. (2011) found that references to sustainability curriculum were 're-stricted and elusive' (35) while Vaughter et al. (2016) noted that in the documents of their sample, discussions of sustainability in the curriculum were vaguely operationalised. Similarly, Lidstone et al. (2015) reported that there is little dis-cussion on how sustainability is to be integrated in the curriculum or what type of learning outcomes are to be achieved. Thus, the coverage of sustainability cur-riculum provision is seen as inadequate across previous studies.

It might be of interest to note that the majority of reports coming from institu-tions with a *Responsible Futures* accreditation appear to cover sustainability cur-riculum provision comprehensively (Codes 3 or 4). *Responsible Futures* promotes the integration of sustainability in the curriculum of tertiary education institutions and two of its mandatory criteria are the inclusion of sustainability in the institu-tion's strategy and the creation of a sustainability curriculum baseline (Responsible Futures 2017). Thus, it is not a surprise to see a reciprocity between the existence of a *Responsible Futures* accreditation and comprehensive coverage of sustainability curriculum in the reports. A similar trend can be observed with institutions par-ticipating in the *UI Green Metric*, a global ranking of universities according to their sustainability performance. One of the *UI Green Metric* criteria asks for the ratio of their sustainability curriculum courses compared to the number of total courses (UI Green Metric 2017). The reports of all institutions that had participated in the *UI Green Metric* presented wide coverage of their sustainability curriculum provision (Levels 3 or 4). A positive correlation appears to exist between comprehensive sustainability curriculum coverage and participation in a sustainability assessment system entailing sustainability curriculum baselining. This finding is aligned with previous literature supporting HE sustainability assessment standards as ways of systematizing and enhancing university sustainability performance and reporting (Fischer et al. 2015; Lambrechts 2015; Ramos and Pires 2013; Disterheft et al. 2012).

In the same vein and looking back at the studies of the Literature Review section (Table 1), one may notice that the studies of STARS created reports (Lidstone et al. 2015; White 2014) record greater percentages of sustainability curriculum coverage compared to studies of non-STARS reports (Fonseca et al. 2011; Lozano 2011). This might be explained by the fact that STARS provides a standardised procedure for reporting on sustainability curriculum, through its Academic Courses (AC1) criterion which encourages institutions to record all taught courses related to sustainability (STARS 2016: 31). Standardisation of information is seen as positively contributing to report quality and credibility (Lock and Seele 2016). Moreover, if the reports do not cover the whole spectrum of sustainability curriculum provision, they are considered insufficient as tools for informing and influencing decision-making (Fonseca et al. 2011). Finally, comprehensive coverage of sustainability curriculum provision by the annual sustainability report can help institutions meet requirements unique to the UK HE sector, like the People and Planet University League, the QAA reviews and in the case of Wales, government regulatory requirements.

5 Limitations

A limitation of the study stems from the fact that it does not explore web-based sustainability reporting. Certain universities not only maintain well-informed and comprehensive sustainability websites but also comprehensively display their undergraduate and postgraduate sustainability courses, in some cases even categorising them by department. Web-based reporting bears great potential (Dade and Hassenzahl 2013) and might become the norm in the future, which turns this limitation into a recommendation for future research. Moreover, having explored a single aspect of sustainability reporting does not allow generalizability of the findings to HE sustainability reporting more broadly. This limitation also justifies the misalignment between this study's records of the existence / non-existence of sustainability reporting and HESA's records of the existence/non-existence of environmental reporting (HESA 2017).

6 Conclusion

The current census of whole-institution sustainability reports among UK HEIs contributes to the under-researched area of university sustainability reporting. Since this type of study has taken place in the US and Canada but not in the UK, the research also covers a geographical gap in the literature. The study argues that the existence of a comprehensive, whole-institution sustainability report is a method of in-house benchmarking of sustainability curriculum provision, which can help universities better communicate and promote their sustainability courses to various stakeholders. Findings indicate that 46 of all UK universities (n=167) issue a whole institution sustainability report and of these 7 cover their sustainability curriculum provision in a comprehensive way. According to the STARS guidelines, recording all sustainability courses constitutes optimal practice in university sustainability reporting. Thus, 7 out of the 167 UK universities perform what is considered by STARS optimal sustainability reporting. The findings are consistent with previous research, where sustainability curriculum is seen as not adequately covered by university sustainability disclosures. A single type of university sustainability disclosure has been analysed, which means that results cannot be generalized to other types of sustainability reporting. As highlighted in the Limitations section, the study does not address web-based reporting with this area providing fertile ground for future research.

According to Evans et al. there are 0.5 million students studying *sustainability related courses* in the UK, who could offer 42.808 years of research time per year (2015: 1). Recording and promoting these courses across multiple channels could further advance the sustainability agenda , granting universities' contribution to sustainability more visibility. In the UK HE sector, reporting on the institution's sustainability curriculum achievements can be aligned with the Teaching Excellence Framework (TEF), which under its *Student Outcomes and Learning Gain* indicator specifies equipping graduates with attributes that allow them to make 'a strong contribution to society, economy and the environment' (TEF 2016: 23). There is thus a demand to equip students with sustainability skills. Part of universities' response to this demand can be the comprehensive reporting of their sustainability curriculum provision. It is hoped that the findings of this study will make a contribution to this direction and encourage more comprehensive reporting of universities' sustainability curriculum provision.

Appendix A

List of UK higher education providers (Higher Education Statistics Agency)

1. Aberystwyth University
2. Anglia Ruskin University
3. Aston University
4. Bangor University
5. Bath Spa University
6. Birkbeck College
7. Birmingham City University
8. Bishop Grosseteste University
9. Bournemouth University
10. Brighton and Sussex Medical School
11. Brunel University London
12. Buckinghamshire New University
13. Canterbury Christ Church University
14. Cardiff Metropolitan University
15. Cardiff University
16. Conservatoire for Dance and Drama
17. Courtauld Institute of Art
18. Coventry University
19. Cranfield University
20. De Montfort University
21. Edge Hill University
22. Edinburgh Napier University
23. Falmouth University
24. Glasgow Caledonian University
25. Glasgow School of Art
26. Glyndŵr University
27. Goldsmiths College
28. Guildhall School of Music and Drama
29. Harper Adams University
30. Heriot-Watt University
31. Heythrop College
32. Hull York Medical School
33. Imperial College of Science, Technology and Medicine
34. King's College London
35. Kingston University
36. Leeds Beckett University
37. Leeds College of Art
38. Leeds Trinity University

(continued)

(continued)

39. Liverpool Hope University
40. Liverpool John Moores University
41. Liverpool School of Tropical Medicine
42. London Business School
43. London Metropolitan University
44. London School of Economics and Political Science
45. London School of Hygiene and Tropical Medicine
46. London South Bank University
47. Loughborough University
48. Medway School of Pharmacy
49. Middlesex University
50. Newman University
51. Norwich University of the Arts
52. Oxford Brookes University
53. Plymouth College of Art
54. Queen Margaret University, Edinburgh
55. Queen Mary University of London
56. Ravensbourne
57. Roehampton University
58. Rose Bruford College
59. Royal Academy of Music
60. Royal Agricultural University
61. Royal College of Art
62. Royal College of Music
63. Royal Conservatoire of Scotland
64. Royal Holloway and Bedford New College
65. Royal Northern College of Music
66. Sheffield Hallam University
67. Southampton Solent University
68. SRUC
69. St George's Hospital Medical School
70. St Mary's University College
71. St Mary's University, Twickenham
72. Staffordshire University
73. Stranmillis University College
74. Swansea University
75. Teesside University
76. The Arts University Bournemouth
77. The British School of Osteopathy
78. The City University
79. The Institute of Cancer Research

(continued)

(continued)

80. The Liverpool Institute for Performing Arts
81. The Manchester Metropolitan University
82. The National Film and Television School
83. The Nottingham Trent University
84. The Open University
85. The Queen's University of Belfast
86. The Robert Gordon University
87. The Royal Central School of Speech and Drama
88. The Royal Veterinary College
89. The School of Oriental and African Studies
90. The University of Aberdeen
91. The University of Bath
92. The University of Birmingham
93. The University of Bolton
94. The University of Bradford
95. The University of Brighton
96. The University of Bristol
97. The University of Buckingham
98. The University of Cambridge
99. The University of Central Lancashire
100. The University of Chichester
101. The University of Dundee
102. The University of East Anglia
103. The University of East London
104. The University of Edinburgh
105. The University of Essex
106. The University of Exeter
107. The University of Glasgow
108. The University of Greenwich
109. The University of Huddersfield
110. The University of Hull
111. The University of Keele
112. The University of Kent
113. The University of Lancaster
114. The University of Leeds
115. The University of Leicester
116. The University of Lincoln
117. The University of Liverpool
118. The University of Manchester
119. The University of Northampton
120. The University of Oxford

(continued)

(continued)

| 121. The University of Portsmouth |
| 122. The University of Reading |
| 123. The University of Salford |
| 124. The University of Sheffield |
| 125. The University of Southampton |
| 126. The University of St Andrews |
| 127. The University of Stirling |
| 128. The University of Strathclyde |
| 129. The University of Sunderland |
| 130. The University of Surrey |
| 131. The University of Sussex |
| 132. The University of the West of Scotland |
| 133. The University of Wales (central functions) |
| 134. The University of Warwick |
| 135. The University of West London |
| 136. The University of Westminster |
| 137. The University of Winchester |
| 138. The University of Wolverhampton |
| 139. The University of York |
| 140. Trinity Laban Conservatoire of Music and Dance |
| 141. University College Birmingham |
| 142. University College London |
| 143. University for the Creative Arts |
| 144. University of Abertay Dundee |
| 145. University of Bedfordshire |
| 146. University of Chester |
| 147. University of Cumbria |
| 148. University of Derby |
| 149. University of Durham |
| 150. University of Gloucestershire |
| 151. University of Hertfordshire |
| 152. University of London (Institutes and activities) |
| 153. University of Newcastle |
| 154. University of Northumbria at Newcastle |
| 155. University of Nottingham |
| 156. University of Plymouth |
| 157. University of South Wales |
| 158. University of St Mark and St John |
| 159. University of Suffolk |
| 160. University of the Arts, London |
| 161. University of the Highlands and Islands |

(continued)

(continued)

162. University of the West of England, Bristol
163. University of Ulster
164. University of Wales Trinity Saint David
165. University of Worcester
166. Writtle University College
167. York St John University

References

Beveridge D, McKenzie M, Vaughter P, Wright T (2015) Sustainability in Canadian post-secondary institutions. The inter-relationships among sustainability initiatives and geographic and educational characteristics. Int J Sustain High Educ 16(5):611–638. doi:10.1108/IJSHE-03-2014-0048

Bloemen O (2013) Learning for change: social responsibility and sustainability courses at the University of Edinburgh 2012–2013. Institute for Academic Development, University of Edinburgh. Edinburgh. http://www.docs.hss.ed.ac.uk/iad/Learning_teaching/Academic_teaching/Resources/Sustainability/SRS_in_UG_Courses_Full_Report.pdf. Accessed 20 Jan 2017

Ceulemans K, Lozano R, del Mar Alonso-Almeida M (2015a) Sustainability reporting in higher education: interconnecting the reporting process and organisational change management for sustainability. Sustainability 7(7):8881–8903. doi:10.3390/su7078881

Ceulemans K, Molderez I, Van Liedekerke L (2015b) Sustainability reporting in higher education: a comprehensive review of the recent literature and paths for further research. J Clean Prod 106:127–143. doi:10.1016/j.jclepro.2014.09.052

Dade A, Hassenzahl DM (2013) Communicating sustainability: a content analysis of website communications in the United States. Int J Sustain High Educ 14(3):254–263. doi:10.1108/IJSHE-08-2011-0053

Disterheft A, Caeiro S, Ramos MR, Azeiteiro U (2012) Environmental management systems (EMS) implementation processes and practices in European higher education institutions: top-down versus participatory approaches. J Cleaner Prod 31:80–90. doi:10.1108/IJSHE-05-2014-0079

Drayson R, Taylor C (2015) The student voice: experiences of student engagement in education for sustainable development. In: Filho W, Brandli W, Kuznetsova O, Paço A (eds) Integrative approaches to sustainable development at university level. World sustainability series. doi:10.1007/978-3-319-10690-8_43

Evans J, Jones R, Karvonen A, Millard L, Wendler J (2015) Living labs and co-production: university campuses as platforms for sustainability science. Curr Opin Environ Sustain 16:1–6. doi:10.1016/j.cosust.2015.06.005

Fischer D, Jenssen S, Tappeser V (2015) Getting an empirical hold of the sustainable university: a comparative analysis of evaluation frameworks across 12 contemporary sustainability assessment tools. Assess Eval High Educ 41(1):1–16. doi:10.1080/02602938.2015.1043234

Fonseca A, Macdonald A, Dandy E, Valenti P (2011) The state of sustainability reporting at Canadian universities. Int J Sustain High Educ 12(1):22–40. doi:10.1108/14676371111098285

Glover A, Peters C, Haslett SK (2011) Education for sustainable development and global citizenship: an evaluation of the validity of the STAUNCH auditing tool. Int J Sustain High Educ 12(2):125–144. doi:10.1108/14676371111118192

Green Gown Awards (2017) Categories and criteria: sustainability reporting. EAUC, Cheltenham. http://www.greengownawards.org/categories-and-criteria#SustainabilityReporting. Accessed 30 Jan 2017

GRI (2015) Sustainability and reporting trends in 2025; preparing for the future. Global Reporting Initiative, Amsterdam. https://www.globalreporting.org/resourcelibrary/Sustainability-and-Reporting-Trends-in-2025-2.pdf. Accessed 30 Jan 2017

Hahn R, Kühnen M (2013) Determinants of sustainability reporting: a review of results, trends, theory, and opportunities in an expanding field of research. J Clean Prod 59:5–21. doi:10.1016/j.jclepro.2013.07.005

HEFCE (2008) HEFCE strategic review of sustainable development in higher education in England by the Policy Studies Institute, PA Consulting Group and the University of Bath Centre for Research in Education and the Environment. HEFCE, London. http://www.hefce.ac.uk/pubs/rereports/year/2008/sdhefcestrategicreview/. Accessed 30 Jan 2017

HEFCE (2014) Sustainable development in higher education: HEFCE's role to date and a framework for its future actions. HEFCE, London. http://www.hefce.ac.uk/pubs/year/2014/201430/. Accessed 30 Jan 2017

HESA (2017) Environmental information by higher education provider. HESA, Cheltenham. https://www.hesa.ac.uk/data-and-analysis. Accessed 20 Jan 2017

KPMG (2012) Sustainability reporting systems: a market review. KPMG Advisory N.V., Amstelveen. https://assets.kpmg.com/content/dam/kpmg/pdf/2012/08/S_CG_5e.pdf. Accessed 15 Jan 2017

Lambrechts W (2015) The contribution of sustainability assessment to policy development in higher education. Assess Eval High Educ 40(6):776–795. doi:10.1080/02602938.2015.1040719

Lidstone L, Wright T, Sherren K (2015) An analysis of Canadian STARS-rated higher education sustainability policies. Environ Dev Sustain 17(2):259–278. doi:10.1007/s10668-014-9598-6

LiFE (2017) Life self-assessment tool higher education version. EAUC, Cheltenham. http://www.eauc.org.uk/life/self-assessment_tool. Accessed 20 Jan 2017

Lock I, Seele P (2016) The credibility of CSR (corporate social responsibility) reports in Europe. Evidence from a quantitative content analysis in 11 countries. J Clean Prod 122:186–200. doi:10.1016/j.jclepro.2016.02.060

Lozano R (2011) The state of sustainability reporting in universities. Int J Sustain High Educ 12 (1):67–78. doi:10.1108/14676371111098311

Parkin S, Johnston A, Buckland H, Brookes F, White E (2004) Learning and skills for sustainable development: developing a sustainability literate society. Guidance for higher education institutions. London: Forum for the Future. https://www.forumforthefuture.org/sites/default/files/project/downloads/learningandskills.pdf. Accessed 30 Jan 2017

PwC (2016) Good practices in sustainability reporting: building public trust awards. PricewaterhouseCoopers LLP., London. http://www.pwc.co.uk/assets/pdf/good-practices-in-sustainability-reporting-2015.pdf. Accessed 30 Jan 2017

QAA (2014) Education for sustainable development: guidance for UK higher education providers. QAA, Gloucester. http://www.qaa.ac.uk/en/Publications/Documents/Education-sustainable-development-Guidance-June-14.pdf. Accessed 30 Jan 2017

Ramos T, Pires SM (2013) Sustainability assessment: the role of indicators. In: Caeiro S, Filho W, Jabbour C, Azeiteiro U (eds) Sustainability assessment tools in higher education institutions; mapping trends and good practices around the world. Springer, London, pp 81–99

Responsible Futures (2017) NUS responsible futures: the criteria. NUS, London. https://sustainability.unioncloud.org/responsible-futures/about/the-criteria . Accessed 10 Jan 2017

Ryan A, Cotton D (2013) Times of change: shifting pedagogy and curricula for future sustainability. In: Sterling S, Maxey L, Luna H (eds) The sustainable university. Routledge, Oxon, pp 151–167

STARS (2016) STARS technical manual; Version 2.1. Administrative Update One. AASHE, Philadelphia. http://www.aashe.org/files/documents/STARS/2.0/stars_2.1_technical_manual_-_administrative_update_two.pdf. Accessed 30 Jan 2017

Sterling S (2012) The future fit framework: an introductory guide to teaching and learning for sustainability in HE. Higher Education Academy, York. https://www.heacademy.ac.uk/system/files/future_fit_270412_1435.pdf. Accessed 10 Jan 2017

TEF (2016) *Teaching excellence framework year two additional guidance*. London: HEFCE. http://www.hefce.ac.uk/media/HEFCE,2014/Content/Pubs/2016/201632/HEFCE2016_32.pdf Accessed 15 Jan 2017

Tilbury D, Ryan A (2011) Embedding sustainability in the DNA of the university. International Research Institute in Sustainability. University of Gloucestershire, Cheltenham. https://www.academia.edu/1409492/Embedding_Sustainability_within_the_DNA_of_Universities. Accessed 10 Jan 2017

Tilbury D, Wortman D (2008) Education for sustainability in higher education: reflections along the journey. J Plan High Educ Soc Coll Univ Plan 36(4):5–16

Tierney A, Tweddell H, Willmore C (2015) Measuring education for sustainable development. Int J Sustain High Educ 16(4):507–522. doi:10.1108/IJSHE-07-2013-0083

UCAS (Universities and Colleges Admissions Service) (2017) Search for courses. UCAS, Cheltenham. https://digital.ucas.com/search. Accessed 10 August 2017

UI Green Metric (2017) University of Indonesia Green Metric: criteria and indicators. University of Indonesia, Depok. http://greenmetric.ui.ac.id/criterian-indicator/. Accessed 30 Jan 2017

Vaughter P, McKenzie M, Lidstone L, Wright T (2016) Campus sustainability governance in Canada. Int J Sustain High Educ 17(1):16–39. doi:10.1108/IJSHE-05-2014-0075

White S (2014) Campus sustainability plans in the United States: where, what, and how to evaluate. Int J Sustain High Educ 3(3):203–220. doi:10.1108/14676371111098320

Wyness L, Sterling S (2015) Reviewing the incidence and status of sustainability in degree programmes at Plymouth University. Int J Sustain High Educ 16(2):237–250. doi:10.1108/IJSHE-09-2013-0112

Author Biography

Katerina Kosta is a funded Ph.D. student researching sustainability reporting in higher education. She has co-led a research project on sustainability assessment and reporting at the EAUC (Environmental Association for Universities and Colleges). She holds an M.Sc. in Educational Research by the University of Exeter and is subject representative for CSR at the International Ecolinguistics Association.

Discomfort, Challenge and Brave Spaces in Higher Education

Lewis Winks

Abstract Examining transformative pedagogical developments which make use of discomfort and criticality, including the suggestion of 'brave spaces', this paper asks how Higher Education might make use of challenge, uncertainty, risk and discomfort in teaching and learning about sustainability. Beginning with an overview of discomfort and uncertainty in relation to the field of sustainability education and transformative learning, the paper moves on to consider the potential for developing a relational education which draws inspiration from the natural (non-human) world aligned with the approaches of ecotherapy. In the second section, the paper discusses the opportunities and assertions arising for Higher Education practice including: making use of 'brave spaces'; encouraging 'challenge by choice'; developing divergent and pluralistic thinking; cultivating a connection in natural environments and considering the 'student satisfaction' agenda in Higher Education in the UK. Finally, the paper concludes with recognition of some of the ethical and institutional barriers to an implementation of discomfort, 'brave spaces' and relational pedagogies. This paper is intended as a provocation for practitioners and aims to signpost the opportunities for developing a more responsive, critical and daring approach to sustainability education.

Keywords Pedagogy of discomfort · Brave spaces · Sustainability education
Relational pedagogy · Higher education · Ecotherapy

1 Introduction

We are living in increasingly turbulent and uncertain times. Much evidence now exists to suggest we face a convergence of social and environmental issues coupled with destructive and failing economic systems, culminating in what Selby has called 'multiple crisis syndrome' (Selby et al. 2015). A series of interconnected

L. Winks (✉)
School of Geography, University of Exeter, Exeter, UK
e-mail: lw425@exeter.ac.uk

© Springer International Publishing AG 2018
W. Leal Filho (ed.), *Implementing Sustainability in the Curriculum of Universities*,
World Sustainability Series, https://doi.org/10.1007/978-3-319-70281-0_7

'wicked problems' now present themselves (Brown et al. 2010), including: conflict; segregation and discrimination; water shortages; loss of biodiversity and habitat; spread of disease; flooding; food shortages and many more, worsened in most cases by climate change (IPCC 2014; Beddington et al. 2012; Franco et al. 2006; Hanjra and Qureshi 2010). While we might agree on the predicaments we face, less clear to us as educators is how to best prepare ourselves and our students for this uncertain future. Although some might take the view that however unknowable the future, the essence of sustainability education and how to go about it is clear (e.g. Kopnina 2015), this paper asserts, alongside others, that a degree of criticality, subjectivity and pluralism are needed for not only structurally addressing these wicked problems, but for transforming and challenging our personal as well as societal norms (Wals 2007, 2010; Wals and Jickling 2002; Huckle 1991).

Much of our socialising occurs within the framework of institutions. This paper considers the role of Higher Education (HE) in cultivating the conditions for radical change and transformative education for sustainability. In particular, this paper considers how sustainability education might be made a centralised component of the curriculum through the use of discomfort as a pedagogical approach. The paper makes the assertion that a truly transformative approach to HE entails a deep questioning of the social and culturally manifested norms which reside within learners. In order to do this, understandings emerging from a pedagogy of discomfort as well as ecotherapy practices which seek communion with the natural world are considered. It is argued that a significant unsettling of the self is required to rupture the hegemonic views which guide unsustainable behaviours, and that the natural world can offer a basis for reconciliation of the self and society. Previous work in this field has made use of a pedagogy of discomfort to confront issues of social and racial justice in the classroom, however, this paper highlights the role of such a pedagogy for sustainability education, especially when combined with the performance of non-human nature. Beginning with an overview of the literature pertaining to discomfort pedagogies and ecotherapy, the paper moves on to consider the challenges and opportunities inherent in such an approach. The paper ends by summarising the discussion and bringing attention to the institutional responsibility within Higher Education in the UK. It is hoped that this paper will be of interest to educators working with principles of discomfort and sustainability education, and that the combining of these pedagogical approaches will provoke further innovative practice within HE.

2   Part I: Learning at the Edge—Breaking the Mould

2.1   Discomfort and Transformation

The concept of a pedagogy of discomfort was forwarded by Megan Boler who suggested that discomfort might operate within educational settings to enable students and educators to "willingly inhabit a more ambiguous and flexible sense of self [and to engage with a] critical enquiry regarding values and cherished beliefs" (Boler 1999, p. 176). We embody a set of culturally and socially manifested beliefs and values which provoke certain behaviours. Unsustainable behaviours, although enacted by individuals are embodied by society and provide a normative set of social conducts and behaviours which become accepted by a social group (Cialdini et al. 1991). Sociologists agree that social norms are key to shifting behaviour, but developing new norms has posed a challenge to educators working with issues of environment and sustainability (Kollmuss and Agyeman 2002).

Transformative learning involves at its heart the challenging of these norms, and provokes a confrontation and eventual shift to what Mezirow (1997) terms 'habits of mind'. Emphasising the significance in challenging our perspectives, Mezirow (1997) comments; "We do not make transformative changes in the way we learn as long as what we learn fits comfortably in our existing frames of reference" (p. 7). The mechanism for creating shifts in habits of mind toward more sustainable behaviours as well as pluralistic conceptions of justice, might be found then in an education which acts to discomfort of our values and cherished beliefs (Cranton and Taylor 2012). The process of working with discomfort in an educational setting entails by its very nature an upset and disruption to the values and beliefs of students. It draws out assumptions and places them under scrutiny. It extols the uncovering of prejudice and avoids simply learning about injustice, and rather attempts to break the mould within which injustice is established. Boler has termed this the act of 'shattering of worldviews' (Boler 1999).

A pedagogy of discomfort has been applied to a variety of issues from gender, class, race, poverty and political conflict in a number of fields including teacher education (Cutri and Whiting 2015), social work training (Coulter et al. 2013; Redmond 2010; Nadan and Stark 2016) and medical education (Aultman 2005; Wear and Kuczewski 2008). Meanwhile, sustainability educators have been growing increasingly aware of the importance of engaging with the underlying worldviews of students, although several potential barriers to implementing such a pedagogy arise.

Not least it is easily argued that placing a learning group into a confrontational situation which entails risk goes against the 'safe spaces' policy of good educational practice, which protects individuals from uncomfortable or risky scenarios. While this guidance might be sound in terms of safeguarding individuals from harm, the use of safe space has been critiqued (Arao and Clemens 2013; Cook-Sather 2016; Rom 1998). Noticing the increasing prevalence of 'safe space' as a metaphor for removing conflict from the classroom (and therefore arguably diminishing prospects of criticality and pluralism), Rom (1998) began a conversation about the appropriateness of such an approach in education. Later, working with issues of social justice in the classroom, Arao and Clemens (2013) came to the conclusion that risk cannot be removed from discourse on such matters, and to suggest so would be counterproductive and disingenuous. Instead they emphasise that open discussion on these pressing issues comes from an acknowledgement that dialogue about them entails a degree of discomfort and risk, and therefore suggest the use of the word *bravery* rather than *safety*. In doing so, Arao and Clemens (2013) propose that students are more prepared to be challenged and confronted in the learning environment. Building upon Arao and Clemens (2013) and highlighting the important role of institutions in supporting learners and educators to step in to such spaces, Cook-Sather (2016) suggest that making use of brave space "implies that there is indeed likely be danger or harm—threats that require bravery on the part of those who enter. But those who enter the space have the courage to face that danger and to take risks because they know they will be taken care of—that painful or difficult experiences will be acknowledged and supported, not avoided or eliminated" (p. 2). This assertion raises with it the importance of institutional support for educators making use of brave spaces, as well and the need for trusting and caring relationships between students and faculty. Furthermore, a pedagogy of discomfort operating in brave spaces raises further questions concerning the significance of those spaces and places of learning, as well as the others with whom we share them. A relational approach to transformative learning for sustainability therefore requires us to look beyond the institutions of which we are commonly a part, and toward the non-human dimensions of the wider environment. It is helpful to look to the practices of ecotherapy for this purpose.

2.2   Building Relationships: The Performance of Non-human Nature

The field of ecotherapy has been growing steadily in recent years and its practitioners are making use of understandings from psychology and sociology to bring participants into communion with the natural world. Ecotherapists propose that non-human aspects of the environment 'perform' on behalf of the needs of participants, allowing for a critical examination of personal and societal narratives (Jordan 2014; Greenleaf et al. 2014; Buzzell and Chalquist 2010). This work is subjective, nuanced and entails an opening of the self in order to illicit these underlying assumptions. Conradson (2005) remarks that "environmental encounters are in part appreciated for their capacity to move us to think and feel differently… [and] in coming close to other ecologies and rhythms of life we may [find] different perspectives upon our circumstances" (p. 105). This approach is more overtly a form of therapy, but the work of ecotherapists requires a relational attitude toward the health and well-being of the self, as well as the health and well-being of the world (Clinebell 2013). A continuous and connected epistemology is sought which brings participants into direct contact with their pain and grief as mirrored in the natural world. The role of non-human nature is fundamental here to the uncovering, deconstructing and healing of the self. Plants, animals and non-sentient nature present themselves and perform in order to create transformative potential and an intersection of 'mind-body-world' (Buzzell and Chalquist 2010). Such practice demands a relational understanding of the world, and a renewed epistemological basis for education which deals with our fundamental connection to our environments and non-human nature. A relational approach has also been taken by proponents of an ethic of care who determine that education should be underpinned with an ethos of interconnectedness and interdependence (Held 2005; Noddings 2013). The relational ontology which emerges from an ethic of care, infused with the practices of ecotherapy foster fertile ground for considering a relational approach to sustainability education within HE.

While these approaches might be concerned with what we could term 'therapy' and for those who are 'sick', to write them off only as such would be to disregard a rich vein of opportunity for our educational institutions grappling with the problem of how to build sustainability education into curricula and pedagogical approaches. We are at a point where the state of society is mirrored in the crises unfolding in the world. It is also clear that these crises are socially and culturally present in each of us, and that many of our behaviours are perpetuating the circumstances in which we find ourselves. Regardless of the diagnosis, whether we term the world in crisis, unsustainable or sick, the immediate prognosis doesn't look good. From this perspective, perhaps therapy is the right term for what is needed. In any case, the examples of practice emerging from a pedagogy of discomfort and ecotherapy prompt us to rethink how we practice sustainability education within universities and colleges. The potential to infuse these institutions with a relational and transformational basis is great. Part II will examine the opportunities and challenges for HE.

3   Part II: Opportunities for Higher Education

Making use of developments of a pedagogy of discomfort and the role of ecotherapy practice in signposting methods for a relational and connected epistemological approach to HE, this section begins to explore some of the opportunities and potential barriers for developing HE practice for sustainability education. First, 'brave spaces' are considered as a potential conversation starter for HE practitioner to begin to work with aspects of discomfort and challenge in their pedagogical approach. Second, the emancipatory and mutualistic preconditions for a brave space to be created are considered. Third, the role of HE institutions as places of artisanal thought and non-conformity is argued. Fourth, the importance of connection with outdoor and natural environments is discussed in light of insights emerging from the field of ecotherapy. Finally, the opportunities and barriers consistent with the 'student satisfaction' agenda in the UK is considered.

3.1   Enabling HE Teachers and Students to Step into 'Brave Spaces' of Learning

Making use of discomfort as an explicit pedagogical tool prompts us to ask questions of the ethical appropriateness of such an approach. While there might be distinction to be made between discomfort as an expectation or a forced part of education, and discomfort through choice—the former being oppressive while the latter might be seen to be emancipatory in nature—to what extent can discomfort be justified within our learning institutions? Much has been written about 'safe space' and safety nets within the educational process. To some this means creating conditions in which no student should feel uncomfortable. To others, this might mean the protection of welfare and personal well-being (Arao and Clemens 2013; Redmond 2010; Rom 1998; Cook-Sather 2016). In response, the term 'brave space' is forwarded as both a challenge and as way of extending the notion of 'safe spaces' when exploring issues pertaining to social justice (Arao and Clemens 2013). The contention is that safe space creates conflation between the topics discussed and wellbeing of the students who felt confronted and put at risk by participation in the discussion. Safe space hints at the removal of risk and hurt from learning, and the dialogue and discussion is therefore limited. Arao and Clemens (2013) contend that "our approach to social justice dialogues should not be to convince participants that we can remove risk from the equation, for this is simply impossible. Rather, we propose revising our language, shifting away from the concept of safety and emphasising the importance of bravery" (p. 136). While 'safety' in learning is challenged here, brave spaces do not claim to leave behind the implications of care. Cook-Sather (2016) explains that brave spaces encourage learners to face danger and take risks in the knowledge that the group and the facilitator are in conscious agreement about the nature of these risks, and will support one another to engage with difficult and risky dialogue. A learning agreement is suggested for such a discussion in order to create a mutuality of understanding and support. Within HE, this approach to discussing issues related to social and environmental justice steps away from the well developed and accepted paradigm of neutrality and security, and instead challenges both educators and students to step into an unknown realm of exploration and risk. However, making use of this opportunity entails a commitment from the learning group to enter into a mutual agreement and acceptance of 'ground rules', as well as a large degree of confidence on the part of the facilitator. For practitioners in HE intuitions, there exists a clear opportunity to begin to work with the principles of brave space, which in turn meshes with the concepts of pedagogical discomfort and uncertainty. The notion of brave spaces helps us to envisage a pedagogical approach which works to confront while at the same time acknowledging the unrest and difficulty which this might provoke in learners. A robust and mutually agreed learning agreement and confidence on the part of the educator is therefore required. In order to achieve this, there exists a necessity to empower learner choice and to develop a culture of emancipatory education resting upon a strong student-faculty relationship.

3.2 'Challenge by Choice': Developing an Emancipatory Education

There are a number of precautions and measures which the educator might adopt in order to become more responsive to the evolving needs of the learner. Boler (2004) has discussed the need for active empathy in understanding the position of the other, even in such circumstances when others views may seem inappropriate or extreme. Her assertion emanates from the need for a compassionate teacher who is not only ready to listen and respond accordingly without judgement, but is understanding, as they also enter into a transformative process alongside their students. A responsive education therefore might be seen to be one in which the path is not predetermined, where right and wrong reactions are not portrayed and where the relationship between spaces of disruption, or brave spaces and safe space is in constant flux. Of importance here then is the emancipatory nature of the discomfort experienced by students, and related closely to this, the choice to place oneself in discomfort—referred to as 'challenge by choice' (Arao and Clemens 2013). An educator who subjects their students to discomfort might be seen as oppressive. An educator who invites their students into discomfort creates the conditions necessary for emancipatory and transformative learning to take place (Mezirow 1990). The ability for an educator to respond empathetically to an individual or group requires a large degree of trust and knowledge that a broad expression of personal opinions will be met with respect and without negative repercussions. Without trust, it would be difficult to gain a true appreciation of the concerns, and ultimately discomfort of the group. In order for empathy to develop within a group the educator must be willing to share and to develop their own acceptance of vulnerability and must become open to emotional exchanges. Many educators reading this will reel at the thought of being emotionally receptive and open with a group of students for a variety of reasons both professional and personal. However, if we are to ask our students to engage with some of the most troubling narratives of our time in a personal and meaningful way, it is likely that they will look to us for guidance on how to respond. A responsive education which champions emancipatory learning and pluralistic thinking should also be willing to provide hopefulness to meet the grief of our own culpability. Boler (2004) provides a distinction between naïve and critical hope; naïve hope being that which can be defined as "those platitudes that directly serve the hegemonic interest of maintaining the status quo … [including]… the rhetoric of individualism; beliefs in equal opportunity; the puritanical faith that hard work inevitably leads to success; that everyone is the same underneath the skin" (p. 128), whereas critical hope is seen as a hope which emerges from a dynamic and constant revaluation of ourselves, resulting in a shift in our relationships with each other, with ourselves and with the world (Boler 2004).

3.3 Pluralising Understandings of Injustice and Sustainability and Visioning HE as a Place of Non-conformity

The writer and English lecturer Nan Shepherd said of her role: "[I consider it] the heaven-appointed task of trying to prevent a few of the students who pass through our Institution from conforming altogether to the approved pattern" (Shepherd 1931). As Wals and Jickling (2002) notes, the concept of sustainability education is messy and difficult. It does not fit into a curriculum box and resists definition. As such, the very notion of instrumentally educating for sustainability has been criticised (e.g. Jickling 1992). If HE is to develop sustainability education as part of its praxis, it must recognise pluralistic understandings and applications of sustainability. Perhaps more fittingly it is necessary to consider the notion of learning *as* sustainability, rather than *for*, and to seek to embed sustainability education at the core of institutional operations and culture (Sterling 2004).

Developing pluralistic thinking demands that we continually question the way in which the world appears to us, make allowances for a diversity of approaches, beliefs and opinions, and that we hold multiple points of view in mind at once (Engberg and Hurtado 2011). The nurturing of pluralistic thinking as part of a HE which also examines our underlying assumptions about the world and our inherited cultural and social narratives places the educator and learner into a position of an 'original thinker'. The artisanal nature of this approach to education also acts to empower learners to seek their own answers to questions, and solutions to problems. A divergence and intersecting of ideas is seen as important for addressing the most significant problems of our time (Brown et al. 2010). The opportunity therefore to see our HE institutions as places of non-conformity should be exciting and essential to sustainable futures.

3.4 Cultivating Connection with the Natural World

Soga and Gaston (2016) argue that increasing disconnection and separation from natural environments, caused by both loss of opportunity and loss of orientation in the natural world constitutes a crisis of experience in young people and adults. Their compelling argument does not stand alone in suggesting that outdoor experiences in natural environments matter a great deal to the development of pro-environmental values concurrent with sustainable behaviours (Kellert 1993; Wilson 1984; Nicol 2014; Dutcher et al. 2007). So too, experiences in nature are seen to be important for mental wellbeing and health (Brymer et al. 2010; Burls 2007). From this perspective comes the assertion that a renewed epistemological basis to sustainability education in HE must invoke regular contact with natural environments. The salutary role of the natural environment, or more specifically 'non-human nature', in 'performing' in order that we might learn something about ourselves has been noted (Buzzell and Chalquist 2010; McGeeney 2016). While Conradson (2005) remarks that by coming closer to non-human nature we might find new ways of examining our own lives, Rust (2014) comments that reconnecting can be frightening and difficult; "withdrawal from an intimacy with the earth has had many consequences: an unfamiliarity with our own wild animal nature, a less embodied life, and a loss of loving exchange and communion with the non-human world" (p. 40). Taking the steps to rebuild such a communion and connection with the non-human world might in the outset simply entail stepping regularly outside of the confines of lecture theatres and seminar rooms. Building intimacy with the non-human environment requires trust and reciprocity and occurs at the juncture of experience spent outdoors and the mutual support required of 'brave spaces' of learning. The fields of Ecotherapy and ecophsychology, (alongside practitioners working in the outdoor and environmental education sector), guide our thinking in terms of how HE might begin to connect its practices with the natural world, local communities and other cultures. The therapeutic practices which attempt to heal disconnection from the natural world might be fundamental to a deeply connected sustainability education.

3.5 Confronting and Appeasing the Student Satisfaction and TEF Agenda Within HE

HE institutions are under increasing pressure to produce high-impact research which draws in further funding, upholds the institution as a place of significant knowledge production and reaches a large range of stakeholders. While the Research Excellence Framework is seen as important in determining the relative standing of one institution against another in terms of research impact, a recent shift in education has seen a similar scrutiny placed upon teaching and learning within HE. The TEF which has been in place since 2015 in the UK acts to examine the attributes which make up a good learning environment for prospective students in what has, by some accounts, become a highly marketised sector (Olssen and Peters 2005; Marginson 2004). For a Higher Education institution considering a turn to uncertainty and discomfort as an aspect of its educational approach, the demands of the TEF sound warning bells and embolden a 'production focused' emphasis on teaching and learning in HE (Frankham 2016). For an institution to score highly on the TEF, it must succeed in demonstrating that its teaching staff are providing engaging and excellent teaching involving 'rigour and stretch' (HEFCE 2016). This is also linked to the student satisfaction survey, undertaken by the NUS which ranks teaching quality based upon; explanation, interest, enthusiasm and stimulation. These scores help to provide a guide for prospective students. An educational approach which prides itself on placing its students into positions of discomfort might struggle to gain approval of an institution subject to such a survey which predicates itself based upon a highly quantitative assessment of teaching excellence, with very little qualitative appreciation of teaching and learning approaches (Berger and Wild 2016). However, this is not to say that student satisfaction and a peda-gogical approach which works with uncertainty, risk and discomfort are totally disputatious. Indeed, the outcome of such a pedagogical practice is likely to be highly emancipatory, and will further the skillsets and abilities of students as result. It is therefore likely that a bold approach to sustainability education would endear institutions to the judgements of the TEF and the NUS survey in the medium to long term [for example, the descriptor of teaching quality reads: "The emphasis is on teaching that provides an appropriate level of contact, stimulation and challenge, and which encourages student engagement and effort" (HEFCE 2016 p. 25)]. However, the problem remains, that these frameworks mostly act to uphold a safe and hegemonic approach to education within HE, and certainly in the short term, it is hard to see a courageous step away from safe spaces of pedagogical engagement and toward brave spaces of learning without further encouragement and institu-tional support.

4 Conclusion

Applying a pedagogy of discomfort to the issue of sustainability education within universities, this paper has highlighted the potential opportunities for HE. The paper has argued that as multiple crises of social and ecological significance take effect, a radical shift in the way that sustainability education is delivered is required. A pedagogy of discomfort lends itself to the radical goals of such an approach as it invites educators and students to step toward vulnerability and uncertainty together. However, to ensure the epistemological basis of such a sustainability education retains a holistic and rounded character, we would do well to listen to, and learn from, the experiences and practices of environmental educators and other practi-tioners who work on the thresholds of our relationship with ourselves and with other-than-human nature. In addition, this paper has suggested that we might begin to think of moving from safe spaces of learning, toward brave spaces in order to bring confrontation, discomfort and uncertainty into the curriculum of HE.

However, several limitations and difficulties present themselves which should be flagged here. Firstly, there are obvious ethical considerations of making explicit use of discomfort and challenge in teaching and learning settings. While the upshot of this approach is to some extent considered by work on brave spaces, there remains a degree of trepidation and risk in taking such an approach within institutions allied to the mantra of safe spaces of learning. Secondly, safe spaces of learning exist alongside an increasingly marketised HE sector in the UK which is in turn con-cerned with student satisfaction. Although this needn't be a direct concern to the focus of this paper, the need to produce satisfaction amongst the student body makes pedagogical approaches championing discomfort and bravery seem less than appealing. While this assertion shouldn't prevent educators within educational establishments from making use of such an approach, perhaps less likely to occur so readily are the important learner-faculty partnerships and institutional support networks for practitioners. Thirdly, alongside the availability of institutional support, a turn to uncertainty in HE demands a host of capabilities from both educators and learners. Amongst these is the ability to create spaces of trust and reciprocity, and as such enter into an agreement to embrace spaces of bravery, entailing danger and risk; eliciting and provoking uncertain reactions, requiring in turn their own response from others. It is beyond the scope of this paper to suggest how educators should manage this, and it is probable that there are no set formula. However, it should not go unsaid that much professional training takes place to equip therapists and councillors with these skills. We would do well to begin with looking to the work of ecotherapists as outlined in this paper. Fourthly, the prompt provided here to make use of the natural world, non-human performance and outdoor spaces for learning comes with it a host of potential barriers to uptake, including: lack of confidence, lack of adequate knowledge and competence of appropriate activities; lack of access to spaces; lack of department provision and so on. There are many other nuanced ways in which educators may feel less than equipped/skilled/supported to take learning beyond the classroom or lecture theatre on a regular basis.

Of course, these limitations should not detract from what this paper has set out to do—that is: to offer a provocation to educators, prompting us to step beyond the security of distant engagement with our students, and away from didactic teaching approaches which are disconnected form the natural world. Instead we are prompted to embrace vulnerability, act with compassion and see ourselves as co-learners. We are given the opportunity to engage with issues of sustainability not as spatially and temporarily distant issues, but as an embodied aspect of the relational self. So too, we are prompted to learn as part of the non-human world which offers alternative perspectives from which we might examine our own lives. From such a standpoint, sustainability education in HE means engaging critically with some of the most difficult problems of our time, enabling the deconstruction of these narratives and worldviews, while reconstructing them with an awareness of how we share responsibility for the state of the world. This paper is an invitation to embrace uncertainty and entertain discomfort as part of our education practice.

References

Arao B, Clemens K (2013) From safe spaces to brave spaces. In: The art of effective facilitation: reflections from social justice educators. Stylus Publishing, Sterling, VA, pp 135–150

Aultman JM (2005) Uncovering the hidden medical curriculum through a pedagogy of discomfort. Adv Health Sci Educ 10:263–273

Beddington JR, Asaduzzaman M, Fernandez A, Clark M, Guillou M, Jahn M, Erda L, Mamo T, Bo N, Nobre CA (2012) Achieving food security in the face of climate change. Final Report ed. Commission on Sustainable Agriculture and Climate Change

Berger D, Wild C (2016) The teaching excellence framework: would you tell me, please, which way I ought to go from here. High Educ Rev 48:5–22

Boler M (1999) Feeling power: emotions and education. Psychology Press, UK

Boler M (2004) Teaching for hope. In: Garrison J, Liston D (eds) Teaching, learning, and loving: reclaiming passion in educational practice. Routledge, UK

Brown VA, Harris JA, Russell JY (2010) Tackling wicked problems through the transdisciplinary imagination, Earthscan, UK

Brymer E, Cuddihy TF, Sharma-Brymer V (2010) The role of nature-based experiences in the development and maintenance of wellness. Asia Pacific J Health Sport Phys Educ 1:21–27

Burls A (2007) People and green spaces: promoting public health and mental well-being through ecotherapy. J Public Mental Health 6:24–39

Buzzell L, Chalquist C (2010). Ecotherapy: healing with nature in mind. Counterpoint

Cialdini RB, Kallgren CA, Reno RR (1991) A focus theory of normative conduct: a theoretical refinement and reevaluation of the role of norms in human behavior. Adv Exp Soc Psychol 24:201–234

Clinebell H (2013) Ecotherapy: healing ourselves, healing the earth. Routledge, UK

Conradson D (2005) Freedom, space and perspective: moving encounters with other ecologies. Emotional Geographies, Ashgate:103–116

Cook-Sather A (2016) Creating brave spaces within and through student-faculty pedagogical partnerships. Teach Learn Together High Educ 1:1

Coulter S, Campbell J, Duffy J, Reilly I (2013) Enabling social work students to deal with the consequences of political conflict: engaging with victim/survivor service users and a 'pedagogy of discomfort'. Social Work Education 32:439–452

Cranton P, Taylor EW (2012) Transformative learning theory: seeking a more unified theory. In: The handbook of transformative learning: theory, research, and practice, pp 3–20

Cutri RM, Whiting EF (2015) The emotional work of discomfort and vulnerability in multicultural teacher education. Teachers Teach 21:1010–1025

Dutcher DD, Finley JC, Luloff A, Johnson JB (2007) Connectivity with nature as a measure of environmental values. Environ Behav 39(4)

Engberg ME, Hurtado S (2011) Developing pluralistic skills and dispositions in college: Examining racial/ethnic group differences. The Journal of Higher Education 82:416–443

Franco A, Hill JK, Kitschke C, Collingham YC, Roy DB, Fox R, Huntley B, Thomas CD (2006) Impacts of climate warming and habitat loss on extinctions at species' low-latitude range boundaries. Glob Change Biol 12:1545–1553

Frankham, J. (2017). Employability and higher education: the follies of the 'Productivity Challenge' in the Teaching Excellence Framework. J Educ Policy 32(5): 628–641

Greenleaf AT, Bryant RM, Pollock JB (2014) Nature-based counseling: integrating the healing benefits of nature into practice. Int J Adv Couns 36:162–174

Hanjra MA, Qureshi ME (2010) Global water crisis and future food security in an era of climate change. Food Policy 35:365–377

Hefce (2016) Teaching excellence framework: year two guidance [Online]. Available http://www.hefce.ac.uk/media/HEFCE,2014/Content/Pubs/2016/201632/HEFCE2016_32.pdf

Held V (2005) The ethics of care: personal, political, and global, Oxford University Press, UK

Huckle J (1991) Education for sustainability: assessing pathways to the future. Austr J Environ Educ 7:43

Ipcc (2014) Climate change 2014–impacts, adaptation and vulnerability: regional aspects. Cambridge University Press, Cambridge

Jickling B (1992) Viewpoint: why I don't want my children to be educated for sustainable development. J Environ Educ 23:5–8

Jordan M (2014) Moving beyond counselling and psychotherapy as it currently is–taking therapy outside. Eur J Psychother Couns 16:361–375

Kellert SR (1993) The biological basis for human values of nature. Biophilia Hypothesis, 42–69

Kollmuss A, Agyeman J (2002) Mind the gap: why do people act environmentally and what are the barriers to pro-environmental behavior? Environ Educ Res 8:239–260

Kopnina H (2015) Sustainability in environmental education: away from pluralism and towards solutions

Marginson S (2004) Competition and markets in higher education: a 'glonacal'analysis. Policy Futures Educ 2:175–244

Mcgeeney A (2016) With nature in mind. In: The ecotherapy manual for mental health professionals. Jessica Kingsley Publishers, London

Mezirow J (1990) Towards transformative learning and emancipatory education. In: Mezirow J (ed) Fostering critical reflection in adulthood. Jossey-Bass, San Francisco

Mezirow J (1997) Transformative learning: theory to practice. New Dir Adult Continuing Educ 1997:5–12

Nadan Y, Stark M (2016) The pedagogy of discomfort: enhancing reflectivity on stereotypes and bias. British J Social Work, bcw023

Nicol R (2014) Entering the Fray: the role of outdoor education in providing nature-based experiences that matter. Educ Philos Theory 46:449–461

Noddings N (2013) Caring: a relational approach to ethics and moral education. University of California Press

Olssen M, Peters MA (2005) Neoliberalism, higher education and the knowledge economy: from the free market to knowledge capitalism. J Educ Policy 20:313–345

Redmond M (2010) Safe space oddity: revisiting critical pedagogy. J Teach Social Work 30:1–14

Rom RB (1998) 'Safe spaces': reflections on an educational metaphor. J Curriculum Stud 30:397–408

Rust M-J (2014) Eros, animal and earth. Self Soc 41:38–43

Selby D, Selby D, Kagawa F (2015) Thoughts from a darkened corner: transformative learning for the gathering storm. Sustainability frontiers: critical and transformative voices from the borderlands of sustainability education, pp 21–42

Shepherd N (1931) RE: Letter from Nan Shepherd to Neil Gunn. Type to GUNN, N

Soga M, Gaston KJ (2016) Extinction of experience: the loss of human–nature interactions. Front Ecol Environ 14:94–101

Sterling S (2004) Higher education, sustainability, and the role of systemic learning. In: Higher education and the challenge of sustainability. Springer, Berlin

Wals AE (2007) Learning in a changing world and changing in a learning world: reflexively fumbling towards sustainability. Southern African J Environ Educ 24:35–45

Wals AE (2010) Between knowing what is right and knowing that is it wrong to tell others what is right: on relativism, uncertainty and democracy in environmental and sustainability education. Environ Educ Res 16:143–151

Wals AE, Jickling B (2002) Sustainability in higher education: from doublethink and newspeak to critical thinking and meaningful learning. Int J Sustain High Educ 3(3):221–232

Wear D, Kuczewski MG (2008) Perspective: medical students' perceptions of the poor: what impact can medical education have? Acad Med 83:639–645

Wilson EO (1984) Biophilia. Harvard University Press, Cambridge

The Teaching-Research-Practice Nexus as Framework for the Implementation of Sustainability in Curricula in Higher Education

Petra Schneider, Lukas Folkens and Michelle Busch

Abstract In the frame of higher education, the Teaching-Research-Practice Nexus (TRPN) considers an equal linking of the subjects to achieve sustainability in applied teaching through a holistic framework, which generally refers to the "research-teaching-practice triangle" according to Kaplan (Account Horiz: 129–132, 1989). The methodology underlines the challenges associated with a Nexus approach: the topics to be linked, their linking mechanisms and the respective communication mechanisms. The present investigation results are based on a data collection through an international questionnaire at institutions of Higher Education. Although there are already courses on sustainability theories at several universities, there is still room for development in terms of the transfer of experiences from projects on sustainable development outside the academic sector gained over the last few years. Governmental agencies, NGO's and the private sector started already implementing practical solutions for sustainable urban planning and development, whose results and experiences should be multiplied in academic teaching. This concerns, for instance, subjects like the ecological footprint, integrated resource management, circular economy and ecosystem services. This contribution discusses the options for the implementation of the TRPN in an international context, and opportunities for universities to become more sustainable by using a balanced proportion of teaching, research, and practice.

Keywords Sustainability implementation · Teaching-Research-Practice Nexus
Sustainability in higher education

P. Schneider (✉) · L. Folkens · M. Busch
University of Applied Sciences Magdeburg-Stendal, Breitscheidstraße 2,
39114 Magdeburg, Germany
e-mail: petra.schneider@hs-magdeburg.de

L. Folkens
e-mail: Lukas.Folkens@outlook.com

M. Busch
e-mail: Michelle.Busch1@web.de

© Springer International Publishing AG 2018
W. Leal Filho (ed.), *Implementing Sustainability in the Curriculum of Universities*,
World Sustainability Series, https://doi.org/10.1007/978-3-319-70281-0_8

1   Introduction

Sustainability awareness plays a crucial role in the practical implementation of strategies for sustainable development in a globalized world. In 2015 the United Nations approved goals for sustainable development (the so-called Sustainable Development Goals, short SDGs) (United Nations 2015). Universities bear a special responsibility in connection with their educational mission with regard to the creation of sustainability awareness, as stipulated in the sustainability objective SDG 4 on university education and lifelong learning. In education, sustainability has different facets. According to the UNESCO's World Action Program "Education for Sustainable Development", this means an education that enables people to think and act for the future (UNESCO World Action Program "Education for Sustainable Development", 2017). In order to convey this knowledge not only in a credible way, but also in an appropriate manner, the academic institution itself has to go hand in hand with sustainable structures.

Starting from a principle of forestry, sustainability has now developed into the guiding principle for the 21st century, based on the leitmotif that society cannot live in the long term at the expense of people in other regions of the world or future generations. In 1713, Hannß-Carl von Carlowitz was the first to formulate the principle of sustainability in his book on the economics of forest culture "Silvicultura oeconomica" (von Carlowitz 2009): "Just cut as much wood as the forest reproduce! As much wood as can grow!" Since Hannß-Carl von Carlowitz, the concept of sustainability has developed far beyond a purely environmental concept, based on the recognition that the environment, the economy and society are mutually influencing each other: While there will be no economic and social progress without an intact environment in the long term, it will not be possible to protect the environment effectively if people struggle for their economic existence (UNESCO World Action Program "Education for Sustainable Development", 2017). Beyond concrete content, it is designed to convey people's competency. Sustainable education goes beyond pure factual knowledge, conveys abilities and values and enables the learning of the forward thinking, interdisciplinary knowledge, acting autonomously as well as the invitation to participate in social decision-making processes. Education for sustainable development thus makes it possible to understand the effects of one's own actions on the world and to make responsible decisions (UNESCO World Action Program "Education for Sustainable Development", 2017).

Although interesting courses on sustainability exist already at numerous universities being implemented both within the framework of various study subjects and by, for example, multi-day summer courses, there is still insufficient knowledge of the experience gained over the last few years. Numerous government agencies and engineers are already implementing practical applications for the implementation of sustainable design solutions, whose results and experience should be multiplied in academic teaching. This concerns, in particular, the areas of integrated water and waste management, ecosystem renaturation and ecosystem services, as well as sustainable land management from land recycling to Water-Energy-Food Nexus (Hoff 2011). The European Commission's communication of 06.05.2013 "Green Infrastructure (GI)—Enhancing European Natural Capital" (European Commission 2013) provides an appropriate planning framework for the content.

An essential didactic feature, which significantly promotes the motivation of students to learn, is that they can understand and also influence practically the meaning of the learning content with regard to their life-world context and their future. In the case of an academic training the claim of being able to pursue later professional practice also scientifically substantiated additionally comes along. This includes elements of the didactic formats of research learning and service learning. Service learning is a teaching and learning strategy that integrates a meaningful non-profit service with lessons and reflection to enrich the learning experience, to teach civic responsibility and to strengthen the communities (Gerholz et al. 2015). Service learning combines cognitive learning with the assumption of responsibility in the learning environment (Seifert and Zentner 2010). Research Learning describes a didactic concept in which the individual phases of the research process are an essential part of a student's learning processes (Tremp and Futter 2012). At universities, this commences already at the bachelor's level (Bartz-Beielstein 2006).

Since the time of the UN Decade of Education for Sustainable Development (2005–2014) started the activities for the development teaching and learning practices on sustainable development all over the world, and on all educational levels (Leal Filho and Salomone 2006; Cambers et al. 2008, and others). For instance, the Organisation for Economic Co-operation and Development (OECD) held a Workshop on 11–12 September 2008 on Education for Sustainable Development, and pointed out the priorities for the teaching on sustainable development: (1) developing strategies for education for sustainable development, (2) devising curricula for education for sustainable development, (3) promoting sustainable schools, and (4) educating for sustainable consumption. In this workshop was also indicated the general curricula framework for education for sustainable development (see Table 1).

From Table 1 can be seen that teaching in higher education as tertiary educational level requires the integration of inter- and transdisciplinary knowledge, including the participatory processes which considers the social aspects as well. The focus of the General Curricula Framework was put on Courses–Concepts–Systems–Measurement–Practices, and highlighted in this way the issues to be considered in teaching for sustainable development. Although the General Curricula Framework was provided already nearly 10 years ago, there are still barriers to innovation and sustainability at universities around the world (Veiga Avila et al. 2017), but there are also samples for successfully implemented curricula (Schneider and Lüderitz 2017).

Even if it is necessary to teach students in the field of sustainable development, though it might be challenging to get their attention for these topics, as these issues often deal with future-related or abstract subjects. The goal of teaching in the field of sustainable development should be to focus on the environmental and social influence of every single individual. From this state, it is possible to raise their

Table 1 General curricula framework for education for sustainable development according to OECD (2008; www.oecd.org/greengrowth/41372200.pdf)

	Courses	Concepts	Systems	Measurement	Practices
Primary school	Single pillars taught broadly in general lessons	(a) Economic (b) Environment (c) Social	(a) Markets (b) Ecosystems (c) Society	(a) Wealth (b) Eco-footprints (c) Voters	(a) Fundraising (b) Eco-schools (c) Citizenship
Secondary school	Integration of two (or more) pillars taught in existing courses (e.g. social studies)	(a) Economic/ environment (b) Economic/social (c) Social/environment	(a) Carbon trading (b) Human capital (c) Transport	(a) Costs of climate inaction (b) Income distribution (c) Measures of well-being	(a) Green entrepreneurs (b) Poverty reduction (c) Fairtrade
Tertiary level	Integration of three pillars taught in stand-alone units (sustainable development studies)	(a) Economic/ environment and social (b) Inter-generational concerns (c) Participatory processes	(a) Sustainable development strategies (NSDS) (b) Sustainable consumption and production strategies (SCP) (c) Education for sustainable development strategies (ESD)	(a) Capital-based indicators (b) Sustainability indices (c) Sustainability impact assessments	(a) Sustainable production (b) Sustainable consumption (c) Corporate responsibility

awareness for instance for the finiteness of resources, social inequalities or climate change. Setting the focus on problem-orientated and case-based teaching and learning, the students will learn how to act interdisciplinary and problem-related to achieve a sustainable solution. As well as the three-bottom-line of sustainability includes social, ecological and environmental issues, the Teaching-Research-Practice Nexus (TRPN) describes the co-equal existence of teaching, research and practice in institutions of Higher Education. As a framework for the implementation of sustainability in Higher Education, the TRPN is intended to lead to the integration of an intensive reference to practice in teaching and research (Schneider et al. 2017). This article aims to discuss the status of the TRPN in an international context, as well as the derivation of methodological approaches to the networking and broadening of the knowledge pool on sustainability in theory and practice. Scope is also to determine the current state of sustainable learning at the University of Applied Sciences Magdeburg-Stendal with special focus on the TRPN. The investigation aimed also on the collection of feasible and applicable approaches and ideas for the implementation of sustainability in Higher Education on international scale.

2 The Teaching-Research-Practice Nexus

At the international level, the term "Teaching Research Nexus" (TRN) has become well-known describing a relationship between the two academic activities, which is mutually beneficial and expresses the fact that both aspects need to be equally integrated into the academic education (Boyd et al. 2010; Magnell et al. 2016). The aim is to equip students on the basis of integrated thinking with skills and the determination for decision-making in order to be able to make practical decisions that are generally acceptable in practice (Locke 2009; Amador et al. 2015). From the point of view of teaching, research shows itself as the basis (research-based teaching), as a linked task (Teaching-Research Nexus), or, for example, as orientation and content (research-oriented teaching, research informed teaching, research enriched teaching). While there are attempts to arrange the different concepts and terms, this subject is confronted with the difficulty that the terms "research" and "teaching" and "Nexus" are used differently (Trowler and Wareham 2008; Visser-Wijnveen et al. 2010). Moreover, this approach is incomplete with regard to the three-pronged approach to sustainability, since the reference to practice is insufficiently considered. The aim of the article is to determine the actual state of the art in terms of sustainable learning at the University of Applied Sciences Magdeburg-Stendal, as well as the derivation of methodological approaches to the networking and broadening of the knowledge pool on sustainability in theory and practice. A practical key focus here is environmental and social management as a preparatory step towards sustainability management.

In the economy, the importance of sustainability management has increased for a considerable time, thanks to the considerably strengthened importance of corporate social responsibility (CSR) in accordance with the "Guidance on Social Responsibility" (ISO 26000). CSR describes the responsibility of an institution towards society and raises the accomplishment of sustainability challenges to a top management task. The German Sustainability Code provides companies with a framework for reporting on non-financial services, which can be used by organizations and companies of all sizes and legal forms.

A sustainable organizational development combined with the integrative consideration of social, ecologic and economic goals and the desire to make a contribution to society may contribute significantly to the reputation of an institution of Higher Education. However, CSR has not yet been transferred to institutions of higher education, although approaches are also being prepared (e.g. the modular system Sustainable Campus BNC). Previous attempts to transfer the German Sustainability Code to higher education institutions, however, have shown its limits, since a much stronger focus in the context of sustainability management must be placed on teaching at universities. In this context, sustainable development means a development which is still relevant and feasible in the future. It's not just about doing the right things, but also about doing things right (Drucker 1964).

While the social component of the approach to sustainability is covered by the education aspect (teaching) and the ecological component by the research aspect (research), the economic component of the TRN is not yet found. To address this deficit, we proposed the concept of Teaching-Research-Practice Nexus (TRPN) as a framework for the implementation of sustainability in Higher Education, which is intended to lead to the integration of an intensive reference to practice in teaching and research (Schneider et al. 2017; see Fig. 1). The general Nexus term refers to a link or set of links that link two or more things or topics. The conceptual definition already underlines the challenges associated with a Nexus approach: Which topics should be linked to one another and using which linking mechanisms? Which communication mechanisms are used as the basis for linking?

Against this background, the TRPN as integrated, interdisciplinary and intersectoral approach, is intended as framework for the consolidation of compulsory sub-areas in a fragmented area of thinking. In addition to the interaction of natural scientists and engineers, this also requires interaction with humanities using qualitative and quantitative methods.

Fig. 1 The triangle of sustainability, related to sustainability in teaching (Schneider et al. 2017)

The TRN can be extended with regard to the three-pronged sustainability approach, in order to take sufficiently account of the relationship to practice. The triangle of sustainability (see Fig. 1) is considered in the literature also as Triple Bottom Line (Elkington 1997), representing an accounting framework with social, environmental (or ecological) and financial aspects. In this regard, Posthumus (2013) proposed the "The Education Triple Bottom Line," which underlines the need for triangularity in the educational context.

Therefore, the term Teaching-Research-Practice Nexus (TRPN) has been proposed in Schneider et al. (2017) as a framework for the implementation of sustainability in applied teaching, referring to the "research-teaching-practice triangle" according to Kaplan (1989). The concept of the TRPN also provides a corresponding answer to the question of the linking mechanisms: Given the fact that the overall objective in higher education is sustainability, it is self-evident that research, teaching and practice must be anchored in equal proportions in higher education, although in reality there is often a lack of appropriate communication mechanisms (Grosu et al. 2015). The underweighting consideration of even one of the three components automatically leads to passing by sustainability. For a practical implementation, the project "Environmental management and environmental certification in theory and practice" was implemented at the University of Applied Sciences Magdeburg-Stendal, in which the students participate with great interest, since it is about their own life-world context of studying with regard to environmental management in their own institution, based on a scientifically sound professional training.

Scope of the research was the investigation on the options for the implementation of sustainability in Higher Education through the implementation of the TRPN and the assessment of the transferability potential on the University of Applied Sciences Magdeburg-Stendal. The contribution discusses the potential approaches for the implementation, and provides a survey on 39 institutions on international scale with focus on Germany.

3 Methodology

Motivation for the methodological background was the collection of feasible and applicable approaches and ideas for the implementation of sustainability under the Nexus in Higher Education on international scale. The approach was tri-fold, following the TRPN concept:

- **Teaching**: TRPN exemplification through presentations (e.g. at workshops at the university and symposia), fostering the transdisciplinary communication, further through publications (e.g. Schneider et al. 2017), and a TRNP implementation survey questionnaire
- **Research**: comprehensive literature review and use of the questionnaire to collect data from higher education institutions on their experiences, ideas and

TRPN implementation development, critical reflection on the own status of TRPN implementation through a SWOT analysis

- **Practice**: analysing the obtained data and collected experiences and derivation of conclusions for the improvement of the sustainability approach at the own institution.

The methodology used for the investigation of the TRPN implementation potential consisted of a background analysis (with a literature review), complemented with the collection of empirical evidence through questionnaires on the TRPN provided to institutions of Higher Education on national and international scale through social networks and direct contact of collaborating institutions. Particular scope of the analysis was to collect information on the variety of approaches and tools for the implementation of the TRPN. The assessment of the results was supported with the tools of a SWOT analysis. The content of the questionnaire focused especially on the two following questions:

- Which topics should be linked under the TRPN using which linking mechanisms?
- Which communication mechanisms are used as framework for the linking?

The questionnaires were prepared using an excel-based form sheet with following type of questions according to Kleber (1992):

- Introductory part on entry into the survey
- Contact details questions on the type of the responding institution and the field of specialisation of the respondent as well as his position in the institution
- General questions on the TRPN and the stage of knowledge of the respondent in terms of the subject
- Questions on the approach of implementing practical aspects in the teaching process of the institutions.

The questionnaire was designed with closed-ended questions, which limit the answers of the respondents to response options, in that case two-point questions. Further, there was the option for putting remarks and explanations for the respondents. The objective was to obtain as clear and complete answers as possible, but in parallel also to give room for the collection of suggestions for sustainability approaches in Higher Education as well as existing experiences. The questionnaires were evaluated in three steps:

- Statistical analysis of the contained numerical information (e.g., number of institutions practicing PC seminars)
- Graphical analysis of the contained numerical information (e.g., type of practical work approaches to be used by the students in the particular institution)
- Evaluation of verbal content.

Furthermore, analytical methods were used to examine which general conclusions can be drawn from the data obtained.

Furthermore, the background analysis was supported by the SWOT analysis tool. The SWOT analysis (**S**trength, **W**eakness, **O**pportunities, **T**hreats) is a tool for position and strategy development (David 1993; Helms and Nixon 2010). The existing strengths, weaknesses, opportunities and risks of a system are compared. The aim is the self-assessment with the aim of deriving strategic decisions, which are derived on the basis of strengths—already included in the system or to be extended—, to reduce weaknesses—, chances, as well as to eliminate risks. The SWOT analysis is used in educational institutions as a tool for analyzing the overall position of work-related learning and its environment. In general, the SWOT analysis is carried after assessment of improvement potentials according to the Deming-Cycle (Deming 1982) out previous of the design phase in order to evaluate deficits as well as improvement potential. Even if the methodology was originally developed for strategic decisions in the economy, the expansion of its application to socio-scientific questions took place in recent years (Hovardas 2015). Using the SWOT approach, the positioning of the University of Applied Sciences Magdeburg-Stendal in the frame of sustainability in Higher Education should be assessed, particularly also in terms of transferability potential of sustainability approaches.

4 Results and Analysis

4.1 Background Analysis: General Linking Mechanisms of the TRPN

To address the deficit of lacking equivalence between the teaching, researching and practicing activities, the concept Teaching-Research-Practice Nexus (TRPN) was proposed as framework for the implementation of sustainability in teaching by Schneider et al. (2017), which naturally leads to the integration of an intensive practice reference in teaching and research. In the way of his methodological approach, the TRNP reflects a practical application of method triangulation of qualitative and quantitative methods in research and teaching, this means the application of mixed methods (Moschner and Anschütz 2010; Ecarius and Miethe 2011; Schneider 2014). The methodological correlation within the components of the TRPN can be summarized as shown in Fig. 2, based on Kinney (1989, modified).

The practical implementation in teaching includes the following activities: Observe (video, excursion), read (article), reflect (key question, self-assessment) and the joint evaluation (in a group). This sequence corresponds to the classical Kolb Learning Cycle (Kolb 1984) and, from a pedagogic point of view, reflects one of the most important functions of the cycle management according to the Deming-Cycle (Deming 1982): feedback loops embedded in the course design.

Fig. 2 Methodological correlation within the components of the TRPN, (own presentation, based on Kinney 1989)

4.2 Background Analysis: General Communication Mechanisms of the TRPN

Feedback loops are the basis of systemic thinking, which supports the understanding of the structuring of operational processes and procedures. Weggeman (2000) transferred the concept of feedback loops to knowledge transfer and developed the knowledge value chain, see Fig. 3. A knowledge value chain in the business sector is a sequence of intellectual tasks by which knowledge workers build their employer's unique competitive advantage (Carlucci et al. 2004) and/or social and environmental benefit. Considering the pillows of the sustainability triangle, the knowledge value chain approach forms a tool to foster sustainability in knowledge management. According to Weggeman (1996) the processes in the knowledge value chain are structured with the help of knowledge management in order to increase the yield and commitment to the production factor knowledge. The knowledge value chain therefore represents a transfer of a technique to teaching which has proven itself in practice. Referring to this, McElroy (2003) highlighted the aspects of the new knowledge management: complexity, learning and sustainable innovation.

With knowledge management, the processes in the knowledge value chain are structured in such a way that the result and the commitment to the factor knowledge. The knowledge value chain defines step by step which processes have to be viewed and optimized repetitive in order to use the knowledge perfectly.

In this context, the teaching of integrated and interdisciplinary thinking is a focus of the teaching approach. According to Biggs and Collis (1982) and Biggs and Tang (2007), the goal of knowledge transfer of integrated and interdisciplinary thinking is the SOLO 4.5-level "deep understanding" and is summarized as a systemic approach (Schneider and Lüderitz 2017):

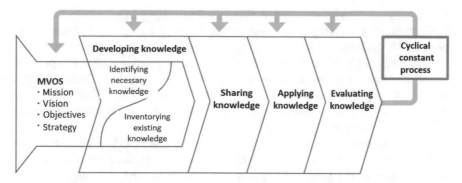

Fig. 3 The model of the knowledge-generating chain according to Weggeman (2000, with kind permission of Prof. Dr. ir. MCDP Weggeman)

(a) The insights of students to develop and sharpen real systems that are typically non-linear, complex and dynamic systems. The application of systemic thinking is aims at:

- an insight into the nature of systems with high feedback rate, their effectiveness and scale,
- mediating the ability to characterize interdisciplinary and interdependent dependencies of resources along material streams,
- providing technical instruments for the identification of the compensation mechanisms of a material flow for the creation or refinement of circular systems,
- imparting awareness of diversity and its central role in creativity, resilience and adaptability, which applies to all dimensions of sustainability.

(b) The overall development of knowledge among students, such as circularity, including the role of natural capital as a vulnerable resource, society as a resource user with economic interests, and governance as a framework for expanding the resilience of the managed systems.

(c) To develop the findings of the students to (a) and (b) with the help of participatory learning approaches as an example for learning with high feedback rate.

4.3 Positioning of the University of Applied Sciences Magdeburg-Stendal

The reference to the TRPN is also reflected in the University Development Plan 2015–2024 (UDP) of the University of Applied Sciences Magdeburg-Stendal, published in November 2014. The university is based on the legal framework and a self-chosen mission statement (University of Applied Sciences Magdeburg-Stendal,

2014): "The University of Applied Sciences Magdeburg-Stendal sees itself as a lively, constantly evolving educational institution, which is geared to research and teaching on the needs of society and business and offers high quality in a wide range of fields. It combines this attitude with the aim of applying science to practice. Due to its facilities at two locations, the university not only has a broad radiance over Magdeburg, but also covers the northern part of Saxony-Anhalt through its location at Stendal, with a special regional policy mandate for the two communes of Salzwedel and Stendal."

The environmental component plays an important role in all fields of action. The motto "studying on a green campus" already suggests an integration of environmental concerns. The fields Energy and Technology as well as Environment and Resources, that are part of the technical profiling, contain environmental concerns as well. This high value of environmental protection at the university means that environmental aspects must also be taken into account in the development of the infrastructure, the control and quality management. The university management would like to raise this challenge, and has consequently adopted a chapter on resource efficiency (Hochschule Magdeburg-Stendal, 2014): "One of the basic conditions for the sustainable development of the university is the responsible and effective handling of resources. Both internal and external measures are developed for this purpose, for example

- the use of an energy manager together with the Otto-von-Guericke-University Magdeburg (OvGU);
- the medium-term implementation of an environmental management system;
- the sustainable use of energy and land resources;
- an expansion of the resource centre, taking into account a strategy of waste prevention;
- the long-term conversion of the university to CO_2 neutrality."

CO_2 neutrality is the long-term goal. However, a complete implementation until 2019 is estimated internally to be too ambitious.

The assessment of the actual state of the University of Applied Sciences Magdeburg-Stendal regarding sustainability and environmental management requirements is based on the SWOT analysis, presented in Table 2. For the development of the SWOT analysis, both internal and external capacities (Ziegler 2015) were used to obtain the most complete and independent picture of the actual situation. Based on the results, potential content and structural opportunities and possible improvements were identified. A special constraint is that, in the field of environmental protection, there are personalised individual activities at the university, both in teaching and in the field of projects, but these are often not interlinked. Networking these activities involves considerable potential leading via the improvement of the visibility of the environmental subject to the expansion of the knowledge pool regarding sustainability in theory and practice in all disciplines.

Currently there exists no closed system of the process description at the university. There are process-oriented approaches in sub-areas (waste management, hazardous material management, energy management, land use) for which

Table 2 SWOT analysis of the actual state of the University of Applied Sciences Magdeburg-Stendal in terms of sustainability

Strengths	Weaknesses
• For the coming years, the UDP provides a strategic framework for the university in which a long-term perspective is given due to the vision of CO_2 neutrality • A number of measures from the UDP, such as the cooperation with the OvGU in energy management, are being implemented. Approaches for the use of renewable energies are being prepared • Energy, technology, environment and resources are two of the five profile shaping of the university, which will soon be supplemented by sustainability • The curriculum in engineering sciences contains a high percentage of practical work, since it includes scientific projects, fieldwork, laboratory and homework • The engineering department has a highly qualified apprenticeship with many years of practical experience, which has been partly acquired in the industrial sector, so that extensive experience with regard to operational material flow optimization can be incorporated • Both the bachelor's degrees and the master's degrees (including MSc in engineering ecology) are recognized as engineering degree in Saxony-Anhalt	• Environmental protection and sustainability have not yet been part of the university's mission statement unlike equal opportunities and family friendliness, although the university is promoting the "Green Campus" • In the perception of most status groups, environmental protection is not a strategic goal of the university, but rather the fulfilling of responsibilities • Due to the fact that environmental protection has not yet been part of the university's mission, there is as yet no organizational framework for a potential environmental certification of the university • In addition, there was no scientific framework for organizing: (a) environmental protection at the university; and (b) integrating environmental protection into the teaching content, in particular, of engineers. In the case of non-technical subjects, the topic has so far only been in the business sector • Environmental protection has not yet been perceived as an interdisciplinary topic, particularly in non-technical subjects, although there is great interest in environmental education
Opportunities	Threads
• The environmental points outlined in the UDP offer concrete docking opportunities for interested actors at the university (i.e. solar thermal energy, environmental education, etc.) • The fields of further education and internationalization offer a suitable platform for environmental issues • Energy management contributes to close cooperation with OvGU in the operational area; further similar uses of synergies are in preparation, e.g. with regard to occupational health and safety • The outlined areas of focus enable cooperation with funding agencies • The open design gives space for a bottom-up process to develop a vision as well as goals and processes	• There is a high dependency on the positioning of environmental protection and sustainability of individuals • There is a risk of the degradation of environmental and sustainability issues due to other priorities or budget cuts • The problem of assigning other than environmental tasks and thus overloading of employees already exists • Environmental protection and sustainability are in competition with other cross-sectional tasks • To improve the state of development of parallel processes, which are not accepted in practice, is already being worked with the aim of the synergetic use of structures

procedures are defined and proven routines are working. However, these are not yet perceived as documented processes. In recent years, the contents of the courses at the University of Applied Sciences Magdeburg-Stendal have been successively updated in order to eliminate the corresponding deficits, which concern the content of the teaching as well as their didactic mediation. In addition, considerable efforts have been made to take explicit account of the developments in engineering practice in teaching. This applies, for example, to the inclusion of the following educational content into the curricula:

- sustainable development and global change
- cycle management with material flow, resource and environmental management
- climate protection, climate change and climate adaption
- environmental design and environmental authorization procedures
- ecosystem services and biodiversity.

4.4   Results of the Questionnaire Survey

A total of 39 institutions of Higher Education responded to the questionnaire, 16 of them being from Germany. Further participating institutions originated from the following countries (in alphabetic order): Bulgaria (1), Canada (1), Colombia (1), Cuba (2), Ghana (1), Hungary (2), Japan (1), Lithuania (1), the Netherlands (1), Portugal (1), Romania (3), Spain (1), South Africa (1), The Czech Republic (1), USA (1), and Vietnam (4). The origin of the respondents in terms of their specialisation was from Humanities (2), Engineering (16), Arts (8), Science (10), and other fields (2). 84% of the respondents responded that they have not yet known about the TRPN, while 71% informed that they understand what TRPN is about. Not all institutions responded to all questions. Even there is a small number of responding institutions, there was obtained a good overview of the TRPN implementation status at the institutions.

Through the respondents, following information was provided (see Fig. 4):

Practical conversion of the provided knowledge takes place inside the institution: 94%
Practical conversion of the provided knowledge takes place externally: 65%
Does practice relation contain study-accompanying courses: 88%
Does practice relation contain active cooperation of students in the private sector: 75%

Details on the type of study-accompanying courses are given in Fig. 5. Regarding the details on the type of active cooperation of students in the private sector, 71% of the respondents mentioned industrial placements, 29% mentioned others, as there are:

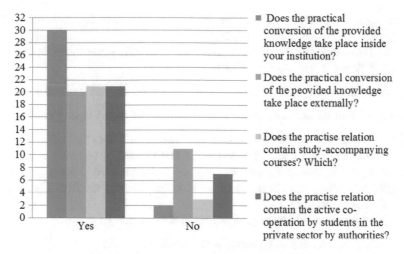

Fig. 4 Summarised results of the questionnaire survey on practice teaching approaches

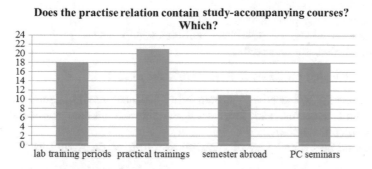

Fig. 5 Details on the type of study-accompanying courses

- internships,
- six to eight week workshops with mentors,
- ring lecture conducted by a practice partner,
- placement to the civil society, NGO's or schools (relevant particularly for teacher's education and didactic specialisation),
- involvement in research projects (relevant for particularly for economics and science),
- sustainability initiatives in the region,
- placement to the NGO's or government (relevant particularly for Life Sciences and Urban Planning).

Four institutions provided reasons why no practical deepening of the knowledge takes place in their institution, as there are:

- inadequate laboratories and equipment, limited demonstration slots on teaching time tables (an institution in Africa)
- lack of financial budget for the equipment for practices (an institution in Asia)
- lack of time (an institution in Germany).

Some institutions provided additional opinions, which are summarised below:

Which topics should be linked to one another and using which linking mechanisms?

Under consideration of the Education for Sustainable Development in Action Learning and Training Tools N°3 of the UNESCO (2012), fostering a multi perspective approach, this approach was proposed to be transferred from school level to Higher Education, namely for example through (Ferreira da Silva Caeiro 2017, personal communication):

In social sciences programs (but not restrictive to …):

- Historical perspective of changes in the world over time
- Geographic perspective of events, problems and issues take on different complexities
- Human rights perspective
- Gender equality perspective
- Values perspective of individuals, cultures and countries
- Cultural diversity perspective.

In natural sciences programs (but not restrictive to …):

- Sustainability perspective—Sustainability balances environmental, social and economic concerns, and focuses on the future to assure the well-being of upcoming generations
- Scientific perspective way of knowing about the world around us.

Which communication mechanisms are used as the basis for linking?

The communication mechanisms cover holistic, collaborative learning methods, approaches and tools (e.g. problem based learning, Triple Bottom Line, Life Cycle Analysis according to DIN ISO 14040, Ecological and Social Footprint, gamefi-cation, on-line collaborative learning etc…). Important is also a re-orientation and training of teachers and professors on the approaches for linking theory and practice, as well as interdisciplinarity and transdisciplinary.

One aspect mentioned by the respondents will be discussed in a bit more detail for reasons of the foundation of the sustainability approach in Higher Education. As the respondents mentioned the Triple Bottom Line as important aspect in the communication mechanisms for linking, there is an approach going beyond this: the Quadruple Bottom Line (QBL) (Roetman and Daniels 2011), which is summarised in "Adding purpose to the mix". The conventional three bottom lines (people, planet, profit) are said to be "transparent". The fourth bottom line, called purpose, is often expressed as spirituality or culture. This approach assumes that sustainable development includes cultural continuity and the development of cultural well-being. QBL provides means to measure, value and assess the addition of

Fig. 6 Conceptual approach of the quadruple bottom line (*source* The local government and municipal knowledge base—encouraging collaboration and knowledge sharing, http://www.lgam. info/quadruple-bottom-line); based on Teriman et al. (2009)

culture, spirituality, and faith in reporting, the Adaptive Quadruple Bottom Line Scorecard (Hadders 2011) and Relational Footprinting, focusing on sustainability is about the quality of relationships (Hadders 2015) (Fig. 6).

Introducing culture through the Quadruple Bottom Line into the sustainability approach provides an upgraded view on the linking concept of Elkington (1997), and has an interface also with the concept of ecosystem services which is applied in ecological engineering. Ecosystem services are the functions of the ecosystem that cause benefit to mankind. The Millenium Ecosystem Assessment (2005) derived provisioning, regulating, cultural and supporting services. Cultural ecosystem services consider nature's contribution to people such as spiritual, recreational, and cultural benefits, e.g. that are making humans happy and give meaning to life. With this approach—introducing ecosystem services into the QBL concept—might be developed the linking between humanities, sciences and engineering in Higher Education. Important is also a re-orientation and training of teachers and professors on the approaches for linking theory and practice, as well as interdisciplinarity and transdisciplinary (Barth and Rieckmann 2012).

4.5 Qualification and Further Education of Teachers and Professors

A sustainable implementation of the TRPN requires as basics teachers and professors who are familiar with the subject of sustainability in Higher Education and innovative teaching methods. According to results of a survey, published on 1st of June 2017 by the common weekly newspaper DIE ZEIT, there are too many postdocs for too few professorships at research universities in Germany. A completely different situation was stated for the Universities of Applied Sciences in Germany, where exists a striking lack of applicants (source: German Centre for Higher Education and Science, DZHW). The Universities of Applied Sciences are hardly able to fill all available professorships. The engineering sciences as well as the legal, economic and social sciences and the professorships in the health care sector are particularly affected. In 49% of the cases, no triple list can be made in the appointing process for a professorship due to lack of feasible candidates. The situation in Baden-Württemberg is particularly difficult: in 48% of the cases, a call for professorship applicants had to be issued more than once. Frequently, the main reason are formal criteria, particularly because the practical experience is missing, which is compulsory at Universities of Applied Sciences, a result which is based on data from 773 appointing procedures for a professorship at 41 Universities of Applied Sciences.

4.6   Transferability Potential of the Questionnaire's Results

The practical reference is of fundamental importance at the University of Applied Sciences Magdeburg-Stendal. It is practiced during the study course through laboratory practice, scientific group work, project days, excursions, scientific projects and specialist internships in companies, whereby a high placement rate for the students after their graduation is given. Irrespective of this, the economy also plays an intermediary role for innovation, which has an effect on the practice of the teaching. The effects of globalization has affected the life cycle of the economy for several years, which is accompanied by increasing internationalization and interdisciplinarity.

An important role in this context plays the qualification work for the successful completion of the studies. In particular, the possibilities offered by the Teaching Research Practice Nexus in practical applications can be seen here. Students will learn how practical interdisciplinary networking on sustainability functions in theory and practice by means of practical topics. The intensifying (international) competition situation has led to a significant increase in the pressure to innovate in companies, regardless of their size. Many companies have added research activities to their portfolio in order to take account of the changed market situation. A system of self-learning knowledge expansion has established itself in the economy in order to keep pace with the latest developments. For the students as well as for the university itself, this is as a rule advantageous because, after completing their internships, they increasingly bring knowledge of current scientific and technical innovations to the university.

Having in view the information collected in the present questionnaire survey, and the ideas and approaches to sustainability in Higher Education provided by the respondents, can be considered that the mentioned tools (e.g. problem based learning, Triple Bottom Line, Life Cycle Analysis, Ecological and Social Footprint, gamefication, on-line collaborative learning etc...) are already practised at the University of Applied Sciences Magdeburg-Stendal, but in a very fragmented manner. Intersectoral cooperation between the departments is practised exceptionally only (for instance through the ring lectures series "Sustainable Development", teaching also about Ecological Footprint. Life Cycle Analysis is teached and practised in the engineering faculty, but might be applicable also for cultural ecosystem services in the social sciences and in the economic accounting in the economic faculty. Teaching in an interdisciplinary and transdisciplinary way forms a challenge also for the professors, as usually they have their field of specialisation. To promote the Quadruple Bottom Line in Higher Education, and to collect information about the fragmented activities in order to link them in a transdisciplinary way, was designed the concept of a "Sustainability Day" at the University of Applied Sciences Magdeburg-Stendal, which takes account of the already taking place activities and actions—mainly performed by students—to celebrate sustainability and in this way to create relationships between people regardless of their nationality, disciplines, and educational status.

5 Conclusion and Outlook

The presented approach is just a first step on the determination of the aspects of the TRPN as framework for the implementation of sustainability in curricula in higher education in order to analyse the current state and to conclude the priorities for further activities. In terms of the questionnaire survey it has to be stated, that 16 institutions from Germany and 23 institutions from all over the world responded to the questionnaire by now. Although this is a small data base, there is enough information for conclusions on the own improvement potentials. Anyhow, this data base is too small to give a representative statement for a whole country or to make comparisons between the countries one by another. However, this survey can afford a general overview about the implementation of the practical dimension of the TRPN at institutions of Higher Education as well as the awareness level of the TRPN itself. It might be considered a constraint that the survey was focused on the practical dimension of the TRPN. This focus was chosen conscious, because one the main goal of the survey was the identification of provided reasons why often a practical deepening of knowledge does not take place in an institution. It has been shown that the main reasons are inadequate laboratories and equipment, limited demonstration slots on teaching timetables, as well as the lack of financial budget for the equipment for practices, or a lack of time. In conjunction with highly qualified practical teaching under the umbrella of an environmentally certified university and the integration into innovative practical research projects, the TRPN represents a future project despite or perhaps precisely because of the extensive challenges. Naturally, a sustainable implementation of the TRPN requires the consideration of two perspectives: that of the teachers as well as of the students.

Thus, against the background of the TRPN, practice presents itself not only as an effective but as a necessary complement to the learning process. In this context, the considerably strengthened importance of Corporate Social Responsibility (CSR) can also be seen (Schaltegger et al. 2007). It describes the responsibility of an institution to society and raises the accomplishment of sustainability challenges to the top management task. In the case of economic enterprises, the company's ideal value has become a tangible asset for a longer time, and the reputation of a company today is more dependent on its environmental and social orientation than ever before. Social awareness of sustainability issues is growing steadily. Although this applies to all social strata and age groups, the young people in particular are emphasised. The awareness of future-oriented issues is often very pronounced with the above, and career decisions are often made on the basis of whether a potential employer contributes to the society and represents values with which the junior staff can identify. This also offers many opportunities for institutions of higher education. A sustainable organizational development combined with the integrative consideration of social, ecological and economic goals and the desire to make a contribution to society can significantly contribute to the reputation of a university or a university.

An instrument that has become an increasingly important focus in recent years is the Sustainability Report, also a CSR or Sustainable Value Report. *"The Sustainability Report is designed to promote the transparency and credibility of sustainability management by providing information on actions and outcomes, both in the past and in the future."* (Berthold and Lingenfelder 2014). This is not just a dialogue tool that improves communication with an institution's stakeholders, but also a tool of continuous improvement and innovation. In this way, a sustainability report can be generated from the management system and, conversely, the management system can be optimized by reporting. The form of a sustainability report is very different in practice. In addition to the integrative consideration of social and environmental aspects in the annual financial reports, the most recent reports have focused mainly on reporting standards, which provide the basic guidelines to be reported. These include in particular the reporting guidelines of the Global Reporting Initiative (GRI) or the German Sustainability Code initiated by the German Council for Sustainable Development. Although independent sustainability reports are hardly ever produced at universities, the combination of interdisciplinary sustainability management and reporting provides great potential for the future. In particular, the coordination and control function, which promotes synergies between faculties and promotes innovation, must be highlighted. Against this backdrop, the challenges for the implementation of the TRPN can also be derived that is nothing less than an integral, interdisciplinary and intersectoral approach, which, in fragmented areas of thinking, can form the framework for the consolidation of conditional parts. In addition to the interaction of natural scientists and engineers, this also requires interaction with humanities scholars using qualitative and quantitative methods. On top of this exists the claim of continuous improvement based on the Deming-Cycle, which draws its philosophy from the Japanese methodology of Kaizen (Kai = alteration, change; Zen = for the better) (Imai 1996).

The networking and expansion of the knowledge pool on sustainability in theory and practice is an important task which will become an increasingly pressing task for universities in the coming years, as well as the development of feasible implementation indicators (Leal Filho et al. 2018). In the student's reality of life, many social questions arise which need to be answered and these answers are often found in sustainability strategies. Sustainability management goes beyond classical environmental management by linking the three pillars of sustainability with regard to all resources. This requires tools for integrated resource management to be developed in the coming years. The TRPN fits seamlessly into the system of the required tools, since it pursues a holistic concept, which is also based on the sustainability principle. The authors are aware of the high requirements with regard to the holistic approach of integrated resource management. Beside this, however, the authors are also convinced that sustainability in education can only arise on the basis of holistic principles.

Acknowledgements The authors are grateful to the respondents of the questionnaires and their willingness to share their experiences.

References

Amador F, Martinho AP, Bacelar-Nicolau P, Caciro S, Oliveira CP (2015) Education for sustainable development in higher education: evaluating coherence between theory and praxis. Assess Eval High Educ 40(6):867–882

Barth M, Rieckmann M (2012) Academic staff development as a catalyst for curriculum change towards education for sustainable development: an output perspective. J Clean Prod 26:28–36

Bartz-Beielstein T (2006) Forschendes Lernen – vom Bachelor zur Promotion in den Ingenieurwissenschaften. In: Berendt B et al (Hrsg.) Neues Handbuch Hochschullehre 3 75 16 04. Griffmarke C 2.36

Berthold N, Lingenfelder M (2014) Wirtschaftswissenschaftliches Studium – Zeitschrift für Studium und Forschung, Heft 8/2014. C.H. BECK und Vahlen Verlag, München, p 2014

Biggs JB, Collis KF (1982) Evaluating the quality of learning—the SOLO taxonomy. Academic Press, New York

Biggs JB, Tang C (2007) Teaching for quality learning at university, 3rd edn. McGraw Hill Education & Open University Press, Maidenhead

Boyd WE, O'Reilly M, Bucher D, Fisher K, Morton A, Harrison PL, Nuske E, Coyle R, Rendall K (2010) Activating the teaching-research Nexus in smaller universities: case studies highlighting diversity of practice. J Univ Teach Learn Pract 7(2):2010

Cambers G, Chapman G, Diamond P, Down L, Griffith AD, Wiltshire W (2008) Teachers' guide for education for sustainable development in the Caribbean published by the UNESCO Regional Bureau of Education for Latin America and the Caribbean OREALC/UNESCO Santiago, ISBN: 978-956-8302-91-7. Available online: unesdoc.unesco.org/images/0016/001617/161761e.pdf

Carlucci D, Marr B, Schiuma G (2004) The knowledge value chain: how intellectual capital impacts on business performance. Int J Technol Manage 27(6/7):575–690

David F (1993) Strategic management, 4th edn. Macmillan Publishing Company, New York

Deming WE (1982) Out of the Crisis. Massachusetts Institute of Technology, Cambridge S. 88. ISBN 0-911379-01-0

Drucker PF (1964) Managing for results: economic tasks and risk-taking decisions. Harper & Row, New York

Ecarius J, Miethe I (2011) Methodentriangulation in der qualitativen Bildungsforschung. Barbara Budrich Publishers, ISBN: 978-3-86649-333-9

Elkington J (1997) Cannibals with forks—triple bottom line of 21st century business. New Society Publishers, Stoney Creek

European Commission (2013) Building a green infrastructure for Europe. doi:10.2779/54125, ISBN 978-92-79-33428-3. © European Union

Ferreira da Silva Caeiro S (2017) Personal communication with Sandra Sofia Ferreira da Silva Caeiro via e-mail on 06.06.2017

Gerholz K-H, Liszt V, Klingsieck KB (2015) Didaktische Gestaltung von Service Learning – Ergebnisse einer Mixed Methods-Studie aus der Domäne der Wirtschaftswissenschaften. bwp@Berufs- und Wirtschaftspädagogik – online. Ausgabe 28:1–23

Grosu C, Almăşan AC, Circa C (2015) Difficulties in the accounting research–practice–teaching relationship: evidence from Romania. Acc Manage Inf Syst 14(2):275–302

Hadders H (2011) The adaptive quadruple bottom line scorecard: measuring organizational sustainability performance. Available online from the Canadian Sustainability Indicators Network: www.csin-rcid.ca/downloads/csin_conf_henk_hadders.pdf. Accessed on 19 July 2017

Hadders H (2015) Relational sustainability: measuring and reporting organizational sustainability performance with relational footprinting, relational academic forum 'relational research in the social sciences: concepts and methodologies'. Cambridge, Available online from researchgate network: https://www.researchgate.net/publication/314096921_Relational_Sustainability_Measuring_and_Reporting_Organizational_Sustainability_Performance_with_Relational_Footprinting. Accessed on 19 July 2017

Helms MM, Nixon J (2010) Exploring SWOT analysis—where are we now? A review of academic research from the last decade. J Strategy Manage 3(3):215–251

Hochschule Magdeburg-Stendal (2014) Hochschulentwicklungsplan 2015 bis 2024 Hochschule Magdeburg-Stendal. https://www.hs-magdeburg.de/hochschule/hochschulentwicklungsplan.html. Aufgerufen am 27 Feb 2017

Hoff H (2011) Understanding the Nexus. Background paper for the Bonn2011 conference: the water, energy and food security Nexus. Stockholm Environment Institute, Stockholm. Available at: http://sei-international.org/publications?pid=1977. Accessed on 27 Feb 2017

Hovardas T (2015) Strengths, weaknesses, opportunities and threats (SWOT) analysis: a template for addressing the social dimension in the study of socio-scientific issues. Aegean J Environ Sci 1:1–12

Imai M (1996) Kaizen. Der Schlüssel zum Erfolg der Japaner im Wettbewerb. 7. Auflage. Ullstein, Berlin u. a. ISBN 3-548-35332-0

Kaplan RS (1989) Connecting the research-teaching-practice triangle. Account Horiz: 129–132

Kinney WR Jr (1989) The relation of accounting research to teaching and practice: a "positive" view. Account Horiz: 119–124

Kleber EW (1992) Diagnostik in pädagogischen Handlungsfeldern: Einführung in Bewertung, Beurteilung, Diagnose und Evaluation. Juventa Verlag, Weinheim; München

Kolb DA (1984) Experiential learning: experience as the source of learning and development, vol 1. Prentice-Hall, Englewood Cliffs

Leal Filho W, Salomone M (2006) Innovative approaches to education for sustainable development. Peter Lang Scientific Publishers, Frankfurt

Leal Filho W, Brandli L, Becker D, Skanavis C, Kounani A, Sardif C, Papaioannidou D, Azeiteiro U, de Sousa L, Raath S, Pretorius R, Vargas VR, Shiel C, Trencher G, Marans R (2018) Sustainable development policies as indicators and pre-conditions for sustainability efforts at universities: fact or fiction? Int J Sustain High Educ 19(2). ISSN 1467-6370 (In Press)

Locke W (2009) Reconnecting the research–policy–practice Nexus in higher education: 'evidence policy' in practice in national and international contexts. High Educ Policy 22: 119. doi:10.1057/hep.2008.3

Magnell M, Söderlind J, Geschwind L (2016) Teaching-reserach Nexus in engineering education. In: Proceedings of the 12th international CDIO conference, Turku University of Applied Sciences, Turku, Finland, 12–16 June 2016. Aufgerufen am 27.02.2017 von www.cdio.org/files/document/cdio2016/68/68_Paper_PDF.pdf

McElroy M (2003) The new knowledge management: complexity, learning and sustainable innovation. Butterworth-Heinemann, Burlington

Millenium Ecosystem Assessment (2005) Ecosystems and human well-being: synthesis. Island Press, Washington. ISBN 1-59726-040-1

Moschner B, Anschütz A (2010) Kombination und Integration von qualitativen und quantitativen Forschungsmethoden in einem interdisziplinären Forschungsprojekt. In: Diethelm I, Dörge C, Hildebrandt C, Schulte C (Hrsg) Didaktik der Informatik - Möglichkeiten empirischer Forschungsmethoden und Perspektiven der Fachdidaktiken. 6. Workshop der GI-Fachgruppe DDI in Oldenburg (S 11–20). Köllen Verlag, Bonn

Posthumus M (2013) The education triple bottom line. Colleagues: 10(1), Article 4. Available at: http://scholarworks.gvsu.edu/colleagues/vol10/iss1/4

Roetman PEJ, Daniels CB (2011) Creating sustainable communities in a changing world. Crawford House Publishing, Adelaide, p 262

Schaltegger S, Herzig C, Kleiber O, Klinke T, Müller J (2007) Nachhaltigkeitsmanagement in Unternehmen – Von der Idee zur Praxis: Managementansätze zur Umsetzung von Corporate Social Responsibility und Corporate Sustainability, Herausgeber: Bundesministerium für Umwelt, Naturschutz und Reaktorsicherheit/ econsense Forum Nachhaltige Entwicklung der Deutschen Wirtschaft e. V./ Centre for Sustainability Management/Leuphana Universität Lüneburg. Aufgerufen am 27 Feb 2017 von www.econsense.de/sites/all/files/nachhaltigkeitsmanagement_unternehmen.pdf

Schneider A (2014) Triangulation und Integration von qualitativer und quantitativer Forschung in der Sozialen Arbeit. In: Mührel E, Bergmeier B (Hrsg) Perspektiven sozialpädagogischer Forschung Methodologien – Arbeitsfeldbezüge – Forschungspraxen, pp 15–24. ISBN 978-3-658-01888-7

Schneider P, Lüderitz V (2017) Integration of ecosystem services as part of the Nexus approach into the applied teaching of ecological engineering. In: Leal W (ed) Handbook of sustainability science and research (in press)

Schneider P, Gerke G, Folkens L, Busch M (2017) Vernetzung und Weiterentwicklung des Wis-senspools zu Nachhaltigkeit in Theorie und Praxis: Umsetzung des Teaching-Research-Practice Nexus an der Hochschule Magdeburg-Stendal. In: Leal W, Diesen R (Hrsg) „Innovation in der Nachhaltigkeitsforschung", Buchreihe "Theorie and Praxis der Nachhaltigkeit" (in press)

Seifert A, Zentner S (2010) Service-learning—Lernen durch Engagement: Methode, Qualität, Beispiele und ausgewählte Schwerpunkte. Eine Publikation des Netzwerks Lernen durch Engagement. Freudenberg Stiftung, Weinheim

Teriman S, Yigitcanlar T, Mayere S (2009) Sustainable urban development: a quadruple bottom line assessment framework. In: The second infrastructure theme postgraduate conference: conference proceedings, 26 March 2009, Queensland University of Technoogy, Brisbane

Tremp P, Futter K (2012) Forschungsorientierung in der Lehre: Curriculare Leitlinie und studentische Wahrnehmungen. In: Brinker T, Tremp P (eds) Einführung in die Studiengangentwicklung. Bertelsmann Verlag, Bielefeld, pp 69–80

Trowler P, Wareham T (2008) Tribes, territories, research and teaching enhancing the teaching-research nexus. The Higher Education Academy, York, p 2008

UNESCO (2012) Exploring sustainable development: a multiple-perspective approach. Education for sustainable development in action learning & training tools N°3. UNESCO

United Nations (2015) Transforming our world: the 2030 agenda for sustainable development, United Nations—sustainable development knowledge platform. 25 Sept 2015

Veiga Avila L, Leal Filho W, Brandli L, MacGregor C, Molthan-Hill P, Gökçin Özuyar P, Martins Moreira R (2017) Barriers to innovation and sustainability at universities around the world. J Clean Prod 164(2017):1268–1278

Visser-Wijnveen GJ, Van Driel JH, der Rijst Van, Roeland M, Verloop N, Visser A (2010) The ideal research-teaching Nexus in the eyes of academics: building profiles. Higher Education Research & Development 29(2):195–210 von Carlowitz HC (2009) Sylvicultura Oeconomica oder haußwirthliche Nachricht und Naturmäßige Anweisung zur Wilden Baum-Zucht. Reprint der 2. Aufl. Leipzig, Braun, 1732. Remagen-Oberwinter: Kessel

Weggeman MCDP (1996) Knowledge management: the modus operandi for a learning organization on increasing the yield of the knowledge production factor. In: Schreinemakers JF (ed) Knowledge management—organizational competence and methodology. Ergon-Verlag, Würzburg

Weggeman MCDP (2000) Kennismanagement: de praktijk. Scriptum, Schiedam

Ziegler M (2015) Erstexploration zur Einführung eines Umweltmanagements an der Hochschule Magdeburg-Stendal, Arbeitspapier: Analyse der Potenziale und mögliche Vorgehensweisen, eidos-Consult, Berlin, 4.6.2015, finanziert im Rahmen Pioneers into Practice-Projektes unter dem Programm Climate-KIC des European Institute of Innovation and Technology (unveröffentlicht)

Education for Sustainable Development: An Exploratory Survey of a Sample of Latin American Higher Education Institutions

Paula Marcela Hernandez, Valeria Vargas
and Alberto Paucar-Cáceres

Abstract Education for sustainable development (ESD) is defined as the knowledge and skills 'needed to work and live in a way that safeguards environmental, social and economic wellbeing, both in the present and for future generations'. Skills for sustainable development include critical thinking, creative thinking, systems thinking and leadership. Over the last decades, there has been efforts across the world to embed ESD into the curriculum. In European Union (EU) countries, some higher education have made efforts to align education strategies with international and national ESD frameworks. A cursory review of the literature seems to indicate that dissemination and implementation of the international ESD frameworks in Latin America has been slow and sporadic. Although there are some signs to implement ESD into curricula of countries such as Brazil and Colombia, these practices have not been substantial or have not permeated higher education sustainable development strategies. This paper aims to explore the developments of ESD in Latin America. As a first step to explore these developments, it intends to survey and map the current ESD processes in eight higher education institutions of four Latin American Countries: Chile, Colombia, Mexico and Peru. The paper also aims to compare ESD developments with some leading EU higher education institutions in ESD with the view to develop a dialogue between the two regions. These will lead to strategies in which ESD processes can be adopted/adapted with benefits in both directions; it will also create, foster and develop mechanisms that

P. M. Hernandez
Processes Engineering Department, Universidad EAFIT, Cra. 49 N° 7 Sur-50,
AA. 3300 Medellín, Colombia
e-mail: phernand@eafit.edu.co

V. Vargas
School of Science and Environment, Manchester Metropolitan University,
All Saints Building Oxford Road, M15 6BH Manchester, UK
e-mail: V.Vargas@mmu.ac.uk

A. Paucar-Cáceres (✉)
Manchester Metropolitan University Business School, All Saints Campus,
Oxford Road, Manchester M15 6BH, UK
e-mail: a.paucar@mmu.ac.uk

© Springer International Publishing AG 2018
W. Leal Filho (ed.), *Implementing Sustainability in the Curriculum of Universities*,
World Sustainability Series, https://doi.org/10.1007/978-3-319-70281-0_9

137

will ensure a sustainable culture of ESD in higher education in both regions. Results of the exploratory survey of a sample of higher education institutions in Latin America are reported.

Keywords Education for sustainable development · Curriculum development Latin America

1 Introduction

In order to help addressing the global challenges that the world is facing, the United Nations have called for the integration of sustainable development at all levels of education in the Agenda 21 Chapter 36 (UN 1992). To support this, the concept of education for sustainable development was refined at the International Conference on Environment and Society held in Thessaloniki—Greece in 1997 (UN 1997). The main message of both of these events was that education should be reoriented towards sustainable development as the means to change the manner society thinks and behave (UNESCO, s.f.). UNESCO has highlighted this by suggesting that current knowledge fails to provide solutions for global challenges. Therefore, they advocated that education must enhance student competences such as critical thinking, imagining future scenarios and work in collaborative ways in order to develop the appropriate responses (UNESCO 2016a, b). Therefore, and for the purposes of this paper, education for sustainable development will be defined as the knowledge and skills 'needed to work and live in a way that safeguards environmental, social and economic wellbeing, both in the present and for future generations' (QAA 2014).[1]

Universities around the world have supported this process by signing declarations of intent since 1988 with the Magna Charta of European Universities followed by the Talloires Declaration in 1990, committing to integrate sustainable development within their activities (Lozano et al. 2013). Whilst declarations provide frameworks for advancing sustainable development at universities, they do not suffice to drive organisational change deeply into processes and disciplinary areas (Bekessy et al. 2007). One of the initial efforts in this process was the integration of environmental education for which there is evidence at higher education courses since 1970 (Alonso 2010) even before the concept of sustainable development was established in the Brundland report in 1987 (UN 1987).

Although the process of integrating sustainable development in universities' activities has included operational issues as well as teaching and learning (Leal Filho et al. 2009), the former has been, to some extent, more emphasised and

---

[1]The concept of education for sustainable development has several meanings and interpretations. UNESCO defines its pillars, but the authors found the concept of QAA clearer for the purposes of this paper.

developed in practice (Lozano 2006). However, in recent years the operational element of sustainable development has been accentuated because of financial support for environmental aspects (Barber et al. 2014), especially in North America. Moreover, there is evidence that there is growing interest in curriculum development issues related to sustainable development from national organisations. Some organisations and agencies that have been active pursuing this aim include the Association for the Advancement of Sustainability in Higher Education—AASHE; the Sustainable Development Solutions Network-SDSN in USA; the Environmental Association of Universities and Colleges-EAUC; and the Higher Education Academy in the UK (HEA 2015; Tilbury 2011).

The process of embedding education for sustainable development in teaching and learning is complex and may include staff development, curriculum review and networking opportunities (Ryan and Tilbury 2016). Although there is a growing amount of literature focusing on learning and teaching aspects of sustainable development at universities, there is a limited number of publications concerning the design and review of curricula (Wals 2014). Methods of assessment have been created and trialled to ascertain the level of integration of sustainable development within universities' activities such as STARS (Sustainability Tracking, Assessment and Rating System), LiFE' (Learning in Future Environments), GASU (Graphical Assessment of Sustainability in Universities) and SAQ (Sustainability Assessment Questionnaire) studied by Ceulemans et al. (2015). Additionally, there are models and tools to assess and better embed education for sustainable development into the curriculum.

Institutions from the Latin American and Caribbean (LAC) region have made attempts to join international forums dealing with ESD. In fact, some universities as Fundacao Universidade Federal de Mato Grosso and Universidad Autonoma de Centro America-Costa Rica and the Colegio de Mexico are original signatories of the Talloires declaration. On the other hand, universities in the Latin American region have been committed and leaders in environmental education (Sáenz and Benayas 2015; Sáenz 2014, 2015); for instance the first conservation courses was offered in 1950 in Colombia (Sáenz 2014). However, there is little information in terms of embedding holistically the aspects of education for sustainable development.

Other organizations working for the promotion, research and implementation of education for sustainable development in LAC include: the Mexican Consortium of University Environmental Programmes for Sustainable Development-COMPLEXUS; the Environmental Committee of the Association of Universities in the Montevideo Group-CA-AUGM; the Argentinian University Network for Sustainability and the Environment-RAUSA; and the Alliance of Iberoamerican Networks of Universities for Sustainability and the Environment-ARIUSA (Sáenz and Benayas 2015).

On the other hand when inspecting ESD strategies deployed by higher education institutions in Latin America, only few models, specifically aim to implement these strategies, have been developed. For example, Geli de Ciurana and Leal Filho (2006) developed the ACES model that includes the complexities to incorporate environmental and sustainable development aspect into the curriculum in higher education. In 2008, the Regional Conference about Higher Education in Latin America and Caribbean Area—CRES signed a declaration where higher education was considered as strategic instrument for sustainable development and inter-institutional and international cooperation, and higher education in Latin America must reaffirm and strengthen the multi-cultural, multi-ethnic and multi-lingual character of the countries and the region (UNESCO-IESLAC 2008). The latest declaration signed in Latin America was the "Declaration of the Americas for Sustainability for and from Universities" signed in Loja-Ecuador in 2011 by 53 universities of 15 countries. Its aims was to organise from the Latin American universities the cultural change necessary to contribute in the mitigations of the social and environmental crisis which is in close relation with poverty and environmental damages, as result of economic growth policies (Mora 2012).

As regards the situation in the EU, research suggests that higher education institutions in Europe have been leading on this area possibly because of more understanding or interest on the topic (Disterheft et al. 2012; Matten and Moon 2004; Lozano et al. 2015). Furthermore, compared to the literature regarding studies in Europe, the research of Latin American cases remains limited.

The main aim of this study is to map curricula in Latin American Universities for their content in terms of education for sustainable development. We plan to continue the study surveying more universities in more detail to then compare these developments with some leading universities based in Europe with the view to develop a dialogue between the two regions. The overall study seeks to support strategies in which education for sustainable processes can be adapted and adopted to benefit both regions.

The paper is organised as follows: after this introduction, in Sect. 2, the methodology used and the survey strategy justifying the convenience sample of four countries is outlined; the limitations of the study are included in this section. In Sect. 3, the initial results of the survey and on-line search is presented. Afterwards Sect. 4 discusses the initial results, and finally, in Sect. 5 conclusions and suggestions for future lines of research to enrich the findings are presented.

2 Methodology

2.1 Survey Research Strategy

2.1.1 Countries and Higher Education Institutions Sampled

Since higher education institutions have different orientations and characteristics to give commonalty to the sample, our sampling strategy was to select countries of the Asia-Pacific Alliance: Mexico, Colombia, Chile and Peru. For each country, the two top universities in the QS ranking (2016) were selected. Eight universities in total including one public and one private were surveyed.

At this exploratory stage, only undergraduate programmes were surveyed. The search done in December 2016, looked at the curricula for all faculties and the survey was made using the universities' web pages. To collect the data, a spreadsheet was made using excel. Each undergraduate programme was checked to find the study plan. From the detailed study plan each semester was reviewed and the subjects that fulfilled the parameters below was incorporated to the matrix.

The subjects selected for the matrix where those related with sustainable development concepts as defined for ESD and sustainable development as: *environment* (e); *social development* (s); *culture* (c); and *sustainable development* (sd) as a holistic concept. Those categories were selected from the UNESCO topic of ESD (UNESCO 2016a, b). All programs that had a publicly access to the syllabus were analysed to guarantee a correct classification. Unfortunately, not all faculties have this information with opened access.

Following the above classification, the subject name of each undergraduate programme was classified into one of the four UNESCO categories. For example, subjects related with biodiversity, climate change, disaster risk, water, pollution, recycling, ecology, or energy were classified as related to the category *environment*. Subjects such as urban planning, social geography, peace, human security, food security, health promotion, gender equity, ethics, peace, and human rights were associated with the category *social development*. Cultural diversity, ancestral knowledge, indigenous knowledge, traditional history, native languages or arts were associated with the *cultural* aspect of sustainable development. Finally, those subjects with the term sustainable or sustainability (i.e. sustainable building) were classified as related to the category of *sustainable development*. Those subjects named or related with ESD skills as systemic thinking were classified as *sustainable development*. Then, university, faculty, undergraduate program subject name, and ESD classification were organized in the spreadsheet.

A total of 422 undergraduate programmes with their respective detailed study plans was classified. The undergraduate programs inspected contained 2868 subjects that were selected and reviewed. The review and classification was done manually: every study plan (i.e. title or content of the unit) was inspected to classify and count subjects related to one of the four sustainable development categories. Table 1 shows the summary of the counting performed. The full spreadsheet with details of the program, subject title and classification assigned is available from the authors by request.

2.1.2 Education Sustainable Development Categories in Higher Education Institutions Undergraduate Programs Sampled

To determine which faculties amongst the sampled universities were more involved with ESD in curricula, similar faculties among the eight universities were identified. The number of occurrence of the categories: *social, environmental, cultural or*

Table 1 Universities sampled according with their position in the QS ranking 2016 (QS 2016)

Country	University	Status	QS Latin America	Total faculties	Total undergraduate programs	Total ESD subjects-undergraduate programs
Chile	Pontificia Universidad Católica de Chile (UCC)	Private	3	18	52	508
	Universidad de Chile	Public	6	14	69	178
Colombia	Universidad de los Andes	Private	8	10	38	217
	Universidad Nacional de Colombia	Public	10	10	51	289
Mexico	Universidad Nacional Autónoma de México (UNAM)	Public	4	15	77	686
	Instituto Tecnológico y de Estudios Superiores de Monterrey (TEC)	Private	7	8	50	274
Perú	Pontificia Universidad Católica del Perú	Private	21	9	52	307
	Universidad Nacional Mayor de San Marcos	Public	70	5	67	406

sustainable development was counted to establish the total of subjects related with ESD in these four aspects, per faculty and per university. It was considered that this would provide a first overview of the faculties that have integrated ESD into the curriculum to then move to a more detail study in the future.

Additionally, the general structure of the faculties and universities was surveyed to validate if the findings per faculty correspond to an institutional agreement in sustainable development or other of its aspects. Those structures included environmental management, university social responsibility, culture and sustainable development statements.

2.2 Limitations of the Survey and Classification Procedure

This is an exploratory study and based on a small sample. This poses obvious limitations to the study. In addition to that, the following limitations to the classification procedure carried out are worth noticing:

- The programmes analysed are only outline documents and they do not represent the complete course aspects.
- The title and general content may or may not describe the programme intentions in relationship to ESD.
- The name of the subject or unit (as appear in the curriculum or course program) is not warranty that is related with the ESD categories used to classify them.
- To review the syllabus of the subject is a proxy to ascertain the content and structure of the subject. However, not all universities have open access to this information.
- When reviewing some subjects (without a syllabus), the classification was relied on the name. But even a fancy name with the word sustainable does not mean that the content or the way is taught is effectively a case of ESD.

3 Initial Results

In this section, results obtained from the search are presented. First, the results of the sustainable development categories in the curriculum of the universities sampled are presented; Secondly, the results by faculty per university associated to the ESD categories proposed by UNESCO are presented. We complement this set of results by presenting a view as to how institutions in the sample have displayed their commitment to environmental and sustainable issues.

3.1 Distribution of Sustainable Development Categories in Undergraduate Curricula for Universities Sampled

Figure 1 shows that *environment* and *social* issues are the commonest subjects related to ESD (blue and orange bars in Fig. 1) followed by cultural ones. Sustainable development or sustainability is explicit in less proportion than subjects related with culture.

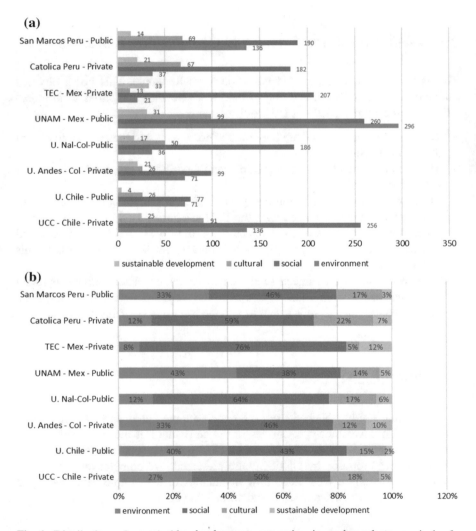

Fig. 1 Distribution of sustainable development categories in undergraduate curricula for universities sampled. **a** Number of subjects per university in each area. **b** Total fraction (%) of EDS topics per university

The two Mexican universities (TEC and UNAM) lead the programmes with curricula directly related with ESD. It is important, also to highlight that, in terms of environmental programmes provision, public universities have a wider offer than private ones.

3.2 Subjects by Faculty Per University Associated to the UNESCO ESD Categories

As it can be seen in Fig. 2, the faculties of engineering, in average, have the highest commitment with environmental aspects and sustainable development. However, the figures seems to suggest that the environmental component is still weak compared with the social aspects of ESD and not being sufficiently addressed.

Also it is worth noticing that the fraction of subjects including *sustainable development* is lower than 30%. The trend in universities, and the composition of programs, is towards the social component of sustainable development. Interestingly, apart from Social Sciences, Design, Architecture and Arts programs are those including social issues in a very significant part of their courses. This may suggest Latin American universities have still much to do in order to guarantee a sustainability culture. However, in universities such as Los Andes, Universidad Nacional Mayor de San Marcos, UNAM and Universidad de Chile (see Table 2), they are becoming, by declaration in their strategic plans, an integration of the environmental and sustainable development component into their academic plans. This could mark significant professional achievements in the pursuit of sustainable development goals by 2030. However, as an example the Social Sciences faculty of Universidad de Los Andes, does not include any environmental neither sustainable development subjects in its academic offer. Social sciences faculties in general have a considerable component of social and cultural subjects as it was expected; however, the environmental component is under 10%.

Although the results (Figs. 1 and 2) presented give an overview of ESD content in the undergraduate programmes and faculties, they do not give in depth information about how programmes address its components. Then, the next phase of this work will be to use a more detailed approach to map sustainable development.

To assess higher education institutions commitment to the operational aspects of sustainable development in general. That is to assess how publically they meant to be committed to engage with sustainable development in their day-today operations. To explore this, a search for evidence of elements of environmental management and sustainable development commitments in their using institutional web pages was made. For university social responsibility, culture and sustainable development, the mission and other institutional declarations were analysed.

All universities have a commitment with social responsibility and/or sustainable practices in campus operation (Table 2). It is important to highlight the universities'

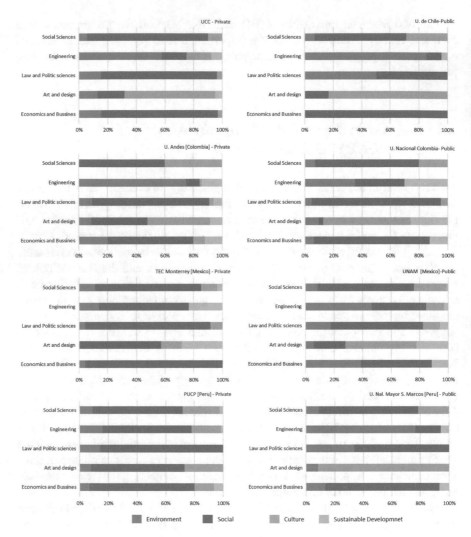

Fig. 2 Number of subjects by faculty per university associated to the UNESCO ESD categories: environment, social issues, culture and sustainable development

compromise with the social and culture promotion, which are considered components of sustainable development.

As it can be seen in Table 2, most of the universities sampled had a University Social Responsibility commitment, and look to contribute with culture in aspects such as cultural heritage preservation and national identity. Six universities, from a total of eight, either have explicitly in their mission or as an institutional programme their awareness and commitment with sustainable development.

Table 2 Evidence of environmental management and sustainable development in universities sampled

University	Environmental management							Sustainable development		
	Solid waste	Transport	Water	Buildings	Energy	Campus	Policy	USR	Culture	Sustainable human development
Pontificia Universidad Catolica de Chile (UC)	x	x	x			x			x	x
Universidad de Chile	x	x	x			x			x	
Universidad de los Andes	x	x	x	x	x	x	x	x		x
Universidad Nacional de Colombia	x	x	x	x	x	x		x	x	
Universidad Nacional Autónoma de Mexico (UNAM)						x	x			x
Instituto Tecnológico y de Estudios Superiores de Monterrey	x	x	x	x	x	x		x		x
Pontificia Universidad Católica del Peru					x	x		x	x	x
Universidad Nacional Mayor de San Marcos	x							x	x	x

4   Discussion

To discuss the results about embedding ESD in the eight Latin American universities e the structure proposed by Gale et al. (2015) (Fig. 3) was used. They found "that different disciplines attract different values" and understandings for sustainable development and it is one impediment to establish ESD in higher education. Their research suggest that business disciplines are more related with the economic dimension of sustainable development because of their view of natural and social capital, while applied life sciences related areas interpret sustainable development as a relation between environment with economic, and ignore the social component. Social disciplines are focused in social justice, equity, poverty among others. Similar trends were found in this exploratory research (see Fig. 3). This model allow us to observe the trends of the different faculties where not equilibrium towards sustainable development can be seen apparently.

Engineering faculties are more concerned towards the environment, maybe because the cost-benefit approach own of the profession; social sciences, law and politic sciences are increasing the relation of resource scarcity and availability with social aspects as inequity, health, peace, human rights, cultural heritage, etc.

As can be seen from the results reported in this section, even business and economics faculties in Latin America, are getting aware of the importance and impacts of environment in economy and society. This analysis suggests that it is necessary an interdisciplinary dialogue towards ESD in Latin America where the view of the university as a competitive commercial enterprise (Tünnermann 2003) needs to be challenged and maybe replaced by a more open and inclusive view that seeks the wellbeing of the different stakeholders including the community where the higher education institution operate.

Regarding the distribution of sustainable development categories in undergraduate curricula for universities sampled, it seems clear that the top universities in the Latin America Asia-Pacific countries have the awareness respect ESD either in the

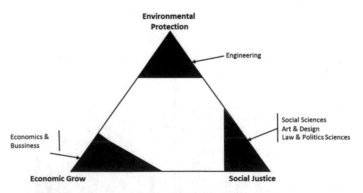

Fig. 3 Trends in ESD by disciplines in Latin America. Adapted from the concept of Conceptions of sustainability in higher education institutions using Connelly's approach (Gale et al. 2015)

social, cultural or environmental aspects of sustainable development. However, the strongest area is the social component. Subjects as ethics, citizenship, rural and social development and cooperation to development are the predominant. It is part of the Latin American universities structure, a combination which evokes the colonial, and religion legacy of the confessional university (Soto 2004) and the search for a more humanistic university in the mid-end of the 20th century (Soto and Forero 2016). However the industrial and commercial purposes influenced by North American universities (Tünnermann 2003), is still prevailing; it can be notice comparing the number of EDS topics and the other subject per undergraduate program. The first ones do not exceed the subjects in professional formation of each program, not even in the social component, and most of them are elective subjects, excepting those programs of the social and human sciences.

Considering that the missions of the eight universities are similar in terms of contributing to social and human development and cultural preservation, it was expected to find more compulsory courses in social development and Latin-American culture. Additionally, the compromise with sustainable development would entail a balance between the social, environmental and cultural components exposed by UNESCO for the incorporation of ESD in higher education which is not reflected in any faculty.

The Universidad Nacional Mayor de San Marcos, declare in their mission that is "Committed to the sustainable development of the country and the protection of the environment; trainer of leading professionals and competent, responsible, courageous and respectful researchers of cultural diversity; promoter of national identity, culture of quality, excellence and social responsibility" (UNMSM 2016). It is contradictory that the only university among the studied universities, who explicitly declares its commitment with sustainable development, has one of the lowest number of subjects in it and just above the average for environmental topics. It suggest as the UNESCO (2012) proposed a clear need to include in the curriculum appropriate subjects for these purposes, a clear own definition in their context of sustainable development, and the creation of its own action plan with their particularities and challenges.

In fact, only one university, UNAM have the total of its studied faculties with all four component of EDS, followed by Universidad de Los Andes with three out of four faculties. Even they are not equity distributed; it partially reflects the findings of Sáenz (2014) about the leadership of Colombia and Mexico in environmental and sustainable development incorporation in higher education. Both Universities have an structured program for sustainable development, for UNAM is called EcoPUMA-University Program of Strategies for Sustainability-(UNAM 2016), for Los Andes there is a complete sustainable structure led by Fenicia Plan (UNIANDES 2016).

Among the subjects found as part of the inclusion of ESD in higher education, it was clear that most of them are part of the elective and flexible curriculum. Although some authors defend the position of including a compulsory subject on sustainable development in the curriculum, the added value of free elective is the interdisciplinary aspect that can emerge in those subjects. Interdisciplinary work is recognised as one of the necessity for champions teachers committed with ESD (Wood et al. 2016). It promotes the understanding and analysis of natural and human phenomena from diverse dimensions.

As mentioned in the introduction, ESD in higher education is approached thought campus operation, institutional policies, integration sustainable development in curricula, staff training, research on sustainable development, and assessment and reporting (Disterheft et al. 2012; Leal Filho et al. 2009; Lozano 2006). Although this research was focused in the curricula, as part of the exploratory phase the exploration in campus environmental management, policies and statements was made to have an idea about the institutional awareness. It is clear, that many universities are getting involved either because of environmental policies in their region or it own conviction about environmental problems as solid waste, water and air pollution, but also concerned about climate change (i.e. energy efficiency, carbon footprint). It is reflected in the eigth Latin American universities surveyed. All of them have some explicitly action about environment. In fact, six of eight universities are participating in GreenMetrics (UI GreenMetric 2017). Pontificia Universidad Catolica del Peru, Pontificia Universidad Catolica del Peru, UNAM, TEC, Universidad Nacional de Colombia, Universidad de Los Andes. This worldwide Metric has five criteria to evaluate campus operation and one to evaluate education and sustainability (Lauder et al. 2015). Thus, Latin American Universities that voluntarily apply to environmental management strategies as it was found.

Lozano et al. (2015) whom found that campus operation is focused, primarily, on waste management, where recycling is the main activity is similar with the findings of this research. As example are TEC and Universidad Nacional Mayor de San Marcos with their paper recycling programms. Although the same authors stated that water management trends to be the least aspect considered by universities, there are important practices in Latin American universities, such as Universidad de Los Andes and TEC, which have wastewater treatment plant in their campus. The EU higher education institutions are highly committed greening the campus (Disterheft et al. 2012) maybe because all the benefits that can it brings (Leal Filo 2015); however, there are no evidence if the Latin American Universities are greening their campus with this knowledge. But, it could be another research topic of Latin American higher education institutions.

5 Initial Conclusions and Further Research

In this paper, an initial approach as to how to ascertain the presence of elements of sustainable development into the curricula of higher education institutions in Latin America was presented. the desirable sustainable development elements as the ones recommended by UNESCO were taken (2016a, b).

This exploratory phase included two of the strands of ESD in higher education the campus operation (e.g. green campus initiatives) and the embodiment of sustainable development elements into the curricula of the higher education institution programs. Ideally, a genuine and lasting commitment to sustainable development principles should be seen and taught for future generations to assimilate and applied these principles.

Based on the results of the Latin American universities sampled s, compared to EU, the Latin-American universities seems to have made some progress. Judging from their activity in regional forums, there is evidence of their commitment to operationalize sustainable development principles. As regards embedding sustainable development elements in the curricula, again there seem to be some progress, especially in the social, and environmental area. Engineering faculties in Latin America seem to be concerned in the environmental issues of sustainable development, whereas social sciences and law and politics sciences showed a compromise with both environment and culture issues for sustainable development. Although all universities have courses explicitly called with the word *sustainable*, the amounts of this courses looks still low compared with the total of social related courses. It suggest that could be relevant to know what sustainable development mean for academic actors (i.e. teacher, students, and other staff) to stablish the relevance in their professions and facilitate the interdisciplinary work that sustainable development is challenging in universities today.

This exploratory research seeks to open possibilities for many universities in Latin America to increase and improve their contributions to Sustainable Development. An important pathway for achieving this goal is related to new curricular projects structured around Education of Sustainable Development. A particular emphasis will be given to Systems Thinking due to the holistic sense this way of structuring social issues, way that fosters the understanding and effective practice of ESD and for the specific characteristics of Latin American students.

As stated before, this is an initial quantitative study based on a small sample. To have a better and complete picture it will be necessary both to increase the sample size and to embark in a qualitative study that is a more in depth study by selecting some representative Latin American universities and extract relevant data about their endeavours to embedded sustainable development in their programs. This can be done maybe by interviewing key staff in each university. Further research also is needed to ascertain other way (this study concentrates on the curriculum content) to see how developments and applications of sustainable development can be seen in the activity of higher education institutions in Latin America.

References

Alonso MB (2010) Historia de la Educación Ambiental. La educacion ambiental en Siglo XX. Asociación Española de Educación Ambiental, España

Barber NA, Wilson F, Venkatachalam V (2014) Integrating sustainability into business curricula: University of New Hampshire case study. Int J Sustain High Educ 15(4):473–49.

Bekessy SA, Samson K, Clarkson RE (2007) The failure of non-binding declarations to achieve university sustainability: a need for accountability. Int J Sustain High Educ 8(3):301–31.

Ceulemans K, Molderez I, Van Liedekerke L (2015) Sustainability reporting in higher education: a comprehensive review of the recent literature and paths for further research. Journal of Cleaner Production Bridges for a more sustainable future: Joining Environmental Management for Sustainable Universities (EMSU) and the European Roundtable for Sustainable Consumption and Production (ERSCP) conferences, vol 106, pp 127–143. doi:10.1016/j.jclepro.2014.09.052

Disterheft A, Caeiro SS, Ramos MR, Azeiteiro U (2012) Environmental Management Systems (EMS) implementation processes and practices in European higher education institutions—top-down versus participatory approaches. J Cleaner Prod 31:80–90. doi:ezproxy.eafit.edu. co:2079/10.1016/j.jclepro.2012.02.034

Gale F, Davison A, Wood G, Williams S, Towle N (2015) Four impediments to embedding education for sustainability in higher education. Aust J Environ Educ 31(2):248–263. doi:10.1017/aee.2015.36

Geli de Ciurana AM, Leal Filho W (2006) Education for sustainability in university studies: experiences from a project involving European and Latin American universities. Int J Sustain High Educ 7(1):81–89. doi:10.1108/14676370610639263

HEA (2015) Education for sustainable development. Retrieved 23 Mar 2015, from The Higher Education Academy: https://www.heacademy.ac.uk/workstreams-research/themes/education-sustainable-development

Lauder A, Sari RF, Suwartha N, Tjahjono G (2015) Critical review of a global campus sustainability ranking: GreenMetric. J Cleaner Prod 108:852–863. doi:10.1016/j.jclepro.2015.02.080

Leal Filho W, Manolas E, Pace P (2009) Education for sustainable development: current discourses and practices and their relevance to technology education. Int J Technol Des Educ 19(2):149–16.

Leal Filo W (2015) Campus greening: why it is worth it. In Leal Filo W, Golda E, Sima M (eds) Implementing campus greening initiatives. Springer, Berlin, pp 359–36.

Lozano R (2006) Incorporation and institutionalization of SD into universities: breaking through barriers to change. J Cleaner Prod 14(9–11):787–79.

Lozano R, Lukman R, Lozano F, Huisingh D, Lambrechts W (2013) Declarations for sustainability in higher education: becoming better leaders, through addressing the university system. J Cleaner Prod 48:10–19. doi:10.1016/j.jclepro.2011.10.006

Lozano R, Ceulemans K, Alonso-Almeida M, Huisingh D, Lozano F, Waas T, Hugé J (2015) A review of commitment and implementation of sustainable development in higher education: results from a worldwide survey. J Cleaner Prod 108:1–18. doi:10.1016/j.jclepro.2014.09.048

Matten D, Moon JJ (2004) Bus Ethics 54:323. https://doi.org/10.1007/s10551-004-1822-0

Mora PW (2012) Ambientalización curricular en la educación superior: un estudio cualitativo de las ideas del profesorado. Profesorado. Revista de Curriculum y Formación del Profesorado, 16 (2), 77–103. Retrieved: 25 Feb 2017, de https://www.researchgate.net/publication/280083228_ambientalizacion_curricular_en_la_educacion_superior_un_estudio_cualitativo_de_las_ideas_del_profesorado

QAA (2014) Education for sustainable development. In Academy TH (ed). Retrieved 20 Feb 2017, from Guidance for UK higher education providers: http://www.qaa.ac.uk/en/Publications/Documents/Education-sustainable-development-Guidance-June-14.pdf

QS (2016) QS Top Universities. (QS Quacquarelli Symonds Limited 1994–2016). Retrieved 25 Dec 2016, from QS World University Ranking: http://www.topuniversities.com/qs-world-university-rankings/methodology

Ryan A, Tilbury D (2016) Learning for sustainable futures: ESD professional development at University of Gloucestershire. Retrieved from University of Gloucestershire: http://www.ue4sd.eu/images/2015/UE4SD-Leading-Practice-PublicationBG.pdf

Sáenz O (2014) Panorama de la Sustentabilidad en las Universidades de América Latina y el Caribe. In Ruscheinsky A, Guerra A, Figueiredo M, Silva Leme P, Lima Ranieri V, Carvalho Delitti W (eds) Ambientalização nas instituições de educação superior no Brasil: caminhos trilhados, desafios e possibilidades. EESC/USP, São Carlos, pp 23–3.

Sáenz O (2015) Diagnosticos nacionales sobre la inclusión de consideraciones ambientales en las universidades de America Latina y el Caribe. AMBIENS. Revista Iberoamericana Universitaria en Ambiente, Sociedad y Sustentabilidad, vol 1, issue no. 1

Sáenz O, Benayas J (2015) Global University Network for Innovation. Retrieved 25 Feb 2017, from Higher Education, Environment and Sustainability in Latin America and The Caribbean. http://www.guninetwork.org/articles/higher-education-environment-and-sustainability-latin-america-and-caribbean

Soto ADE (2004) La reforma del Plan de Estudios del fiscal Moreno y Escandón. 1774–1779. Centro Editorial Universidad del Rosario, Bogotá

Soto ADE, Forero RA (2016) La Universidad Latinoamericana y del Caribe en los desafíos del Siglo XXI. Rev Hist Edu Latinoam 18(26):279–309. http://dx.doi.org/10.19053/01227238.4375

Tilbury D (2011) Higher education for sustainability: a global overview of commitment and progress. High Educ World 4:18–2.

Tünnermann C (2003) La Universidad Latinoamericana ante los retos del siglo XXI. México D.F, Coleccion UDUAL

UI GreenMetric (2017) UI GreenMetric. Retrivied 26 Feb 2017, de Paticipants 2016: http://greenmetric.ui.ac.id/participant-2016/

UN (1987) Report of the world commission on environment and development: our common future. Retrieved 24 Feb 2017. http://www.un-documents.net/our-common-future.pdf

UN (1992) Agenda 21-Chapter 36. Promoting education, public awareness and training. United Nations Conference on Environment & Development (pp 320–328). Rio de Janerio, Brazil, 3 to 14 June 1992: United Nations. Retrieved 15 May 2016. https://sustainabledevelopment.un.org/content/documents/Agenda21.pdf

UN (1997) United nations educational, scientific and cultural organization educating for a sustainable future. Retrieved 25 Feb 2017, from a transdisciplinary vision for concerted action: http://www.unesco.org/education/tlsf/mods/theme_a/popups/mod01t05s01.html

UNAM (2016) Ecopuma Universidad Sustentable. Retrieved 12 Dec 2016; Universidad NAcional Autóma de México: http://ecopuma.unam.mx/

UNESCO (2012) Exploring Sustainable development: a multiple-perspective approach. Education for Sustainable Development Section (ED/PSD/ESD), Paris

UNESCO (2016a) Educación. Retrieved 24 Apr 2016, from Educación para el Desarrollo Sostenible: http://www.unesco.org/new/es/education/themes/leading-the-international-agenda/education-for-sustainable-development/education-for-sustainable-develo

UNESCO (2016b) Education. Retrieved 2 Apr 2016, from Education for Sustainable Development: http://www.unesco.org/new/en/education/themes/leading-the-international-agenda/education-for-sustainable-development/

UNESCO (s.f.) Education for sustainable development. Retrived 20 Feb 2017. http://en.unesco.org/themes/education-sustainable-development

UNESCO-IESLAC (2008) Conferencia Regional de la Educación Superior en América Latina y el Caribe (CRES). Declaraciones y plan de acción. Retrieved 25 Feb 2017, from www.unesco.org.ve/documents/DeclaracionCartagenaCres.pdf

UNIANDES (2016) Sostenibilidad. Retrieved 3 Jan 2017, de Universidad de Los Andes: https://campusinfo.uniandes.edu.co/es/sostenibilidad

UNMSM (2016) Universidad Nacional Mayor de San Marcos. Retrieved 27 Dec 2016, de Mision: http://www.unmsm.edu.pe/home/inicio/mision
Wals A (2014) Sustainability in higher education in the context of the UN DESD: a review of learning and institutionalization processes. J Cleaner Prod 62(1):8–15. doi:10.1016/j.jclepro. 2013.06.007
Wood BE, Wood BE, Cornforth S, Cornforth S, Beals F, Beals F, Tallon R (2016) Sustainability champions? Academic identities and sustainability curricula in higher education. Int J Sustain High Educ 17(3):342–36.

Biorefinery Education as a Tool for Teaching Sustainable Development

Ari Jääskeläinen and Elias Hakalehto

Abstract Savonia University of Applied Sciences, Finland works closely with economic life and makes studies authentic by using Open Innovation Space learning. In this spirit, ABOWE biorefinery project and the related educational links served as a concrete tool for teaching sustainable development. Savonia invested in a mobile biorefinery pilot plant, which is based on the research of Finnoflag Oy, Finland. In this biorefinery plant solution, biodegradable wastes are valorized to platform chemical and energy products with the aid of microbes and their enzymes. The process is carried out in the same way as the circulation is taking place in the Nature. However, skilled personnel are required for conducting the process runs. Therefore, the engineering, construction and testing of the mobile biorefinery pilot plant served as an excellent opportunity for educating engineers to encounter the basics of biological industrial processes. The interdisciplinary project also provided means for learning from each other in a practical way. ABOWE has been positively evaluated for its multisector approach and for involving students and personnel from many educational organizations in the cooperation. Sustainable development was clearly a through-cutting principle during the project. This chapter aims to give an example of how a multidisciplinary development project can act as an educational tool for teaching sustainable development for the participant students as well as a part of continuing education for the personnel.

A. Jääskeläinen (✉)
Environmental Engineering Department, Savonia University of Applied Sciences,
P.O. Box 6, 70201 Kuopio, Finland
e-mail: ari.jaaskelainen@savonia.fi

E. Hakalehto
Finnoflag Oy, P.O. Box 262, 70101 Kuopio, Finland
e-mail: elias.hakalehto@gmail.com

E. Hakalehto
Department of Agricultural Sciences, University of Helsinki, Helsinki, Finland

E. Hakalehto
Faculty of Science and Forestry, University of Eastern Finland, Joensuu, Finland

© Springer International Publishing AG 2018
W. Leal Filho (ed.), *Implementing Sustainability in the Curriculum of Universities*,
World Sustainability Series, https://doi.org/10.1007/978-3-319-70281-0_10

Keywords Biorefinery · Open innovation space · Constructivism
Sustainable development · Interdisciplinary project

1 Introduction

1.1 Savonia and OIS

Savonia University of Applied Sciences, Finland (Savonia) offers education in seven
fields or disciplines and its campuses are situated in Kuopio, Iisalmi and Varkaus, all
in the Northern Savonia region. Savonia has almost 6000 students and 460 employees.
In its education Savonia implements Open Innovation Space (OIS) learning, in which
the learning is combined with development, research, and teaching.

In the OIS learning the key elements are working life proximity, authenticity and
cross-sectoral activities. The students work actively and collaboratively with their
learning guided and supported by the teachers, other staff and representatives from
working life. The forms of cooperation include various assignments rising from
various practical real-life needs as well as an outcome of the development projects in
which new products and services are produced for different customers. Thus the
activities in the field as well as theoretical knowledge are well integrated in the studies.

The working life adjacent OIS learning is reflected in Savonia campuses by diverse
learning environments, as well as shared learning planning and implementation.
Various learning practices take place largely in expert networks and in virtual environments. Hence students learn practices and gain contacts and know-how that are
needed in the working life (Savonia University of Applied Sciences 2017).

1.2 Constructivistic View on Learning and Action Research

The constructivistic view on learning emphasizes the learner's own responsibility for
the success of the learning process. According to this view, the learner is active, controls
one's operations and solves problems in a self-directed way. Experiments, problem
solving, thinking, innovating and understanding the background have a central value in
the learning. Moreover, learning should simulate the reality at a level that is complex
enough. Consequently, complex content in natural situations inspires and motivates the
thinking of students and teachers in a sufficiently demanding level. A learner usually
gets motivated from the clear objectives and favorable operational preconditions
(Patrikainen 1999; Rauste-von Wright et al. 2003; Liukkonen et al. 2002).

Factors promoting motivation in communities are, according to (Räsänen 2008):

- a powerful challenge or a noble mission that leads to the best contribution from
  all,
- strong group loyalty with group members caring about and taking care of each
  other,

- diversity of talent in the working group being a good basis for innovative action,
- trust and selfless cooperation with persons having the ability and willingness to help others,
- willingness to focus on the things and passion for working,
- an amusing and rewarding task that motivates as such,
- allowing creativity in the group giving room for renewal and enthusiasm.

The pedagogic research need behind this chapter was to obtain impact on practical level among students and personnel in learning sustainable development. The main tool for the learning process was the biorefinery concept and the target was to obtain an example for industry and academia about the new opportunities that are in line with sustainable development to the core. Students and personnel were giving their contribution to the development work. At the same time they were learning new skills all the time, which was in line with Open Innovation Space learning frame.

The method used in the analyses resemble somewhat to action research (see e.g. Sagor 2000), i.e., development work and observation of the learning are being done at the same time. Other sources for information regarding students' learning were their theses which showed their own reflection of the learning process.

1.3 ABOWE Project Topic and Goal

Sustainable development is a wide and diverse concept, which may remain rather blurry in students' minds if not concretized to practical level. Renewable energy, energy efficiency and waste recycling are clear and well known themes or principles within the framework of sustainable development, but if many sectors could be integrated into a novel system which promotes at the same time e.g. climate change abatement, bioenergy and biomaterials production as well as waste management, we start to be in the core regarding the future systems and solutions. And if students from various fields join their forces, in an organized way during this developmental project, we can well talk about teaching sustainable development. This approach includes the use of microbial cultures and communities, as well as their enzymes, in the processes.

In this chapter the ABOWE biorefinery project and the related educational links are presented as concrete tools for teaching sustainable development: The multi-sector ABOWE project involved tens of students from various fields, levels and educational institutions from Finland, Poland, and Sweden in the biorefinery pilot project, and also from Germany, Lithuania and Estonia with respect to another subproject on the biogas production.

The ABOWE project (Implementing Advanced Concepts for Biological Utilization of Waste, 12/2012–12/2014) belonged to the EU Baltic Sea Region Programme 2007–2013. It was initiated and led by Savonia as an extension project for REMOWE project (Regional Mobilization of Sustainable Waste to Energy Production 9/2009–12/2012) to pilot two promising technologies in a semi-industrial scale.

Within the ABOWE project, Savonia invested in a mobile biorefinery pilot plant, which is based on the novel biorefinery concept innovated and developed by Adjunct Professor Elias Hakalehto (Finnoflag Oy, University of Eastern Finland and University of Helsinki). Another technology piloted in ABOWE was dry digestion technology, with Ostfalia University of Applied Sciences, Germany investing in a mobile dry digestion pilot plant, in the lead of Professor Thorsten Ahrens. In this chapter we concentrate on the mobile biorefinery pilot plant.

The goal of the ABOWE project and the mobile biorefinery pilot plant was to provide "proof of concept" on the ways, how biowaste and waste biomasses could be used as raw materials for microbial bioprocesses. The idea was to study the combination of gaseous, liquid and solid phases in the pilot plant's bioreactor in order to produce biofuels, organic chemicals, fertilizers and nutrients, or their raw materials. These products are to be produced in an economically feasible way. The raw materials could include, besides various biomass or biowaste sources, also different degradable mixed wastes (Hakalehto and Jääskeläinen 2017).

One biochemical product alternative, 2,3-butanediol, could be used for producing plastic monomers, synthetic rubber, textiles, cosmetics, anti-icing chemicals and other commodities. Ethanol, hydrogen and various organic acids are also potential fermentation products from the biorefinery.

The leading principle in the Finnoflag Oy's biorefinery technology is the implementation of the degradative and recycling functions of the Nature's microbiota into industrial applications. This requires understanding on the interactions between the members of natural flora, and the added strains and enzymes (Hakalehto et al. 2016a).

The production exploited results by the PMEU enhanced cultivation unit (Portable Microbe Enrichment Unit), and in larger vessels in the Finnoflag laboratory since 1997. As the products are produced faster, the full scale production facility can be smaller in size and the initial investment lower. Moreover, the end product concentrations can be increased and the total duration of the process shortened by this acceleration of the biological reactions.

Novel production principles were tested in Finland, Poland and Sweden with various raw materials. Downstream techniques for product recovery were developed in cooperation with Ostfalia University of Applied Sciences, Germany, under the supervision of Professor Thorsten Ahrens (Hakalehto et al. 2016a).

This novel sustainable technology innovation environment offered students a unique setting to develop their skills in a multidisciplinary project where the biocatalysts were integrated with the machinery, electrical steering of the process and under other control and follow up. The special features of the biological materials were also learnt as a group.

Sustainable development was the leading principle throughout the project. This chapter aims to give an example of how a multidisciplinary development project can act as an educational tool for teaching sustainable development for the participant students as well as a part of continuing education for the personnel.

2 ABOWE Biorefinery Pilot Plant Overall Design, Engineering and Construction

A draft lay-out of the biorefinery pilot plant was designed for ABOWE kick-off meeting in February 2013. Overall designing of the novel biorefinery pilot plant's bioprocess was conducted by Adjunct Professor Elias Hakalehto (Finnoflag Oy, University of Eastern Finland and University of Helsinki). Moreover, he brought the microbiological and biotechnical process know-how and point of view in the engineering of the various technical sectors as well as in the construction of the pilot plant.

The pilot plant was engineered and constructed by the ABOWE Engineering team. The team consisted of Finnoflag Oy experts and Savonia's engineering teachers, project engineers and engineering students. The project team was built in such a way that teaching was integrated as widely as possible already at the planning stage of the process. Versatile knowledge of process and instrumentation, layout, mechanical, electrical, automation, IT, environmental and manufacturing engineering was combined.

Savonia constructed the pilot plant and several locally operating industrial enterprises were supplying its components. Altogether more than 50 persons were involved in the pilot plant engineering and construction efforts. Savonia students from many educational programmes participated in the engineering, construction and testing phases. Also numerous trainees from Savo Vocational College, Finland participated in the pilot plant construction and testing.

Part of the Engineering team is shown in Fig. 1.

An empty used 12 m (40 feet) long sea freight container was received in Mid-Summer 2013. Autumn 2013 was a very intensive time for various

Fig. 1 Part of the ABOWE Engineering team. *Photo* Aku Tuunainen

Fig. 2 The mobile biorefinery pilot plant in its construction site in front of Savonia's educational workshop in Kuopio, Finland *Photo* Ari Jääskeläinen

installations. First, mechanical process equipment was installed. Heating, ventilation and air conditioning (HVAC) comprised major parts of the installment and here was needed a local engineering company as well as local installation companies.

The biorefinery pilot plant was constructed outside, in front of Savonia's educational workshop in one of its campuses in Kuopio, Finland. The site, neighbored by three educational institutions and a sports hall, was very lively with several hundred of passers-by daily during seven months (Fig. 2).

In January 2014 the pilot plant was completed and was ready for the testing series in three ABOWE countries. The pilot plant was moved in the coldest day of that Winter, −25 °C.

3   ABOWE Biorefinery Pilot Plant

The description of the Finnoflag biorefinery technology was the basis for the proof of technology (Hakalehto et al. 2016a).

There are four main tanks in the pilot plant biorefinery process:

1. HOMOGENIZER is equipped with a biomass crusher and effective mixing. In the homogenizer various biomasses are being mechanically broken in micro- and macroscale. Their dry weight and total masses of solid and liquid raw materials are measured.
2. HYDROLYZER is a thermostatic and pH controlled reactor for producing, maintaining and adjusting the optimal conditions for chemical and/or enzymatic hydrolysis of the macromolecules in the raw material biomasses. Main

parameters are the water content, fill in level, temperature, pH of the biomass, viscosity and the hydrolysis time.

3. BIOREACTOR (Fig. 3) metal works have been manufactured by Brandente Oy in Kuopio according to the instructions of the innovator Adjunct Professor Elias Hakalehto (Finnoflag Oy, University of Eastern Finland, University of Helsinki) and Senior Lecturer Anssi Suhonen, Savonia. The patented design is based on numerous bioprocess runs in Finnoflag Oy's laboratory projects preceding the ABOWE project. Different homogenized and hydrolyzed biomass fractions are processed in adjustable gas conditions in order to produce biofuels, gases and chemicals by the metabolic activities of several bacterial strains and other micro-organisms. During the process runs, temperature, pH, dissolved oxygen, total volume and the gas mixing and measurement are adjusted by the central computer control together with real time operating activity by the personnel on site and connected via 3G network to the pilot plant. The gaseous products are recorded from the volatile outflow of the bioreactor prior to the stabilization.

The microbiological and biotechnical process control was designed by Finnoflag Oy. The microbiological inocula for the bioreactor are produced first in the

Fig. 3 The bioreactor of the pilot plant. *Photo* Ari Jääskeläinen

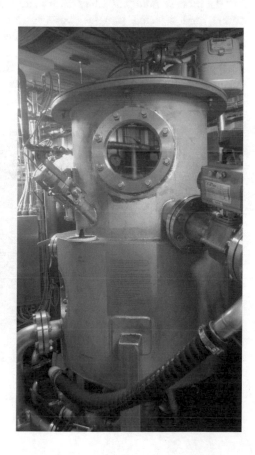

PMEU equipment (Portable Microbe Enrichment Unit) (Samplion Oy, Siilinjärvi, Finland) (Fig. 4), and then in the seed fermenters connected to the main bioreactor. By the PMEU device it is possible to get homogenous cultures in the same active growth phase in a few hours of cultivation (Hakalehto and Heitto 2012). The PMEU was originally planned for environmental, clinical, food and other hygienic control purposes (Hakalehto 2012, 2013).

4. STABILIZER is a cooled collection unit of the bioprocess fluid containing liquid (and possibly solid) products. There the temperature is lowered to 15–18 °C from the usually much higher production temperatures in order to avoid losses in the product concentrations after the process.

Bioreactor is the sole entirely novel big tank in the pilot plant. All the other three tanks have been recycled and modified. Hydrolyzer, for example, originated from the Finlayson Oy cotton factory in Tampere, which is the city in Southern Finland where Finnish metal engineering and other industrialization began almost 200 years ago. There the forthcoming bioreactor tank was used for staining textiles before the tank was modified in Kuopio into a crucial part of the upstream process of biorefining waste biomasses in the pilot plant, during the era of modern reindustrialization.

The process fluid was analyzed on-site with Gas Chromatography by a laboratorian, and selected samples were further analyzed with the NMR (Nuclear Magnetic Resonance) spectroscopy at the University of Eastern Finland, Kuopio by Professor Reino Laatikainen.

Figure 5 presents the Process room of the pilot plant and the Fig. 6 has been taken from the Laboratory and Control room of the pilot plant.

Fig. 4 PMEU—Portable Microbe Enrichment Unit. Manufactured by Samplion Oy, Siilinjärvi, Finland

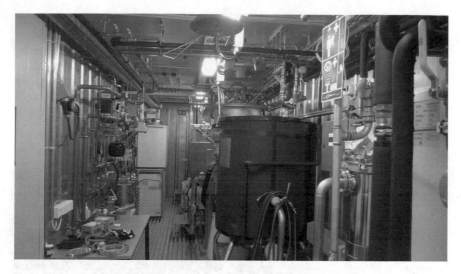

Fig. 5 Process room of the mobile biorefinery pilot plant. *Photo* Ari Jääskeläinen

4   ABOWE Biorefinery Pilot Plant Testing and Public Exhibiting

ABOWE testing partners were in charge of organizing the testing periods in the three countries. Finnoflag Oy brought microbiological and bioprocess know-how in the test runs in Finland and in Sweden and consulted the Polish test runs. Finnoflag Oy also planned the tests and led the testing operations in Finland and Sweden and had a major participation in the Polish tests. They had crucial role in analyzing the test data and in the reporting.

The User's manual was compiled by the Engineering team with help from environmental engineering students. Training for the representatives of Polish and Swedish testing teams took place in February 2014, during the Finnish testing period.

Savon Sellu Oy cartonboard factory's wastewater treatment plant in Kuopio was the first testing site, for February–March 2014. Finnoflag Oy operated the pilot plant tests, with Savonia's support. An environmental engineering student from Savonia and two process operator students from Savo Vocational College participated in the test runs (Hakalehto et al. 2016b).

The pilot plant was transported to ZGO Gac Ltd.'s waste management centre in Lower Silesia, Poland for tests during May to early July 2014 with potato waste from chips production and with municipal biowaste. Wroclaw University of Technology operated the pilot plant tests together with Finnoflag Oy. The Polish testing team included 24 students from Wroclaw University of Technology, in the lead of Associate Professor Emilia den Boer and with the help from five other staff members.

In Sweden the pilot plant was tested at the Hagby's bird farm in Enköping with chicken manure and slaughterhouse waste during August-early October 2014.

Mälardalen University operated the pilot plant tests together with Finnoflag Oy and Savonia. In the Swedish testing team participated a female environmental and energy technology student from Mälardalen University.

National and international teams learned to co-operate well in this milieu where biological components (biomass, microbes and enzymes) met with metal hardware, sensors and computerized control.

The ABOWE biorefinery pilot plant fulfilled its purpose in giving a proof of concept on the biorefining of various biowastes into useful products. The results are presented in (Hakalehto et al. 2016a, b, c; Hakalehto and Jääskeläinen 2017; den Boer et al. 2016a, b; Schwede et al. 2017) as well as in the ABOWE test reports (www.abowe.eu).

Fig. 7 International ABOWE project team visiting the mobile biorefinery pilot plant at its Swedish testing site, the ecological bird farm Hagby's. *Photo* Ari Jääskeläinen

Based on testing results and the proposed biorefinery value chain, investment memos were compiled to promote commercialization of the technology within ABOWE. The target was that after ABOWE there would be industry-driven continuation projects targeting full scale implementation of the technology. A Finnish engineering company was commissioned to pre-engineer a full scale biorefinery plant and its investment costs. RDI Specialists from Savonia Business School coordinated investment memo preparation in each project region.

The project held two national events in Finland in 2014, with the pilot plant publicly exhibited: in April in the Finnish Science Centre Heureka and in October at Helsinki University Viikki campus and Gardenia Tropical Garden. In Poland there was an open seminar in Wroclaw and an exhibition at ZGO Gac Ltd waste management center in July 2014.

Figure 7 shows the international ABOWE project team visiting the Swedish testing site.

5 Student Participation and Their Views on Learning within ABOWE Project

Here are presented various student cases as well as views from some of them based on their theses or other correspondence. The overall view was that the students were glad of the opportunity to receive a valuable working experience within a multi-sector, international and meaningful research, development and innovation (RDI) project.

5.1   Participation in the Pilot Plant Engineering and Construction

Interaction with students began already before the ABOWE project got approved as a team of four young male environmental technology students of Savonia joined in a project assignment. Their task was to investigate technical properties of target wastes as well as existing patents elsewhere regarding the pretreatment technology. Both aspects were helpful for engineering the pilot plant's pretreatment part.

A young male electrical engineering student participated in the Engineering team for eight months as an electrical and automation engineer. He got electrification of the plant as the original task, including the engineering and assembly of the main distribution board and the design of general electrical wiring. Moreover, he participated in device selection and procurement. His task enlargened to cover also other engineering sectors, and supervising electrical and automation installations. He also led the commissioning and related measurements of the pilot plant. Moreover, he took a major responsibility in the technical documentation such as updating process and instrumentation diagram (P&ID). He kept the memos of the Engineering team meetings.

The student in question learnt the theory of bioprocess and biotechnology as well as the automation choices of the process. He learnt that it is essential to know the process thoroughly when planning and drawing the P&ID (Hirvonen 2015). He was a very unprompted and multiskilled member of the Engineering team. He took numerous new tasks which often went even beyond his study field without prejudices. He took care of the tasks in a responsible and competent way. As the team leader of the electrical and automation installment team he showed good manager and problem-solving skills.

Another young male electrical engineering student co-operated in part of these tasks during two months during his practical training period.

Two electrical installer students from Savo Vocational College took their practical training period within the ABOWE pilot plant electrical installments for autumn 2013. One German upper secondary school (with technological emphasis) student joined them for two months. They performed the practical installments, in the supervision of the earlier mentioned electrical engineering student. A senior laboratory engineer from Savonia Electrical Engineering Department supervised the electrical engineering student, in turn. This electrical installment team was practically in charge of the electrical and automation installments in the ABOWE pilot plant.

The senior laboratory engineer reported that in the engineering of instrumentation and automation, students needed to widely familiarize themselves to various alternatives. Concerning sensoring many issues needed to be solved: process demands, maintenance, hygiene and strain to the instruments from cleaning of the process equipment. The biological needs of the microbes needed to be taken into consideration in the structure of the equipment. Intelligent sensors, control devices and motor drives were integrated in the system (Jääskeläinen and Hakalehto 2014; Jääskeläinen et al. 2016).

The modification of recycled vessels into a new use as part of the pilot plant's process as well as mechanical assembly work and the routings of piping were hand made by the construction team of Savonia. Two adult sheet iron worker—welder students from Savo Vocational College and one mechanical engineering student from Savonia took part in the mechanical assembly works.

A Russian young male IT student from Savonia participated the ABOWE Engineering team for couple of months. He participated in gas mixing system investigation and some other technical sectors.

Eight ICT students from Savonia joined the ABOWE pilot plant construction in terms of telecommunications. They engineered practical solutions for internet-connection, back-up copying, antenna etc. Their teacher guided them with help from the Engineering team's process control and automation system engineer.

Three young female environmental engineering students did their project assignment course in ABOWE by participating in the pilot plant User's manual compilation in co-operation with the experts. For example, they adopted a dry matter meter since its arrival and compiled instructions for its use.

5.2 Participation in the Pilot Plant Testing

A shortly graduated young female laboratorian participated ABOWE for about eight months. She adopted a gas chromatography system since its arrival and compiled a manual for its use, in guidance of Savonia's chemist. She brought the system to the pilot plant's laboratory room and independently was in charge of its use during the Finnish tests. She trained the Swedish and Polish testing personnel in its use, as well as the pretreatment of samples, both in theory and practice.

A young male environmental engineering student participated in ABOWE for nearly one year with a varying intensity. He participated in compiling the User's manual for the pilot plant and in running the pilot plant during the Finnish testing period.

Two young male process operator students from Savo Vocational School joined the ABOWE test runs in the Finnish testing site. One of them reported that he started to understand why it would be rational to start producing biofuels and chemicals from sewage sludge. At the same time he learnt what the processed products could be used for. He did not know beforehand how the wastewater is treated at Savon Sellu Oy and how much potential the sludge still contains. Further it came clear for him how much sludge material is all the time going for nothing and how in the future a plant like the ABOWE biorefinery pilot plant could significantly help in utilizing the sludge waste (Auriola 2015). Later on he started environmental engineering studies within Savonia.

A young male laboratorian student participated ABOWE by conducting laboratory analyses of samples from the Finnish pilot plant test runs, in the guidance of a project engineer from Savonia.

One Ph.D. student from Mechanical Engineering Department and one M.Sc. student from Environmental Technology Department of Wroclaw University of Technology, Poland, participated in the training period in Kuopio, together with their associate professor. In turn, they have participated in educating other students when the pilot plant was in Poland.

Two Polish environmental technology students that were interviewed during the Polish exhibition event reported that they were really happy of the opportunity and had experienced the operation with the pilot plant very interesting and useful for them.

5.3   Participation in the Investment Memos and the National Events

A young male environmental engineering student participated in ABOWE by gathering information and drafting an investment memo of ABOWE biorefinery based on the Finnish tests. An outline for this formed the framework for investment memo compilation in the all ABOWE regions. He took part in open ABOWE seminars in Lithuania and Finland. He participated also in representing ABOWE during one international conference in Finland.

Independent and responsible working in an international and multisector environment was the student's personal learning objective. He commented that the training opened up his eyes regarding bio economy and he also got absorbed in the EU climate policy objectives and how these will be implemented in practice. As a whole he experienced that the practical training was different than his earlier training periods in which there had been more supervision and the tasks had been mainly practical duties. He was happy that he had opportunity to be a part of ABOWE project (Vehviläinen 2014).

A female adult business student joined ABOWE for nine months with the primary task to perform economical calculations with the cost data based on the Finnish test runs. This gave example of corresponding calculations also for Poland and Sweden. A secondary task was to take care of business model work with Savonia's InTo and Extended Business Model Canvas tools. As a result was created a starting point for a business plan based on the ABOWE test runs in Finland. She also had a major role in the practical arrangements, together with an RDI specialist from Savonia Business School, as well as the project manager, of the two national events that were held in Helsinki area in 2014.

The business student reported that she was lucky as she could conduct a diploma work that supports global sustainable development and is a multinational one. She has viewed waste based business as a very promising one. The diploma work process was interesting and very educating in her opinion (Kauppinen 2014).

6 Conclusions

The purpose of this chapter was to give an example of how a multidisciplinary development project can act as an educational tool for teaching sustainable development for the participant students as well as a part of continuing education for the personnel.

During the project many challenges were faced in designing and constructing the mobile biorefinery pilot plant which would be both a real production process as well as a research laboratory. There were not available many previous models or practical experience about the effects of various functions. Both the several Engineering team members and numerous students bravely threw themselves in the common working and interaction as well as in the empowering flow of learning. Sustainable development, that the novel piloted biorefinery concept is straight promoting, was clearly one of the key motivating factors.

Students were an integral part of the ABOWE team throughout the project. They really took responsibility in many individual parts and phases. Without their contributions ABOWE project would not have been possible to implement. A general comment from all students was that they experienced that they had got more responsibility than they had expected. Many students have viewed their participation as a very educating experience. This motivated them a lot and they grew during the project both as persons and as professionals.

Also the staff members became enthusiastic about the ABOWE goals and the mobile biorefinery pilot plant and it clearly kindled them. The process control and automation system engineer commented in an informal discussion during the project, that ABOWE was a "self-motivating project". Personnel got freedom to implement the agreed sectors and they tried their best, thus the action clearly matched with the concept of internal entrepreneurship.

ABOWE team received the first place in the national "Kärjet" (Spearheads) competition among the universities of applied sciences in Finland, in the series Applied research knowledge and innovations. In the assessment ABOWE was positively evaluated for its multisector approach and for involving students and personnel from many educational organizations. ABOWE project facilitated international educational and scientific cooperation around the themes and goals of sustainable development.

It can be supposed that each student and staff member who took part in ABOWE has learnt sustainable development in practice, and is better aware to apply the related principles in one's activities in the future. Moreover, all the publications made on the basis of the project are spreading biorefinery education and thus contributing to the teaching of sustainable development internationally.

Acknowledgements ABOWE project was funded by the European Union (European Regional Development Fund). Construction of the mobile biorefinery pilot plant was co-funded by the Ministry of Employment and Economy, Finland and the Regional Council of Pohjois-Savo, Finland.

References

Auriola M (2015) Jäteveden puhdistuksen lietteen hyötykäyttö (Utilization of wastewater treatment sludge). Final Project. Savo Vocational College. Degree Programme in Process Operating (in Finnish)

den Boer E, Łukaszewska A, Kluczkiewicz W, Lewandowska D, King K, Jääskeläinen A, Heitto A, Laatikainen R, Hakalehto E (2016a) Biowaste conversion into carboxylate platform chemicals. In: Hakalehto E (ed) Microbiological industrial hygiene. Nova Science Publishers Inc., New York

den Boer E, Łukaszewska A, Kluczkiewiczc W, Lewandowska D, King K, Reijonen T, Kuhmonen T, Suhonen A, Jääskeläinen A, Heitto A, Laatikainen R, Hakalehto E (2016b) Volatile fatty acids as an added value from biowaste. Waste Manag 58:62–69

Hakalehto E (ed) (2012) Alimentary microbiome—a PMEU approach. Nova Science Publishers Inc., New York, USA

Hakalehto E (2013) Enhanced mycobacterial diagnostics in liquid medium by microaerobic bubble flow in portable microbe enrichment unit. Pathophysiology 20:177–180

Hakalehto E, Heitto L (2012) Minute microbial levels detection in water samples by portable microbe enrichment unit technology. Environ Nat Resour Res 2:80–88

Hakalehto E, Jääskeläinen A (2017) Reuse and circulation of organic resources and mixed residues. In: Dahlquist E, Hellstrand S (eds) Natural resources available today and in the future: how to perform change management for achieving a sustainable world. Springer, Germany

Hakalehto E, Heitto A, Suhonen A, Jääskeläinen A (2016a) ABOWE project concept and proof of technology. In: Hakalehto E (ed) Microbiological industrial hygiene. Nova Science Publishers Inc., New York

Hakalehto E, Heitto A, Niska H, Suhonen A, Laatikainen R, Heitto L, Antikainen E, Jääskeläinen A (2016b) Forest industry hygiene control with reference to waste refinement. In: Hakalehto E (ed) Microbiological industrial hygiene. Nova Science Publishers Inc., New York

Hakalehto E, Heitto A, Andersson H, Lindmark J, Jansson J, Reijonen T, Suhonen A, Jääskeläinen A, Laatikainen R, Schwede S, Klintenberg P, Thorin E (2016c) Some remarks on processing of slaughterhouse wastes from ecological chicken abattoir and farm. In: Hakalehto E (ed) Microbiological industrial hygiene. Nova Science Publishers Inc., New York

Hirvonen T-P (2015) Siirrettävän biojalostamon prosessi ja sähköistys (Process and Electrification of Mobile Biorefinery). Bachelor's Thesis. Savonia University of Applied Sciences. Degree Programme in Electrical Engineering (in Finnish)

Jääskeläinen A, Hakalehto E (2014) ABOWE-hanke integroi tekniikan opetuksen ja huippututkimuksen (ABOWE project integrated technology education and top research). Automaatioväylä (in Finnish)

Jääskeläinen A, Rissanen R, Jakorinne A, Suhonen A, Kuhmonen T, Reijonen T, Antikainen E, Heitto A, Hakalehto E (2016) How does modern process automation understand the principles of microbiology and nature. The 9th Eurosim congress on modelling and simulation. 12–16 September, Oulu, Finland

Kauppinen M (2014) Uuden Liiketoiminnan luominen asiantuntijaorganisaatiossa. Case: Biojalostusteknologiaosaamisen tuotteistaminen Savonia-ammattikorkeakoulussa (Creating new business in a professional organization. Case: creating services based on biorefinery technology knowhow at Savonia University of Applied Sciences). Bachelor's Thesis. Savonia University of Applied Sciences. Degree Programme in Business and Administration (in Finnish)

Liukkonen J, Jaakkola T, Suvanto A (2002) Rahasta vai rakkaudesta työhön. Likes- ja työelämäpalvelut Oy, Jyväskylä

Patrikainen R (1999) Opettajuuden laatu. Ihmiskäsitys, tiedonkäsitys ja oppimiskäsitys opettajan pedagogisessa ajattelussa ja toiminnassa. PS-kustannus, Jyväskylä, Finland

Rauste-von Wright M, von Wright J, Soini T (2003) Oppiminen ja koulutus. WSOY, Helsinki

Räsänen J (2008) Notes from the lecture at Savonia University of Applied Sciences Developing Days in August 7th and 8th. Koli, Finland

Sagor R (2000) Guiding school improvement with action research. Association for supervision and curriculum development

Savonia University of Applied Sciences (2017) Web-pages. http://portal.savonia.fi/amk/en

Schwede S, Thorin E, Lindmark J, Klintenberg P, Jääskeläinen A, Suhonen A, Laatikainen R, Hakalehto E (2017) Using slaughterhouse waste in a biochemical based biorefinery-results from pilot scale tests. Environ Technol 10:1275–1284

Vehviläinen M (2014) Assessment of the investment memo of ABOWE Pilot A Finland. Bachelor's Thesis. Savonia University of Applied Sciences. Degree Programme in Environmental Engineering

Author Biographies

Ari Jääskeläinen M.Sc. (Industrial Management, with technological orientation in environmental protection technology) and B.Sc. (Social Sciences), serves since 2005 as Project Engineer/RDI Specialist and part-time Lecturer in the Environmental Engineering Department of Savonia University of Applied Sciences. He has managed or been involved in several international and regional EU-projects within environmental technology and business field. In ABOWE he managed and coordinated the overall project from the original project idea through preparation and implementation to closure. To his responsibilities included e.g. overall project coordination and administration, budget planning and finances, communications, coordination of the Engineering team and organization of events. He would like to thank all the students, personnel, companies and partners who have taken part in the ABOWE project for each one's important contributions for the common goal.

Elias Hakalehto serves as an Adjunct Professor of Microbiological Agroecology for the University of Helsinki, and as an Adjunct Professor for Biotechnical Microbe Analytics at the University of Eastern Finland, Kuopio. He has participated with his R&D company, Finnoflag Oy, in more than 100 specific investigation tasks in industrial hygiene monitoring, environmental protection, diagnostics, bioprocess development, as well as research on clinical microbiology and probiotics. He is the principal author and editor of the "Alimentary Microbiome—A PMEU Approach", as well as in the ongoing edition of the series 1. Microbiological Clinical Hygiene (2015), 2. Microbiological Food Hygiene (2015), 3. Microbiological Industrial Hygiene (2016), 4. Microbiological Environmental Hygiene (to be published in 2017–2018). He has contributed also to several other books, scientific articles and patents in the fields of microbiology, biotechnology, microbiomes and biorefinery technology. Validations of his inventions regarding the PMEU (Portable Microbe Enrichment Unit) technologies have been carried out e.g. in water quality control, and in the surveillance of contamination in specific industries. Adj. Prof. Hakalehto has been the principal technology provider in the six nation EU Baltic Sea Biorefinery project ABOWE 2012–14. He is interested in sustainable bioindustries, environmental health and monitoring of microbial metabolism in real-time. Dr. Hakalehto has given numerous lectures and conducted teaching sessions in domestic and international conferences, fairs and courses. In summer 2017 he served as the Chair of the 22nd International Conference on Environmental Indicators in Helsinki, Finland.

Reflections on Using Creativity in Teaching Sustainability and Responsible Enterprise: A First and Second Person Inquiry

Helena Kettleborough, Marcin Wozniak and David Leathlean

Abstract This paper is a response to the need for teachers in higher and further education to better engage students in learning about the urgent but complex issues of sustainable development and responsible business practice and applying that learning in a range of disciplines. The potential of higher education (HE) to address these challenges through creativity is highlighted alongside emerging challenges for the HE sector. Undertaking creativity in teaching in practice is explored through the 'I Love Learning' Project (I♥L), an interdisciplinary project aiming to encourage creativity in teaching as a means to motivate and inspire students, supported by the Centre for Excellence in Learning and Teaching at Manchester Metropolitan University (MMU). The contribution of first and second person action research methods to the development of creativity in higher education is examined. Drawing from practice, the paper offers two models employing extended ways of knowing to categorise creative interventions and suggests a **'creativity map'** as a model to identify different elements of learning, presenting feedback from students on how the initiatives were received and providing some reflections on quality in the creativity initiatives undertaken. The paper argues for a participatory action research model to grow creativity in higher education through a self-organising structure, imagining **'creativity greenhouses'** as a model. The paper draws the threads together to highlight creativity's role in teaching sustainability and responsible enterprise and offers practice based insights. The authors are colleagues from different disciplines, working together in a co-operative inquiry to identify ways to make learning more creative.

H. Kettleborough (✉) · M. Wozniak
Manchester Metropolitan University Business School, All Saints Campus,
Oxford Road, Manchester M15 6BH, UK
e-mail: h.kettleborough@mmu.ac.uk

M. Wozniak
e-mail: m.wozniak@mmu.ac.uk

D. Leathlean
Manchester Fashion Institute, Manchester Metropolitan University,
Righton Building, Cavendish Street, Manchester M15 6BG, UK
e-mail: d.leathlean@mmu.ac.uk

© Springer International Publishing AG 2018
W. Leal Filho (ed.), *Implementing Sustainability in the Curriculum of Universities*,
World Sustainability Series, https://doi.org/10.1007/978-3-319-70281-0_11

Keywords Creativity · Learning · Action research · Sustainable development
Responsible enterprise · Extended ways of knowing

1   Introduction: A New Eaarth

Teaching sustainability and responsible enterprise is gaining ground within UK
universities and globally, evidenced through such networks as PRME (2017) and
the Inter-University Sustainable Development Research Programme (IUSDRP
2017). The reason for this is to be found in the urgent challenges facing the planet.
Indeed McKibben argues eloquently that the planet we are living on, which he
renames Eaarth, is completely different from the Earth our human civilisation
developed on (2010: 45). As the Millennium Eco-Assessment identified, human
beings depend on ecosystems for their economic survival (2005). In 2009,
Rockström et al. identified nine planetary boundaries, of which three—Loss of
Biosphere Integrity, Climate Change and Altered Biogeochemical Cycles—had
been breached, to which Steffen et al., added a fourth breach—Land System
Change—in 2015.

Whilst human development levels have been rising in terms of numbers lifted
out of extreme poverty (MDG 2012), the challenges facing the more-than-human
world[1] continue to increase in difficulty. For example, 'Loss of Biosphere Integrity'
is captured in a recent WWF report:

> Species populations of vertebrate animals have decreased in abundance by 58 per cent
> between 1970 and 2012 (2016:14).

Further, reporting on a major conference at the Vatican:

> Biologists think 50% of species will be facing extinction by the end of the century (McKie
> 2017)

Despite this, Kolbert argues that even critically endangered species can be saved but
it is necessary to have the resources and people power to save them (2014).

Attracting more political and media attention, 'Climate Change' is a daunting
threat both to humanity and to the more-than-human world. 2016 was the third
hottest year on record (Graham-Harrison 2016) and 16 of the 17 warmest years
since records began occurring since 2001 (NASA 2017). As an indication of how
fast ice in the world is melting, the Global Ark for seeds built in Norway in 2001 is
now threatened because the permafrost is melting (Graham 2017).

In the Tutorial room or seminar, however, it is not always immediately obvious
how to take these imperatives forward within or alongside the curricula and contact
time of the diversity of disciples where the students are neither expert in the science
nor predisposed to see the connections between these urgent issues and their chosen

---

[1]More-than-human-world was coined by David Abrams to describe the "sensuous world in which
our techniques and technologies are all rooted" (1996: x).

subjects. The purpose of this paper is to present the authors' exploration in creative approaches to the teaching of sustainable development and to present two practical models with the advantage that they can be developed and applied to teaching by individual teaching staff without requiring institutional buy in or leadership, frequently a barrier to progress (Leal Filho 2015).

Teaching on sustainable development and responsible enterprise can be traced from its beginnings with a variety of 'issue based' education initiatives and with environmental education (Bullivant 2010: 9). The UN Decade for Education for Sustainable Development 2005–2010 (UN 2005) provided a major impetus. Alongside this the private sector has increasingly been placing emphasis on corporate social responsibility (Laasch and Conaway 2015: 86). Examples of teaching for sustainable development appear in a variety of the literature, for example, from the Development Education Centre (2003), Gadsby and Bullivant (2010) considering the secondary sector, specific examples in the higher education sector (Grinsted 2015) and HE sector reviews (Leal Fihlo et al. 2014).

2   Universities and Sustainability

What has been be the response of universities to these challenges? Universities cover the entire globe and have an important role in tackling these issues. Berry identified universities "as one of the four basic establishments that determine human life in its more significant functioning" which he identified as "the political, religious, intellectual and economic establishments" (1999: 72). Two decades later, Leal Filho re-enforced that importance:

> universities train millions of graduates each year—across all the disciplines and are hence in a strategic position to foster a broader awareness of what sustainable development is (2015: 5).

Given the importance of the challenges facing human society, Berry identified "…it is the time for universities to rethink themselves and what they are doing" (1999: 85).

Homer-Dixon (2006) argues that students need to be flexible in the face of unknown challenges and gain a 'prospective mind' in addressing the great challenges facing humanity. Seeking thinking which can help develop solutions, Craft (2011) wants students to acquire skills for discovering 'future possibilities'. In an important study, Botkin, Elmandjra and Malitza argue for 'anticipatory learning' whereby humanity engages with the challenges and works out appropriate responses rather than learning by reacting to crisis and disaster (1979). Philosopher Gregory Bateson sees that teaching is predominately within linear and silo approaches, whilst the planet operates in holistic and systematic ways (1972). For Berry, universities need to teach that humanity is part of the planet and wider universe, and take responsibility for acting, rather than feeling fear for the future (1988: 97).

Fig. 1 Main approaches
taken by universities towards
sustainability (Leal Filho
2015: 5). With permission of
Springer Nature

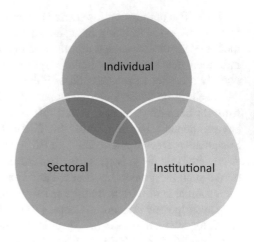

Sadly, Leal Filho (op. cit.: 5–6) concludes that, whilst individual members of staff and certain faculties and departments have taken this message on board: "Unfortunately only a small proportion of Universities adopt the institutional approach…where there is a commitment from the whole university towards sustainable development…which shows that much still needs to be done". Bateson and Berry's calls have yet to be wholeheartedly taken up (Fig. 1).

3 Creativity

Creativity in education is a developing field. For Perkins: "(a) a creative result is a result both original and appropriate (b) a creative person—a person with creativity is a person who fairly routinely produces creative results" (Perkins cited in Starko 2010: 5). Starko distinguishes between "Creativity with a big 'C' that changes disciplines and creativity with a little 'c', the more common place innovations of daily life" (ibid: 5) to which Beghetto and Kaufman added the concept of 'mini-c' creativity, which "focuses on the novel and personally meaningful interpretation of experiences, actions and events" (2010: 195). Creativity can be considered subject specific, as covering the whole institution, and as engaging with activities outside the learning institution; a "lifewide curriculum" (Jackson 2010: xvii). Examples of creativity in informal educational settings are, for example, extensively explored in Playing for Time (Neal 2015).

Teaching can be creative in its methodology or creativity can be taught (Cropley 2001; Reid and Petocz 2010). However creative teaching requires thought and effort, and where innovative creative techniques are introduced, may face opposition. Higher Education Institutions can be ambivalent about creativity (Jackson 2013; Beghetto and Kaufman (2010). The latter suggest that education claims to value creativity but teachers often see creative students as 'disruptive, impulsive or wayward' (op.cit: 454–5). Equally, in the authors' experience, not all students will welcome active creative processes, at least initially. Using creativity well requires wisdom in teaching practice and an ethical approach (Craft et al. 2008). It was in response to these challenges that the authors proposed and conducted the 'I Love Learning' project from 2015 onwards.

Creativity can be seen as belonging to a range of initiatives to embed excellence within teaching provided in Higher Education. Manchester Metropolitan University, in its submission to the UK Government's Teacher Excellence Framework (HEFCE 2017), cites how it seeks to "build a culture for teacher innovation" through "supporting and disseminating creative and effective approaches to teaching and learning", including the CELT Scholarship fund, the CELT website 'good practice exemplars' and the CELT Academic Journal (ibid: 6–7). The I♥L project participated in these initiatives. The same submission makes a reference to how initiatives aimed at teaching excellence often arise from "discipline and practice specific experience" (ibid: 15) lending force to the I♥L project which focuses on bottom up creativity in teaching responsible enterprise and sustainability (see below). This paper seeks to pull out the practice and learning from embedding creativity through specific practice in teaching for teaching in sustainability and responsible business.

4 'I Love Learning' Project

'I Love Learning' (I♥L) was a Scholarship of Teaching and Learning project funded by the Centre for Excellence in Learning and Teaching (CELT) at Manchester Met. The authors were academic and support staff who worked together collaboratively and across our different disciplines—Business Responsibility & Sustainability, Student Support and Fashion Promotion—to explore how to develop and embed creative approaches within our practice and the institution. It focused on encouraging creativity in teaching and learning support to motivate and encourage students.

We considered how to motivate students to help them be successful in their studies though innovative and creative approaches across teaching and support. Our aim was to support students' learning in a creative and imaginative ways, to deliver value in education that is not just about money, but also inspiration and creativity and which helps them develop key competencies for lifelong learning. These key competencies (ec.europa.eu 2017) combine knowledge, skills and attitudes needed for personal fulfilment, active citizenship, social inclusion and employability. Such competencies include communication, science and technology, social and civic competence, a sense of initiative and entrepreneurship, cultural awareness and expression, digital competence and learning to learn. These key European Union defined competences are mutually dependent and rely on critical thinking, finding creative solutions, taking initiative, problem solving, risk assessment, decision taking and management of feelings. I♥L's starting point was that incorporating creative approaches within HE can help students cultivate such competencies and so to become lifelong learners to help them transform their lives and those of others.

5 Putting Creativity into Practice

During the I♥L project, we introduced various creative approaches within teaching and learning, such as the use of stories and metaphor and bringing students into new environments inside and outside of the university to enhance their motivation, engagement and empowerment.

Marcin Wozniak, in his role as Student Experience Support Tutor, used a metaphor of a growing plant, nurturing and cultivating its roots in a context of professional development. During a 'Social Enterprise Ideas Workshop' that aimed to engage students in extracurricular activities, develop ideas and 'grow' them into projects or social enterprise, students planted a seed representing their developing idea. Later they nurtured their growing plant and documented the progress both of the plant and the idea via social media with peers and staff, to motivate one another. The metaphor of nurturing of a growing plant, supported by the making process of the seed planting and aftercare activities, aimed to enhance memorisation of the activity and of the message conveyed within the workshop (Low 2008). The experience is presented in Wozniak et al. (2017) by the present authors. Owen (2001: xvi) suggests that metaphor can help to "externalise abstract thinking and translate it into a sensory-based tangible representation".

Within the I♥L project, we explored different environments that allowed us to bring creativity into our teaching and to take learning outside the classroom including the university Library's Special Collection, the foyer of a theatre, art galleries, typical classrooms which were filled with sounds, colour and creative materials, and an unused Greenhouse on campus—an active project space and meeting point for students adjacent to an Ancient Woodland and a space for reflection and fresh air. As James and Brookfield (2014:4) suggests, 'for learning to "stick" … the fullest range of our imaginative faculties needs to be engaged'. Such experiences of various environments led to subsequent development within David Leathlean's Manchester Fashion Institute Foundation Year Programme, where newly introduced 'Active Learning Weeks' encourage student participation in their learning outside of the classroom. The student cohort experienced visits to easily accessible Manchester Art Gallery and Manchester City Centre's retail district, which they perceived to be fun, interesting and exciting. New or changed environments make it easier for students to be inspired, creating an energy that would not have been achieved otherwise (Wozniak et al. 2017). Clark (2002: 9), comments on "the positive impact [*on learning*] of changing the environment". This simple change in offering students a different location can enthuse students' approach and awaken their creativity as well as helping learning to 'stick'.

5.1   Creativity and Responsible Enterprise

Tutorial teaching to second year undergraduates on the core Responsible Enterprise Unit on the University's Business School degree courses—a core unit for over 500 students—was used as a case study for I♥L.

> *Responsible Enterprise*
> *The Unit examines the nature and shape of business structure resulting from the responsible business initiatives and approaches enshrined in corporate governance/legal frameworks and implied by the growth/rise of CSR and sustainability consciousness.*
>
> *Term 1 focuses on understanding the context and drivers behind responsible enterprise. Specifically, we consider the key global challenges we face, how "responsible enterprise" might help address those challenges & why greater collaboration between business, government and non-profits' might be necessary in order to do so*
>
> *Term 2 focuses on the responsible enterprise in action. Specifically, we consider how organisations respond to the challenges identified in Term 1, including RE as a competitive opportunity, Strategy as a tool for more responsible enterprise, Responsible enterprise across the supply chain and Addressing human rights & other ethical challenges* (Manchester Metropolitan University 2017).

In all, twenty four weeks of tutorials with two cohorts were used to explore creative interventions developed through I♥L project inquiry by Helena Kettleborough.

6   Growing Out of Action Research

The methodology chosen for the project inquiry was action research, utilising characteristics drawn out in this definition:

> A participatory, democratic process, concerned with developing practical knowing in the pursuit of worthwhile human purposes, grounded in a participatory worldview...it seeks to bring together action and reflection, theory and practice, in participation with others, in the pursuit of practical solutions to issues of pressing concern to people, and more generally to the flourishing of individual person and their communities (Reason and Bradbury 2001: 6)

Quality issues which arose for the authors included the complex nature of the language and the extensive number of approaches in the literature which were challenging to learn (Bradbury 2015a, b). Quality issues are a major concern in the action research literature (Reason and Bradbury 2006, 2008; Marshall 2004, 2016). The authors were aware of critiques raised by these and other sources relating to the process of first person enquiry, interpreting the data, making sure the enquiry was sufficient in extent and depth and awareness of issues of diversity and power, the latter a particular issue in the teacher–student relationship (Kettleborough 2014).

The researchers used first and second person action research methodology (McNiff and Whitehead 2011; Bradbury 2015a, b) in the setting up of the project and in undertaking the research, outlined in further detail in the following two sections.

7   Categorising Creative Interventions

Studies of the literature showed various ways to categorise creative interventions, for individuals, for groups, by content area, for group working and team working (Cowley 2005; Craft 2011; Starko 2015). In the I♥L project, teaching and learning support were undertaken between the cycles of action and reflection. In project meetings the authors would talk about practice, what was being taught, how it was being taught, and the creative interventions being made. In these reflections, ideas for different or alternative ways forward emerged, sometimes from authors' experience, sometimes linking back to ideas from the CELT CPD Creativity for Learning course, sometimes from reading undertaken and sometimes through discussion. Over time, and due to the extensive and varied experience in the group, a large number of creative processes emerged. The authors decided to use extended ways of knowing as a helpful way to categorise interventions and enable understanding of creativity to link into the learning undertaken.

8   Towards a New Model for Creative Learning 1: A 'Creativity Map' Using Extended Ways of Knowing

A characteristic of action research is an extended epistemology, suggesting that there is not one way of knowing but many. One way of defining an extended epistemology is through the work of Heron who explores four ways of knowing (1996).

First: <u>experiential knowing</u>, including the knowing of experience, life, living.
Second: <u>presentational knowing</u>, of story, art, music, expression.
Third: <u>propositional knowing</u>, the knowing of academic literature, policy, science, left brain knowing.
Fourth: <u>practical knowing</u>, taking action out in the world to create a better place, building on the other three forms of knowing as well. (ibid: 32–34).

These ways of knowing could be seen as circular or as forming an 'up hierarchy', in the shape of a triangle, with practical knowing at the top (ibid: 33).

The following are examples of creative interventions understood through extended ways of knowing which were all used in the Responsible Enterprise Unit modules to encourage learning in the Tutorials.

Creative interventions understood through experiential knowing by influencing the environment of the seminar room, both visually and through sound, for example through using coloured pens, putting out creative arts and crafts material and starting the session with music. Sharing pieces of music selected by students changed the mood in the room, through bringing in feelings and emotions (Boyce-Tilman 2016). Another example of this was taking students out of one visual environment into an entirely new one, which David did in his Active Learning weeks (Wozniak et al. 2017) or in Responsible Enterprise, bringing outside environments into the classroom through video clips.

Creative interventions understood through presentational knowing. Students were shown some line poems from Richard Long about the earth and encouraged to have a go writing their own, using instructions provided by university colleague Sam Illingworth (2017). Case studies used within the course curriculum are stories, but generally presented in a concise, factual way. A creative approach was to seek to become a storyteller, for example recounting conservationist Aldo Leopold's epiphany moment, of seeing the 'fierce green light dying' in the mother wolf's eyes as he shot her (1968). Such a story helps illustrate how we need to see the earth as living to be shared rather than a resource to be exploited. These approaches helped explore how knowledge can become embodied within the student, rather than as a detached observer (Seeley and Reason 2008).

Creative interventions understood through propositional knowing. The theoretical model of the self-organised learner was explained to students, seeking to empower them and believe in themselves (Zimmerman 2002). A further model of co-learning was offered to students, suggesting that in the tutorial there are three sources of learning—the tutor, the individual students and their peers—and encouraging them to make best use of all three. Students were encouraged to get to grips with reading an academic text or understanding the elements of their assignment through using colour to separate the text into sections, thus helping to organise and structure the text.

Creative interventions understood through practical knowing. These interventions involved making things to deepen knowledge. Drawing representations of strategic ideas using colour and creative materials supported students' understanding of concepts. For example, the Natural Step Business Strategy, which began as a dry, conceptual text, became a visual funnel shape, drawn by students or made from pipe cleaners, that the world needs to pass through (Robert 2003). The drawing or making process allowed for better understanding of the concept by each student. The 'closed loop economy' turned into a drawing of a cherry tree, as everything the tree produces is recycled and used again: the tree has no waste which is discarded (McDonough and Braungart 2002). Using creative materials to make a sustainable supply chain, such as smiley faces, stickers, pompoms, and coloured pens turns the students into creators, moving from being consumers of knowledge to makers of knowledge (Kumar 2016). Other techniques employed included free fall writing [writing what comes into participants mind and neither censor or cross out (Turner-Vesslago 2013)], using some of the skills of dreaming, from appreciative inquiry (Zandee and Cooperrider 2008; Berry 1988) learning the skills of forecasting, used in the Natural Step models (Robert: 2003: 245) and by Jonathon Porritt in imagining how humanity created a sustainable world (2013).

9 'A Creativity Map' of Interventions

I♥L project participants came to see they had created a 'map' or box of ideas, which linked each of the way of knowing to an example (Table 1).

Table 1 Growing a 'creativity map' using Heron's extended ways of knowing

Extended ways of knowing	Extended ways of knowing as used with the I♥L project	Examples of creativity used in learning environments
Experiential	Changing the environment to allow for different experiences in learning	• Coloured pens on tables; art and crafts materials out. • Teaching in a different environment: outside • Playing music in the session • Using social media (Instagram, Twitter, Facebook, Snapchat) to explore new environments and connect with others
Presentational knowing	Learning through use of story, art, poetry, literature, music	• Creating stories • Using creative space with art materials • Creating line poems • Showing short video clips
Propositional knowing	Developing academic skills	• Sharing academic models which empower students (e.g. models of reflection) • Sharing techniques to visualise or dream • Encouraging students to see peers as part of the learning process—collaboration
Practical knowing	Learning through doing and including all the others	• Creating an academic poster using creative approaches: drawing images, stories, to explain and develop ideas • Exploring academic ideas using drawing, stories, making
	Developing reflection	• Using free fall writing • Working in silence • Writing down dreams for the future • Using writing to develop reflections on progress made

Source Kettleborough et al. (2017)

10   Reflections on Quality and the Creative Interventions Undertaken

The literature points to a variety of criteria for quality evaluation, including the feedback from the students and how the students have learned. In using particular creative methods, we reflected on particular quality issues. Mead asks whether stories have helped the listener pass over a threshold to gain wider understanding (2011). For Illingworth and writing line poetry, an important aspect is to encourage the students to have a go (2016). In free fall writing, facilitator Helena wanted the students to get started and write, and not be judgemental towards their 'standard' of writing (Murray 2013).

As part of the second person inquiry the authors identified their own indicators of the degree to which the project delivered on its aims. For Marcin, getting students involved and encouraging them to be creative themselves as part of university life was important. This was reflected in his work with a group of students who set up a 'Sustainability and Growth Society' and a 'Student Academic Innovation Journal'. For David, it was the visual or creative quality of the work produced. For Helena there was the quality of silence when students are engaged with reflecting or free fall writing and there is the contented buzz when people are creating happily in groups. Other indicators of quality emerged through practice when students remember what you have taught them through the creative methods used, for example the visual shapes of the sustainability strategies for business. Helena requested anonymous, freeform feedback from students on the Tutorial at the beginning and end of each term. In a number of Tutorials posters and artefacts were produced and students were encouraged to take photographs not only of their own work but also of other students' work, to share their learning. At the end of the year, students were asked to give prompted anonymous feedback against checklist headings.

Finally Helena, nominated by her students was awarded Outstanding Teacher in Sustainability in 2016 by the MMU Students Union and shortlisted following her students' nomination for Outstanding Teacher in Innovation in 2017 which she directly attributes to the work undertaken in the I♥L Project (Table 2).

The academic literature identifies many ways in which learning can be evaluated. In reflecting on student feedback the authors reflected on how creativity interventions helped deepen the learning undertaken, akin to Jackson's idea of the life deep learning (2010), how it helped students engage with the learning, starting on the road to lifelong learning (Longworth 2003) and simply the love of learning implied by some of the feedback, a humanist impulse particularly valued by Helena (Hutchins 1968).

Table 2 Feedback from students related to the 'I ♥ Learning' creativity project, 2016

Checklist headings	Written anonymous feedback from students
Free fall writing	Free fall writing is a good method that stimulates creativity I appreciated free fall writing exercises, as it simply encouraged me to write, especially when starting essays Free fall writing is a good method that stimulates creativity
Use of colour	When we use coloured pens, it is not highlighted only on a paper but in my mind Coloured pens are useful for memory retention in assignments; same for essays I like colour—images help me memorise
Drawing	When I drawing a poster I manage to visualise the idea Creativity helps you memorise a lot more of the learning than just paper and ink Drawing and posters, free fall writing: I enjoyed these and found them useful at encouraging participation and engaging the class, as they were creative Drawing hot posters and interacting with one another has helped us become a learning community
Music at the start of the session	Music in the tutorial was creative and fun when we could pick the songs to listen too, good icebreaker Music in the tutorial makes me wake up and relax Music can be influential if used properly, i.e. uplifting and powerful
Using visual illustrations of subject matter	The newspaper cuttings regarding environmental issues were relevant—a good creative way to make subject topics more real and relevant and brought home real life examples Cuttings for biodiversity loss great knowledge
Empowerment	When I think about spaceship earth I recognise that the problems faced by the earth and resources are very much my own issues and not just for someone else to deal with Positive attitude to students work The Unit helped me understand how to be more responsible as an individual and how take the best of it. How to combine the sustainable with the business ideas
General feedback	I enjoy this Unit more than any other, as we are able to be more creative and interactive with our learning By bringing positivity and creativeness into the classes, it is more uplifting and motivational to get involved and try new things Interactive (not sitting and listening) I like having more creative tutorials, as makes the content more interesting and memorable Level of interaction among groups
Critique	Creativity plays a significant role in the Tutorials, whilst for some this is beneficial, for others it can be distracting. So maybe less creativity activities, and more traditional teaching, or a half medium Do not enjoy free fall writing/don't benefit from it

11 Towards a New Model for Creative Learning 2: Kick Starting Creativity in Teaching and Learning— Growing 'Creativity Greenhouses'

Over the three years they have worked together the authors reflected on how creativity might be further encouraged in higher education through a whole institution approach, adopting creativity at all levels, or as the individual teachers' responsibility (Leal Filho et al. 2014).

For Jackson, as a keen advocate of creativity in higher education, the challenge is how creativity can be more deeply embedded:

> If we want learners to be creative, we have to foster their will to be creative and help them develop the confidence, knowledge and capabilities to be creative. Imagine inventing an education system that has (this) as its core value and purpose (Jackson 2010: xix).

The authors see the adoption of the whole institution approach as powerful and a worthy end goal. They were encouraged by the Creativity Greenhouse group run by CELT, a group of likeminded people who meet and explore creative approaches in teaching. However, they recognised the many barriers in the way of creativity, such as time, lack of confidence and resources. Marcin was keen to collaborate with students and staff to develop the derelict greenhouses, a place for growing plants but also growing ideas and creativity with students and had some successes in this area (Wozniak 2016). His approach was to engage students through creative approaches that enthuse and encourage them, that lead to the development of new projects and collaboration between students and staff and further expand creativity, to motivate and inspire others. The authors developed this idea through the I♥L project, seeing that staff can build their own 'creativity greenhouses' which in turn can foster creativity within tutorials, lectures and across programmes, departments and HE institutions.

12 'Grow Your Own 'Creativity Greenhouse'—A 'Can Do' Approach

The first quality of 'Creativity Greenhouses' is that they are self-organising. They do not need the approval of the HE institution to get started. The model has a potential number of distinct elements which can be re-arranged at will:

- *Getting started: external or internal impetus* An initial impetus for the creativity learning needs to be found, either external to or internal to the participants. We argue that developing creativity in learning requires a small group of companions traveling together. External impetus can be, for example, through CPD opportunities, research scholarship opportunities and conferences. For the I♥L project the initial input to drive the project was the role of the Centre for

Excellence in Learning and Teaching. The learning provided aimed to take participants out of their comfort zones and into new ways of experiencing learning (Cobb et al. 2016; Kettleborough 2015). The second way is an internal impetus, when a group of staff committed to creativity meet over coffee and agree to work together to promote creativity in learning. Such a commitment arises from the knowledge that to implement creativity in learning on one's own is lonely and will not result in synergy between group members.

- *Growing out of an action research methodology* As indicated earlier, the methodology chosen for the project was action research (Reason and Bradbury 2006, 2008). In engaging with such a methodology, the authors felt that using elements of a co-operative inquiry, of equal co-authors inquiring together to meet a shared challenge over cycles of action and reflection, was appropriate to the project (Torbert 2001; Torbert and Taylor 2008). In the I♥L project, the shared challenge was how to motivate and inspire students through using creativity in learning. The characteristics of action research as drawn out by Marshall et al. (2011) and further developed by Bradbury (2015a, b) allow for an emergent process within an overall frame.

- *Faithful adherence to cycles of action and reflection throughout the project* In meeting together in regular cycles of action and reflection, firstly to discuss and undertake the research (2015/16) and during 2016/17 to process and write up, the emergent nature of the inquiry came to the fore, which can be traced in other accounts of co-operative inquiry practice (Maughan and Reason 2001; Yorks 2015; Kettleborough 2015). In developing their work, the authors used elements of both first and second person inquiry. First person inquiry involved the development of ideas 'through inner and outer arcs of attention' in individual teaching sessions and in supporting students (Marshall 2001). In coming back to inquire together, second person inquiry methods were used.

- *Growing skills in creativity through learning together* 'Creativity Greenhouses' need input in terms of ideas, theory and practice in creativity and learning. At Manchester Met, these ideas were initially developed through the Creativity for Learning course at CELT in which the authors participated. Ideas were also learnt through peer learning and knowledge exchange during cooperative inquiry groups where staff from across roles and faculties were invited to trial and discuss creative approaches. In developing these ideas in teaching, it is only through practice that the relevant skills can be developed. Such practice and reflection occurred with the use of music, free fall writing, developing active learning and using skills in social media with students (Wozniak et al. 2017).

- *Using the characteristics of action research to deepen inquiry* As well as using first and second person inquiry, the group used some of the characteristics of action research to help structure what they were doing. Clear targets were established for the project but in following the characteristics of action research and using the notion of emergence, of allowing events and understanding to unfold, the detailed shape of what was to be done was not set in stone. The shape of the project was therefore allowed to emerge as the weeks went by and it could vary as the situation changed. The group explored the academic literature they brought to the process and in so doing started to work with the concept of knowledge in action and action in knowledge (Friere 1970/96.). One of the main characteristics of action research developed as central to the project was the notion of extended ways of knowing, which was explored above.

- *Keeping going: continuing light touch support* Plants in greenhouses need water, fertilizer and sunshine. Mentoring for Helena was provided by Liz Walley, who introduced sustainability into the curriculum of the Business Management Degrees where were the processor of the Responsible Enterprise Unit (Christian and Walley 2015). Liz supported the introduction of creative ideas into the tutorial learning on an ongoing basis. Further support came from CELT, through informal sessions with Chrissi Nerantzi, course leader for Creativity for Learning (CELT 2017).

13 Strengths of the Model

The model created mutual peer support and encouragement, which enabled participants to learn new skills and take risks in their thinking and teaching. Trying something new can often require courage. The model took participants out of their comfort zone in a secure way, providing a structure for trying out new skills. The model was rooted in a strong theoretical tradition, with plentiful examples of practice in the literature (Marshall et al. 2011; Mead 2011; Richardson 2000).

14 Self-organising, from the Bottom up

The beauty of such an approach was that it did not require any radical paradigm shift in thinking in any part of the organisation. The I♥L project was self-organised and able to develop as participants saw fit. The model allowed the development of

Table 3 Elements of 'Creativity greenhouses' to grow creativity in HE

Stages	Examples of Elements
Getting started	Inputs (external or internal) to kick start the process
Growing out of action research	Using definitions of action research
Faithful adherence	Using cycles of action and reflection research
Growing skills in creativity	Learning with practice and support
Deepening inquiry	Using characteristics of action research
Keeping on	Light touch support
Utilising strengths	Peer support and encouragement Self-organising, from bottom up

Source Kettleborough et al. (2017)

approaches and ideas which cut across disciplines and did not need approval of the hierarchies to go forward. As such it emerged as a bottom up approach to developing creativity in the organisation, small steps on the way to Jackson's creative learning institution (2010) or Berry's self-organising university (1988: 107). The jury is still out, however, as to the extent to which the 'creativity greenhouses' are adopted more widely across the university (Table 3).

15 Conclusions

This paper started with the challenges facing the planet and the role of Universities in engaging with them, before exploring creativity and action research as ways forward in the classroom developed through co-operative learning over cycles of action and reflection. Drawing on the I♥L experience, the authors present two models to bring creativity into teaching and learning for sustainability and responsible enterprise. Their next step will be to create a metaphorical 'chocolate box' of ideas, techniques and inspiration for colleagues and students drawn from this work.

The paper shares the authors' ideas on how creativity in learning might be taken forward, through offering the two models. The first of these is to develop a '**creativity map**' using extended ways of knowing to help categorise the interventions made and maximise the understanding of the students and staff. The second offers a model to grow creativity in learning in HE using '**Creativity Greenhouses**' on a self-organising model. We see that these two models may be applied together, complementing each other to develop the skills of creativity within higher education. It is not necessary for either model to have a 'top down' permission or drive.

The paper also aims to develop practice in teaching and learning through experimenting with a range of creative interventions and receiving positive feedback, presenting the experience of the project.

In our work, we were mindful of the constraints and limitations of the project. Firstly, concern about using creativity in the teaching context and in relation to a business degree including responsible enterprise/sustainability. The models we propose are a way to overcome and resolve initial perceived limitations and tackle assumptions about the challenges of using creativity. Secondly, that creativity in teaching sustainability and responsible enterprise requires constant feedback with students to ensure that all learners are catered for, which was attempted in the I♥L project, but was sometimes frustrated by the constraints of the curriculum, the fact that students were not always comfortable with this process and time limitations on the authors. Thirdly, we are aware that work in one university cannot necessarily be assumed to be valid in other higher education institutions, cultures and countries. The authors would like to see future research include working with students and staff further in a global range of campuses on their feedback to co-create greater learning and creativity in their academic studies. In so doing, learning for sustainability and responsible enterprise takes on the characteristics of 'future possibilities' suggested by Craft (2011) and the anticipatory learning of Botkin et al (1979). In such a way, we believe, we can open the doors to staff and students working together on making creativity happen in lectures and tutorials and to deepening understanding of sustainable development and responsible business practice across disciplines and in the world beyond the campus.

The authors would like to thank Phil Barton and Liz Walley. From Manchester Metropolitan University they would also like to thank: participating students from Responsible Enterprise, the Cheshire Campus and the Manchester Fashion Institute, Chrissi Nerantzi and Charles Neame (CELT), Helen Wadham & staff team of Responsible Enterprise and Tom Scanlon (Business School).

References

Abrams D (1996) The spell of the sensuous. Vintage, New York

Bateson G (1972) Steps to an ecology of mind. University of Chicago Press, London

Berry T (1988) The American college in the ecological age. In: The dream of the earth. San Francisco: Sierra Club

Berry T (1999) The great work, our way into the future. Bell Tower, New York

Beghetto R, Kaufman J (2010) *Nurturing* creativity in the classroom. Cambridge University Press, Cambridge

Botkin J, Elmandjra M, Malitza M (1979) No limits to learning, bridging the human gap: a report to the club of Rome. Pergamon, Oxford

Boyce-Tilman J (2016) Experiencing music, restoring the spiritual. Oxford: Peter Lang

Bradbury H (2015a) How to situate and define action research. In Bradbury H (ed) Sage handbook of action research, 3rd edn. Sage, London, pp 1–14

Bradbury H (2015b) Sage handbook of action research (3rd edn). Sage, London

Bullivant A (2010) Global learning: a historical overview. In: Gadsby H, Bullivant A (eds) Global learning and sustainable development. Routledge, London

CELT (2017) Creativity for learning CPD, http://www.celt.mmu.ac.uk/cpd/accredited/unit_details. php?unit id=93. Last accessed 24th June 2017

Christian J, Walley L (2015) Termite tales: organisational change—a personal view of sustainable development in a university—As seen from the tunnels. In: Leal Filho W (ed) Integrative approaches to sustainable development at university level. Springer International Publishing, Switzerland

Clark H (2002) Building education: the role of the physical environment in enhancing teaching and research. Institute of Education, University of London, London

Cobb S, Leathlean D, Kettleborough H, Wozniak M (2016) I love learning: innovating for creativity. Creative Academic Magazine, 4a, pp 86–88. http://www.creativeacademic.uk/uploads/1/3/5/4/13542890/cam_4a.pdf

Cowley S (2005) Let the buggers be creative. Continuum, London

Craft A (2011) Creativity and education futures: learning in a digital age. Trentham Books, Stoke-on-Trent

Craft A, Gardner H, Claxton G (2008) Creativity, wisdom and trusteeship, exploring the role of education. Sage, London

Cropley A (2001) Creativity in education and learning, a guide for teachers and educators. Kogan Page, London

Development Education Centre (2003) Lessons in Sustainability: What an earth is happening? How do we respond?. Development Education Centre, Bristol

EC.Europa. EU: https://ec.europa.eu/education/consultations/lifelong-learning-key-competences-2017_en. Last accessed 28 July 2017

Fisher R (2008) Teaching thinking, 3rd edn. Continuum International Publishing Group, London

Freire P (trans. M. Ramos) (1970/1996) Pedagogy of the oppressed (Penguin edn). Penguin Books, London

Gadsby H, Bullivant A (2010) Global learning and sustainable development. Routledge, London

Graham C (2017) 'Doomsday Artic Seed Vault breached', *Telegraph* http://www.telegraph.co.uk/news/2017/05/20/doomsday-arctic-seed-vault-breached-permafrost-melts/. Last accessed 15 June 2017

Graham-Harrison E (2016) 'July confirmed as the hottest month ever', *the Guardian*, WEditionesday 17 Aug 2016, p 3

Grinsted T (2015) The matter of geography in education for sustainable development: the case of the Danish university. In Filho L (ed) Transformative approaches to sustainable development in universities. Springer, Switzerland

Heron J (1996) Co-operative inquiry, research into the human condition. Sage, London

HEFCE: Manchester Metropolitan University UKPRN: 10004180Year Two Provider Statement http://www.hefce.ac.uk/media/HEFCE,2014/Content/Learning,and,teaching/TEF/TEFYearTwo/submissions/TEFYearTwoSubmission_10004180.pdf. Last accessed 2 Aug 2017

Higher Education Funding Council for England (2017) The Teaching excellence framework: http://www.hefce.ac.uk/lt/tef/. Accessed 2 Aug 2017

Homer-Dixon H (2006) The upside of down, catastrophe, creativity and the renewal of civilisation. Souvenir Press Ltd, London

Hutchins R (1968) A learning society, a Britannica perspective. Penguin, London

Illingworth S (2016) Using line poetry in teaching (personal email)

Illingworth S (2017) https://thepoetryofscience.scienceblog.com/author/thepoetryofscience/. Last accessed 19 June 2017

IUSDRP (2017) https://www.haw-hamburg.de/en/ftz-nk/programmes/iusdrp.html. Last accessed 15 July 2017

Jackson N (2010) Forward. In Nygaard C, Courtney N, Holtham C (2010) Teaching creativity—creativity in teaching. Libri Publishing, Faringdon

Jackson N (2013) Tackling the wicked problem of creativity in higher education, http://www.creativeacademic.uk/uploads/1/3/5/4/13542890/. Last accessed 19 June 2017

James A, Brookfield S (2014) Engaging imagination: helping students become creative and reflective thinkers. Jossey-Bass, San Francisco

Kettleborough H (2014) Joining self and community to earth and cosmos, a first person inquiry of discovery and understanding, unpublished PhD thesis, Lancaster University

Kettleborough H (2015) Using play in HE—reflections on participation in a creativity for learning HE course. In Nerantzi C, Jones A (eds) Exploring play in higher education, creative academic magazine, No. 2A

Kettleborough H, Wozniak M, Leathlean D (2017) Reflections on using creativity in teaching sustainablity in responsible enterprise: a first and second person inquiry. In Leal Fihlo W (ed) Implementing sustainablity in the curriculum of universities, world sustainablity series. Springer, Switzerland

Kolbert E (2014) The sixth extinction, an unnatural history. Bloomsbury, London

Kumar S (2016) Soil, soul and society. Leaping Hare Press, Lewis

Laasch O, Conway R (2015) Principles of responsible management. Cengage, Stamford CT

Leal Filho W, Manolas E, Pace P (2014) The future we want: Key issues on sustainable development in higher education after Rio and the UN decade of education for sustainable development. Int J Sustain High Educ 16(1):112–129, https://doi.org/10.1108/IJSHE-03-2014-0036 Permanent link to this document: https://doi.org/10.1108/IJSHE-03-2014-0036

Leal Filho W (2015) Education for sustainable development in higher education: reviewing needs. In Leal Filho W (ed) Transformative approaches to sustainable development at universities, working across disciplines. Springer Nature, Switzerland

Leopold A (1968) A sand county almanac. OUP, Oxford

Longworth N (2003) Lifelong learning in action: transforming education in the 21st century. Kogan Page, London

Low G (2008) Metaphor and education. In Gibbs Raymond W (ed) The Cambridge handbook of metaphor and though. Cambridge University Press, New York, Cambridge, pp 212–231

Manchester Metropolitan University (2017) Teaching excellence framework, year two provider statement, http://www.hefce.ac.uk/media/HEFCE,2014/Content/Learning,and,teaching/TEF/TEFYearTwo/submissions/TEFYearTwoSubmission_10004180.pdf. Accessed 2 Aug 2017

Manchester Metropolitan University (2017) (Internal, online) Course Handbook, Responsible Enterprise. Last accessed on 19 June 2017

Marshall J (2001) Self reflective inquiry practices. In Reason P, Bradbury H (eds) (eds) Handbook of action research concise paperback edition. Sage, London, pp 335–342

Marshall J (2004) Living systemic thinking: exploring quality in first-person action research. Action Res 2(3):305–325

Marshall J (2016) Living life as inquiry, an action research approach. Sage, London

Marshall J, Coleman G, Reason P (eds) (2011) Leadership for sustainability: an action research approach. Greenleaf Publishing, Sheffield

Maughan E, Reason P (2001) A co-operative inquiry into deep ecology. ReVision 23(4):18–24

Mead G (2011) Coming home to story: storytelling beyond happily ever after. Vala Publishing Co-operative, Bristol

McDonough W, Braungart M (2002) Cradle to cradle. North Point Press, New York

McNiff J, Whitehead J (2011) All you need to know about action research, 2nd edn. SAGE Publications, London

McKibben B (2010) Eaarth: making a life on a tough new planet. Times Books, New York

Millennium Ecosystem Assessment http://www.mil enniumassessment.org/en/index.aspx. Last accessed on 19 June 2017

McKie R (2017) Biologists fear 50% of all species will be extinct. https://www.theguardian.com/environment/2017/feb/25/half-all-species-extinct-end-century-vatican-conference

Millennium Development Goals (2015) http://www.un.org/millenniumgoals/2015_MDG_Report/pdf/MDG (last accessed 15 June 2017)

Murray R (2013) Writing for academic skills. McGraw-Hill, Maidenhead

NASA & National Oceanic & Atmospheric Administration, 18th January 2017, https://www.nasa.gov/press-release/nasa-noaa-data-show-2016-warmest-year-on-record-globally . Last accessed 29 July 2017

Neal L (2015) Playing for time, making art as if the world mattered. Oberon, London

Owen N (2001) The magic of Metaphor: 77 stories for teachers, trainers & thinkers. Crown House Publishing, Carmarthen

Porritt J (2013) The world we made. Alex McKay's Story. Phaildon, London

PRME (2017) http://www.unprme.org/about-prme/history/index.php. Last accessed 29 June 2017

Reason P, Bradbury H (2006) Introduction: inquiry and participation in search of world worthy of human aspiration. In Reason P, Bradbury H (eds) Handbook of action research (Concise Paperback Edition). Sage, London, pp 1–14

Reason P, Bradbury H (2008) Concluding reflections: whither action research? In: The sage handbook of action research, participative inquiry and practice (2nd edn). Sage, London, pp 695–707

Reid A, Petocz P (2010) Diverse views of creativity for learning. In: Nygaard C, Courtney N, Holtham C (eds) Teaching creativity—creativity in teaching. Libri Publishing, Faringdon

Richardson L (2000) Writing: a method of inquiry. In: Denzin N, Lincoln Y (eds) Handbook of qualitative research, 2nd edn. Sage, Thousand Oaks, CA, pp 923–948

Robert K (2003) The natural step story: seeding a quiet revolution. New Society Publishers, Gabriola Island

Rockström J, Steffen W, Noone K, Persson Å, Chapin III FS, Lambin E, Lenton TM, Scheffer M, Folke C, Schellnhuber H, Nykvist B, De Wit CA, Hughes T, van der Leeuw S, Rodhe H, Sörlin S, Snyder PK, Costanza R, Svedin U, Falkenmark M, Karlberg L, Corell RW, Fabry VJ,

Hansen J, Walker B, Liverman D, Richardson K, Crutzen P, Foley J (2009) Planetary boundaries: exploring the safe operating space for humanity. Ecol Soc 14(2):32. http://www. ecologyandsociety.org/vol14/iss2/art32/. Last accessed 12 July 2017

Seeley C, Reason R (2008) Expressions of energy: an epistemology of presentational knowing in Action Research. In: Liamputtong P, Rumbold J (eds) Knowing differently: arts based and collaborative research. Nova Science Publishers, New York

Steffen W, Richardson K, Rockström J, Cornell S, Fetzer I, Bennett E, Biggs R, Carpenter S, de Vries W, de Wit C, Folke C, Gerten D, Heinke J, Mace G, Persson L, Ramanathan V, Reyers B, Sörlin S (2015) Planetary boundaries: guiding human development on a changing planet. Science 347(6223):736–748

Starko A (2010) Creativity in the classroom: schools of curious delight, 3rd edn. Routledge, London

Torbert W (2001) The practice of action inquiry. In Reason P, Bradbury H (eds) Handbook of action research concise paperback edition. Sage, London, pp 207–208

Torbert W, Taylor S (2008) Action inquiry, interweaving multiple qualities of attention for timely action. In Reason P, Bradbury H (eds) The Sage handbook of action research, participative inquiry and practice (2nd edn). Sage, London, pp 239–251

Turner-Vesslago B (2013) Freefall: writing without a parachute. Vala Publishing Company, Bristol

United Nations Decade for Education for Sustainable Development (2005) http://unesdoc.unesco. org/images/0014/001486/148654E.pdf. Last accessed 1st August 2017

WWF (2016) Living Planet Report, Risk and Resilience in a New Era, http://awsassets.panda.org/ downloads/lpr_living_planet_report_2016.pdf. Last accessed on 24 June 2017

Wozniak M (2016) Use of creative Spaces within Higher Education to generate a motivational spark moment [Online]. Last accessed on 4 Aug 2017. https://cpdmarcinwozniak.wordpress. com/flex1/flex-2/flex-3/flex-4/

Wozniak M et al (2017) Working together collaboratively and in partnership to explore 'I ♥ learning' creative approaches to motivate and inspire students', CELT Journal of Teaching and Learning: http://www.celt.mmu.ac.uk/ltia. Last accessed on 25 July 2017

Yorks L (2015) The practice of teaching co-operative inquiry. In Bradbury H (ed) The Sage handbook of action research, pp 256–264

Zandee P, Cooperrider D (2008) Appreciable worlds, inspired inquiry. In Reason P, Bradbury H (eds) The sage handbook of action research, participative inquiry and practice (2nd edn). Sage, London

Zimmerman B (2002) Becoming a self-regulated learner. Theor Pract 41(2):64–70

A SDG Compliant Curriculum Framework for Social Work Education: Issues and Challenges

Umesh Chandra Pandey and Chhabi Kumar

Abstract Goal 4 of the Sustainable Development Goals (SDGs) calls for Inclusive and equitable quality education and promoting lifelong learning opportunities for all. We have enough reasons to believe that education has significant bearing on others SDGs. This brings the issue of curriculum development to the core of sustainable development. Among all the disciplines, Social Work carries specific importance, primarily because the competence of Social Work Practitioners will have direct impact on the pursuit of SDGs. Hence there has been a realization that curriculum planning for social work education need to be suitably aligned to the requirements of Sustainable Development Goals. It will equip the Social Workers with required skills and competencies to work with the target communities. Furthermore, SDGs have created new opportunities for social workers to advance their professional pursuits in a global perspective. There is need to identify such new opportunities for social work professionals, incorporate them in their professional curriculum and create pedagogical tools to connect to new target groups of Social Work Education. It will equip the social workers to effectively deal with the concerns of SDGs while working with target communities. This study reviews the existing research literature and identifies the new ingredients for curriculum planning for social work education. An indicative framework has been presented for the curriculum development for Social Work profession.

Keywords Curriculum development · Sustainable development goals
Education sustainable development

U. C. Pandey (✉)
Evaluation Centre, Indira Gandhi National Open University, 3rd Floor, Sanchi Complex, Shivaji Nagar, Bhopal 462047, Madhya Pradesh, India
e-mail: ucpandey@ignou.ac.in

C. Kumar (✉)
Indira Gandhi National Open University, Jabalpur 482001, Madhya Pradesh, India
e-mail: chhabidubey@gmail.com

© Springer International Publishing AG 2018
W. Leal Filho (ed.), *Implementing Sustainability in the Curriculum of Universities*,
World Sustainability Series, https://doi.org/10.1007/978-3-319-70281-0_12

1 Introduction

As Post 2015 development agenda progresses, Education has come to a much sharper focus primarily due to its potential to achieve and influence all the Sustainable Development Goals (SDGs). Education is seen at the heart of development and as a potent tool to develop and transfer capacities within and between communities (UNESCO 2016). National Governments are under obligation to make policy interventions and investments for equitable and inclusive quality education at all levels (UNESCO 2017). It is well realized and understood that Education enables individuals to transform and improve their lives by empowering them to make informed decision making. Thus, education has a pivotal role to play in the era of knowledge driven economies relying heavily on skills, ideas, research and knowledge acquired by individuals. Higher Education is therefore gaining the attention never seen before (Mohamedbhai 2015).

The sector of Higher education is perhaps the most vital which has great bearing on the achievement of SDGs (OECD 2006). It is at this level that individuals are prepared to enter the labor market and are trained with the necessary skills and knowledge to assist them in contributing positively to the knowledge economy. The Kyoto Declaration on Sustainable Development adopted in 1993 by the International Association of Universities (IAU) highlights the role of Universities and other seats of higher learning in the promotion of sustainable development. The IAU has proclaimed its resolve to stay committed to the cause of sustainable development as an active partner to further the cause. Thus, Universities and other institutions of higher learning have become inseparable from the development sector as these have the ability to deliver the tools of adaptation, resilience and transformation of individuals and communities alike (Rwomire 2011).

Social Work has tremendous potential to address the SDGs primarily due to the nature and scope of this discipline. Its core principles are intimately linked to the practice of sustainability (UNRISD 2017). Therefore it is felt that this profession is ideally positioned to address the Sustainable Development Goals. However, the curriculum of Social Work Education is yet to be aligned with the changing developmental context across the world. Social workers need to be equipped with skills and competencies required to handle the concerns of SDGs while they work with target communities. Therefore teaching on the issues of SDGs carries a pivotal role. In this perspective, the curricula of social work profession and pedagogical methods to impart training are of tremendous strategic importance and need to be carefully planned.

This chapter seeks to explore a SDG Compliant Curriculum for the social work at University level and a presents an indicative framework for its effective delivery. The existing research literature has been examined and new perspectives for curriculum for Social Work Education have been explored. The chapter is divided into four parts. In the first part we have described the role of Higher Education sector for Sustainable Development Goals. The international concerns in this regard have been described. These developments have been divided in two parts namely the developments till 2015 and further the developments after 2015. In the second part rationale for curriculum restructuring in Social Work has been highlighted. It is followed by Part 3 where an indicative framework for SDG compliant framework has been given. It contains relevant developments which have crucial significance for curriculum restructuring in Social Work Education. The recommendations for an indicative framework for curriculum planning have also been presented in this section. Lastly, the fourth part brings out the discussions and conclusions.

2 International Concerns for Higher Education

The need for integrating concerns of sustainable development in education is well realized and understood (SDELG 2006). However it is essential to develop a holistic understanding of the sustainability and to help the students to evaluate their action in relation to requirements of sustainability (UWE, Bristol). Universities need to include the concerns of sustainable development in their curricula so that students can relate the context of their discipline to intricacies of Sustainable Development. The extent of this relation will be different depending upon the nature of the discipline.

2.1 Developments till 2015

The concerns of the Universities for the issues of Sustainable Development had started coming up during nineties. The International Association of Universities (IAU) has been very proactive on this matter. It also adopted a policy statement in 1993 named as Kyoto Declaration on Sustainable Development which gives a comprehensive outline to promote Universities' role for promotion of sustainable development across the world. The document calls upon the Universities to seek, establish and disseminate a clearer understanding of sustainable development. The IAU continued to work for networking among the Universities and sharing of their activities through online portal on Higher Education (Mohamedbhai 2015).

UNESCO's report prepared at the end of the decade of education for sustainable development gives an account of the challenges which educational systems have to face in performing their roles. The report highlights the progress made by the institutions of Higher Education to support sustainable development through the development of new programmes, teaching and Research. In fact the most important role of Universities during the decade came up in the form of networks of educational institutions in different parts of the world—MESA in Africa, ProsPER. Net in Asia-Pacific, COPERNICUS Alliance in Europe, ARIUSA in Latin America and the Carribean. These networks have significantly advanced the issues of sustainable development through capacity building, sharing of experiences and influence of education. United Nations have taken steps to bring the Universities together and thereby promote the sharing of their experiences, activities and scope for collaborations so that concerns for sustainability can be addressed. United Nations Environment Programme (UNEP) made a significant initiative through GUPES to promote partnerships of 370 universities across the globe to implement environment and sustainability practices into their curricula. However, the impact of such initiatives have been limited primarily due to the compartmentalized approach of University functioning which leaves little scope for interdisciplinary interactions and curriculum development. The capacities of the faculty and staff responsible to transform the curriculum have been limited and the existing systemic practices do not encourage the required staff development activities. Besides there have been little effort to realign the pedagogical practices to meet the requirements of target groups. Within the higher educational institutions the administrative practices have not been very supportive for change and there have been little coordination among the various disciplinary groups to develop curricula as per the requirements of Sustainable Development.

Despite the rising concerns, the effective roadmap for sustainable development did not take shape by the end of decade. However, based on extensive consultations with various stakeholders UNESCO came up with Global Action Programme and an indicative roadmap to explore role of education in sustainable development. The roadmap highlights for the need for mainstreaming education for sustainable development policies, integrating sustainable development principles in daily activities, building capacities in educators and trainers, empowering and mobilizing youth; and to explore solutions local and community levels (Flor 2016).

The end of Decade of Education for Sustainable Development concluded with two major conferences organized in Aichi-Nagoya in Japan in November 2014. The first conference came to an consensus that Higher Education Institutions need to follow a 'whole-institution approach' including transformative leadership, encouraging capacity development and undertaking an assessment of the institution for sustainability (Mohamedbhai 2015). It further highlighted that there is a need for institutional engagements in diverse knowledge areas, policy issues and critical community groups such as youth and the private sector. The Nagoya Declaration which followed the conference emphasized on essential role and responsibility of higher education institutions for creating sustainable societies. This international conference was followed by another major conference organized by UNESCO and the government of Japan which was attended by nearly 1000 participants (Mohamedbhai 2015). A declaration on Education for Sustainable Development at the end of conference emphasized on implementation of the GAP priority actions. The end of the year 2015 marked the beginning of the historical Sustainable Development Goals of United Nations which recognized the importance of Higher Education. There have been unprecedented responsibilities and policy thrust laid on systems of Higher Education across the world. Post 2015 period has seen a great deal of sensitization at the policy making levels in Higher Education sector across the world to integrate basic tenets of Sustainable Development in their teaching, research, community engagement and campus operations.

2.2 Post 2015 Developments

However, High-Level Policy Forum 26th ICDE World Conference Pretoria, South Africa in 2015 came up with five challenges that are being faced by the systems of Higher education across the nations in realizing their resolve to work for sustainable development. These challenges relate to ensuring equity in access and outcomes, making the curriculum relevant to the present needs and situation, strengthening and expanding alternate modes/systems of education dissemination, supporting lifelong learning experiences of individuals and communities and finally strengthening collaboration and partnerships across sectors for sustainable development.

The Goal 4 along with its 10 associated targets is explicitly related to the 'development of education' and 'education for development' (Lewin 2016). These targets serve as global signposts expressing global commitments that need to be contextualized and reflected in respective national education policy priorities for member countries (UNESCO 2015). The National Governments are expected to integrate the global commitments of SDGs into their national education development plans. It is beyond doubt that the international goals and objectives must resonate with the respective national priorities, else they are likely to lose their relevance. Thus, contextual orientation is the hallmark for the SDGs as envisaged by the UN.

3 Rationale for Curriculum Change

The specialized methods and techniques of social work enables the social work professionals to work in a wide variety of settings, both institutional and community based and work with individuals, groups and communities to achieve the desired objectives. Social Workers are therefore, seen as the linking pin to associate various stakeholders in the field of development. This makes them ideally positioned to work for the most disadvantaged and marginalized groups, thereby making inclusive and humane development a reality (Jaysooriya 2016). Social Work training emphasizes on developing an understanding for the various forms, mechanisms and reasons for oppression, conflict and discrimination leading to poverty, human rights violation and inequality. The social workers are expected to develop humility and respect for cultural and social diversities and contexts. Thus, the training is expected to play a unique role in enabling the social work professionals to educate the masses, advocate on behalf of the oppressed and the marginalized and plan required intervention after a careful and scientific process of need assessment. Thus, the curriculum for the social work training and education has a direct bearing on the effectiveness and capacities of social work professionals to contribute and partner for SDGs.

4 Social Work Education: A Major Concern

The changing context of developmental priorities across the world in post 2015 era has given rise to a much more meaningful role for the social work professionals (UNRISD 2017). Post 2015 developmental paradigm has given unprecedented focus on collective responsibilities across the political boundaries, partnerships, social justice and human rights. These priority areas have already been the underpinning principles of the social work profession. It has led to following developments at the level of curriculum planning and outreach of social work programmes:

(a) Social work has gained much more attention as a tool to accomplish SDGs (UNRISD 2017). Hence curriculum planning came up as a major issue. It was well realized that the curriculum of social work has to be aligned with the concerns of SDGs if it has to serve a meaningful role.

(b) The concerns for Sustainable Development Goals have created a conducive atmosphere for social workers to connect to their counterparts in different countries, look for synergies and enter into partnerships in developmental projects across the political boundaries (Faruque et al. 2013). There is widespread belief that prosperity of one nation cannot be seen in isolation with others. It requires a collective international action to achieve sustainable economic, social and environmental wellbeing (UNRISD 2017). It is a challenge for the social workers to acquire competencies for such new development paradigms.

(c) The changes are required not only at the level of curriculum but also how it needs to be transacted, how it can dynamically evolve with the societal needs and how well these programmes are accessible to the target groups. The delivery structure of these programmes through Universities has to be articulated in such a way that it has maximum impact. It is in this context that curriculum for Social Work Programmes, Pedagogy and Outreach of the Universities needs to be addressed (Kumar 2017).

(d) The rising concerns of Environmental Degradation have posed multitude of challenges for the practice of social work. The impact has been different on diverse groups of populations across the world. It makes the job of social workers more complex, requires social workers to acquire new competencies and perform new roles (Drolet et al. 2015).

The social work education has to dynamically evolve with time, strategize its curriculum, develop suitable pedagogy and outreach strategies as per changing requirements. As the concerns for Sustainable Development Goals are embedded in Social Work profession these emerging issues cannot be ignored in the emerging discourse for sustainability (Dominelli 2012). However these issues have not been sufficiently dealt in the research literature

5 Ingredients for SDG Responsive Curriculum

There has never been a realization so strong that most trying issues of mankind popularly known as "Sustainable Development Goals (SDGs)" can be addressed through a time bound action based on basic principles like Social Justice, Human Rights, Collective Responsibilities and Respect for Diversities. Though these basic principles have been the cornerstones of Social Work profession, we now have global commitments to address these issues. Therefore, Global Developmental Agenda of United Nations and Social Work are based on the identical doctrines which make this profession a powerful instrument to accomplish the "Sustainable Development Goals (SDGs)". SDGs aim to explore collective responsibilities of nations to resolve some of the most outstanding problems of the world.

5.1 Curriculum to Be Dynamically Evolving

The Higher Education Institutions should engage not only in what is transacted in classrooms but also how well the Universities are able to relate to experiences of the students who go back to real life conditions and put the structured knowledge acquired in classrooms into practice. There has largely been a disconnect between the practitioners of Social Work and Educators of Social Work. There has not been a shared platform through which they could interact and exchange knowledge. Hence, theory and practice could not benefit from each other. Moreover, Higher Education Institutions do not facilitate such interactions. The curriculum planning has largely remained confined to the faculty of the University and there has been little participation of those who practice the profession. Blending of the knowledge earned by the practitioners could have been utilized by the practitioners to discover new paradigms of the theoretical foundations of the discipline. Higher Education Institutions have not done enough to take benefit of such tacit knowledge which practitioners acquire while working in the field. Hence, the linkages of the Universities to real life conditions are important. Social Work as a discipline needs such linkages much more than any other discipline. The tacit knowledge acquired by such practitioners is vital for further enriching the theoretical foundation of the social work. This ongoing interaction between the structured knowledge and tacit knowledge acquired by its practitioners will generate new knowledge which needs to be captured, suitably incorporated into existing pool of knowledge and utilized for upgrading the competencies of social work professionals (Kumar 2017). Hence, curriculum development, pedagogy and involvement of stakeholders in curriculum planning is vital. The theoretical foundations of the discipline should dynamically evolve to match the requirements of SDGs.

5.2 Identifying the Synergies with SDGs

The social work professionals need to acquire competencies to identify synergies of their operations with SDGs. Existing curriculum does not capacitate them to adjust with changed context and realities. This is particularly true for several developing countries like India, where the content, approach and methodology need to be refined and reoriented according to the changing field realities (Shodhganga 2016). There are a large number of gaps which exists in the current curriculum framework of social work education in developing countries where Social Work interventions are immediately needed. First and foremost, a large body of literature in social work theory and practice has been borrowed from the west, particularly from United States which is often considered as the land of its origin, especially in the current form. Thus, the curriculum is deprived of contextual relevance which leads to deprivation of students from getting a contextual and holistic understanding of the field realities existing in the society and the social work approach that needs to be followed. There is a dire need to update and revise the content of social work curriculum and embed the various SDG concerns within the framework of the social work education and training. This is particularly required to create awareness on the global concerns at the local levels and ensure their participation for collective action.

5.3 Need for Regulatory Structure

An important lacuna in the social work curriculum is the lack of standardization of curriculum in programmes and trainings offered by different institutes mostly in developing countries. This leads to disparities in the education scenario and compromise on the overall quality of training/research on the subject. There is a need for strong regulatory system which should govern the professional training of social workers. Unfortunately such a regulatory system is not in place in most of the developing countries. This has prevented the growth of standardized curriculum.

5.4 Need for Interdisciplinary Courses

One of the major issues in Higher Education Institutions is the lack of connections between various academic disciplines. It has hampered free flow of knowledge across these different areas and prevented growth of academic disciplines. Concerns for Sustainable Development have brought these issues to the forefront. We need to have an integrated approach to address the issues of sustainability (Hopkinson et al. 2008). SDGs are composite goals and therefore any SDG compliant framework will require curriculum development in several interdisciplinary areas within the larger ambit of social work. Such interdisciplinary courses are required to sensitize and enable the students to acquire knowledge and information on certain specific aspects of development. These courses may be related to NGO Management, Psychosocial Counseling, Gerontological social work and Care giving, Development Management, Project Management, Human Rights Education, Climate Change education and Environment Conservation, Disaster Management and Mitigation, gender studies among others. These specialized courses would assist in making the curriculum more responsive and relevant to the changed realities post SDGs. This is all the more important as the discipline of social work gives most weighage to the practice aspect and hence there is a critical link between the theoretical aspect and the practical training in social work.

5.5 Developing Global Competence

The SDGs call for creating partnerships at all levels for promoting coordination and collaboration for sustainable development. The similar case can be built for social work education and training where the increasing interdependence between the nations and growing demand for increased collaboration has resulted in international concepts and global competence within the social work profession (Faruque and Ahmmed 2013). The challenges and problems arising due to globalization and other humanitarian crisis situations developing as a result of global terrorism, climate change shocks and global warming require immediate humanitarian aid and response and services of volunteers and professionals across the borders. This situation also requires a relevant response in the curriculum.

5.6 Beyond Political Boundaries

The SDGs have given rise to new opportunities for developing collaborative Social Work projects for Regional Cooperation among countries. It is no longer a matter of choice but a compulsion for developing countries (Prodhan and Faruque 2012). The issues pertaining to SDG target groups cannot be seen within specific political boundaries. They need to be addressed holistically and therefore necessitates the social work projects beyond political boundaries. This type of international diplomatic environment is very conducive to bring about sharing of best practices, collective efforts and developing partnerships among the social work organizations across the political boundaries. It can further be a soft diplomatic tool for Regional Peace and Cooperation. Countries world over are resorting to this diplomacy to bring about greater cooperation, peace and regional power balance.

Social Workers need to develop professional competencies to take advantage of this new emerging world order. Thus, developing an international outlook in the curriculum part of the discipline would certainly provide leverage to the discipline and aid in furthering the cause of international social work (Kumar 2017).

5.7   Need to Enhance Outreach

The social workers practicing in developing countries are largely untrained and lack professional skills to perform their roles. It is adversely affecting the concerns of SDGs (Mohapatra 2014).There is a dire need for the Universities to connect to such groups, build up their capacities and thereby advance the progress on SDGs. It will require innovative and Flexible methods through effective use of ICTs (Gulati 2008). The Universities are yet to realize full potential of ICTs to create flexible and innovative learning environments for prospective clientele working in remote areas.

6   Discussions and Conclusion

Education for Sustainable Development implies acquiring the knowledge, skills, attitudes and values which are indispensable for the creation of a sustainable future (Lewin 2016). Post 2015 development agenda therefore calls for complete aligning of Social Work Curriculum with the requirements generated by SDGs. This reform process should involve total structuring of academic activities which teachers and Students get involved in. Such a curriculum would include key sustainable development issues like climate change, disaster risk reduction, biodiversity issues, poverty reduction, sustainable livelihood opportunities and sustainable consumption. It is also expected to promote a participatory teaching and learning methods for creating a learner centric environment.

Such a system of education would entail promoting competencies like critical thinking, questioning and empowering communities to make informed decisions in a collaborative way (Lewin 2016). Focus on effective, coherent and relevant learning would require review of existing teaching and learning contents, current pedagogical orientation, materials and prevailing classroom teaching practices, assessment frameworks, teacher's training and professional development among others. Curriculum development related to achieving the Sustainable Development Goals has to be located within the framework of national systems of education. It must be based on their specific histories, preferences, pedagogy and aspirations depending on their unique situation or context. The curriculum development for achieving SDGs need to be conceived as change oriented and responsive to the emerging development scenarios. This is essential for making the curriculum relevant and resilient. The various patterns of curriculum development have been categorized as adoption, adaptation and local development (Lewin 2016). Adoption refers to picking up material from a foreign source without making significant changes to it. Adaptation refers to changing curricula to reflect one's own context and requirement. Local design, on the other hand occurs when curriculum development is based on characteristics of a particular system in consultation with the local community and is not adopted from materials developed elsewhere. Needless to say, that all the three patterns of curriculum development have their own sets of advantages and disadvantages. It is also not possible to prefer one over the other as the choice of the pattern to be followed would depend upon the need and situation of a country.

This chapter has dealt with unique traits and roles of Social Work which has made it a powerful instrument to achieve SDGs. The concerns arisen in the wake of Post 2015 developmental agenda give a range of new perspectives for curriculum planning for social work education. These emerging new perspectives should be embedded in the curriculum of this discipline. It will require appropriate structural and procedural changes in the present system of curriculum development. The following trends have crucial significance for curriculum development for the Social Work Education.

(a) There is a widespread acceptance of the fact that basic issues of poverty, climate change, human rights, education etc. cannot be addressed within the confines of National Boundaries (Global Agenda for Social Work and Social Development 2014). The contemporary world order has created crisis situations which quite often transcend political boundaries. These issues certainly have global ramifications and collective solutions have to be articulated by the national governments. It is due to this reason that Social Work profession has attracted attention across the world. This important trend can give rise to new paradigms of Collaborative Developmental Projects, International Diplomacy and Regional Cooperation. The social work response for the SDGs will therefore depend upon how well the curriculum of the Social Work Education has been aligned to the concerns of global consensus on developmental issues. It will give opportunities to Social Workers to connect to their counterparts beyond the political boundaries to realize the mandate of the profession more effectively. It is an unprecedented trend for social work profession as the National Governments will encourage the collaborations among the social work organizations, allocate budgetary allocations for them and will ensure their participation in the global development agenda. Thus, the curriculum needs to be designed in such a way that social workers are equipped with competencies required to perform such roles beyond national boundaries.

(b) Furthermore the social workers are confronting with another professional challenge as the global environmental crisis is affecting the diverse target groups in a different way. For example Climate change would affect different regions of the world in different ways. It calls for contextualized solutions for such target groups with a global outlook.

(c) There are certain challenges faced today by countries which are likely to create situations which would impact the social fabric of the society like the refugee crisis across countries and the problem of displacement. This would require interventions even beyond the political boundaries.

(d) Regional cooperation for development necessitates the participation of all stake holders in the process including the civil society of which the social workers are a part. Diverse socio cultural environments require a suitable orientation of the social workers of the same so that they are able to work with people coming from different cultural and ethnic backgrounds. It's a challenge to the social workers as they require a deeper understanding across the sectors. SDG concerns have given rise to many such challenges and opportunities for the Social Workers. Hence the curriculum of the training programmes planned at the level of Universities should be oriented to meet such requirements. This study advocates for dynamic interaction between Educators and Practitioners, Cross border interaction between Social Work Organizations to suitably align the Social Work curriculum to SDGs and finally more involvement of Social Workers in policy planning for Regional Cooperation. Social work curriculum needs to enhance for meeting such professional requirements. It will inculcate the skills and competencies required by social work professional to respond to professional challenges.

(e) It finds much more relevance in the context of developing countries where most of the SDG target groups live (Prodhan and Faruque 2012). Over the years Social Work as a discipline has made significant impact in developmental arena of developing countries which is evident from the fact that the number of NGO working in India is more than number of schools in India (Mahapatra 2014). The Government involves such NGOs in large numbers for the implementation of their developmental programmes. The curriculum of the professional training of these social workers is therefore a serious matter of concern. However there is no regulatory system exists for the professional training of these social workers. In this context the curriculum and delivery strategies of professional training of social workers need to be examined through the lens of sustainability.

(f) Further, there is a need for a continuous updation, improvement and inclusion of context specific content in the curricula at all levels of the educational and training in Social Work. It is essential that the themes and concepts related to sustainable development are embedded within all areas of teaching and learning in social work. Moreover the rich tacit knowledge of social work practitioners should be captured and blended with structured knowledge of social work to evolve new paradigms. It will help to dynamically modify the curriculum as per developmental needs.

These new paradigms of curriculum planning would capacitate the social workers to take the challenges of SDGs which will ultimately hasten the process of problem solving at the local level, making social work education more relevant, practice oriented and one that promotes multi stakeholder's participation at all levels. Further it would be a good idea to create international platforms to capture emerging opportunities for social workers, to develop an international understanding for collaborations and thereby address the concerns for SDGs through social work. However the prescriptions given here just give an indicative roadmap and need to be contextualized as per the national level policies and priorities.

Refrences

Dominelli L (2012) Green social work. Polity Press, Cambridge. https://books.google.co.in/books? id=EXTjAQAAQBAJ&printsec=frontcover&dq=Green+Social+Work.+Cambridge:+Polity+Pre ss.&hl=en&sa=X&ved=0ahUKEwiJ3_WM_rzVAhUBWbwKHej_BwMQ6AEIJTAA#v=onepa ge&q=Green%20Social%20Work.%20Cambridge%3A%20Polity%20Press.&f=false. Accessed 04 Aug 2017

Drolet J, Wu H, Taylor M, Dennehy A (2015) Social work and sustainable social development: teaching and learning strategies for 'Green Social Work' curriculum. Soc Work Edu 34(5). http://www.tandfonline.com/doi/full/10.1080/02615479.2015.1065808?scroll= top&needAccess=true. Accessed 3 Aug 2017

Faruque CJ, Ahmmed F (2013) Development of social work education and practice in an era of international collaboration and cooperation. J Intl Soc Issues 2(1):61–70. https://www.winona. edu/socialwork/Media/JISI_Faruque_Ahmmed.pdf. Accessed 25 July 2017

Flor A (2016) The promotion of ASEAN regional integration through open and distance higher education. file:///C:/Users/my/Downloads/001_Flor.pdf. Accessed 26 July 2017

Global Agenda for Social Work and Social Development (2014) First report. Intl Soc Work 57 (S4):3–16. http://journals.sagepub.com/doi/pdf/10.1177/0020872814534139. Accessed 04 Aug 2017

Gulati S (2008) Technology enhanced learning in developing nations: a review. Intl Rev Res Open Distance Learn 9(1):1–16

Hopkinson P, Hughes P, Layer G (2008) Education for sustainable development: using the UNESCO framework to embed ESD in a student learning and living experience, policy and practice: a development education review. https://www.developmenteducationreview.com/ issue/issue-6/education-sustainable-development-using-unesco-framework-embed-esd-student- learning. Accessed 06 Aug 2017

IFSW (2012) Poverty eradication and the role for social workers, International Federation of Social Workers. http://ifsw.org/policies/poverty-eradication-and-the-role-for-social-workers/. Acces sed 26 July 2017

Jayasooriya D (2016) Sustainable development goals and social work: opportunities and challenges for social work practice in Malaysia. J Hum Rights Soc Work 2016(1):19–29

Kumar C (2017) Strategizing social work response for sustainable development. In: Pandey UC, Verlaxmi I (ed) Open and distance learning initiatives for sustainable development, IGI Global Publications, USA. https://books.google.co.in/books?id=pjsqDwAAQBAJ&pg=PA245&lpg= PA245&dq=STRATEGIZING+CURRICULUM+DEVELOPMENT+FOR+SOCIAL+WOR K+EDUCATION+SUSTAINABLE+DEVELOPMENT&source=bl&ots=eRXXvnrAFo&sig =oAoOvmhVtxXzyy1y13lJAOpbAbU&hl=en&sa=X&ved=0ahUKEwiw9L3Y0brVAhUFu4 8KHdsNBuwQ6AEINjAB#v=onepage&q=STRATEGIZING%20CURRICULUM%20DEVE LOPMENT%20FOR%20SOCIAL%20WORK%20EDUCATION%20SUSTAINABLE%20D EVELOPMENT&f=false. Accessed 25 July 2017

Lewin (2016) Curriculum assessment and education for sustainable development: does the emperor have new cloths. In: CIES conference held at Vancouver on March 7th 2016. http:// keithlewin.net/wp-content/uploads/2016/03/CURRICULUM-INNOVATION-AND-SDGsv3. pdf. Accessed 14 July 2017

Mahapatra D (2014) India witnessing NGO boom, there is 1 for every 600 people. Times of India. http://timesofindia.indiatimes.com/india/India-witnessing-NGO-boom-there-is-1-for-every- 600-people/articleshow/30871406.cms. Accessed 23 July 2017

Mohamedbhai (2015) What role for higher education in sustainable development, Issue 349. http:// www.universityworldnews.com/article.php?story=20150108194231213. Accessed 25 July 2017

OECD (2006) Higher education for sustainable development, based on research carried out by Dr. Andy Johnston, seconded to the OECD from Forum for the Future, in October 2006–March

2007. https://www.oecd.org/education/innovationeducation/centreforeffectivelearningenviron mentscele/45575516.pdf. Accessed 25 July 2017

Prodhan M, Faruque CJ (2012) The importance of social welfare in the developing world. J Intl Soc Issues 1:1 II-21

Rwomire A (2011) The role of social work in national development. Soc Work Soc Intl Online J 9:1

SDELG (2006) Sustainable development education in a curriculum of excellence. file:///C:/Users/ Admin/Downloads/SDELG.pdf. Accessed on 2 Aug 2017

Shodhganga (2016) http://shodhganga.inflibnet.ac.in/bitstream/10603/23970/9/09_chapter%201. pdf. Accessed 24 July 2017

UNESCO (2015) http://www.un.org/sustainabledevelopment/blog/2015/08/transforming-our-world-document-adoption/. Accessed 06 Aug 2017

UNESCO (2016) Unpacking sustainable development goal 4 education 2030. http://unesdoc. unesco.org/images/0024/002463/246300E.pdf. Accessed 24 July 2017

UNESCO (2017) Education for sustainable development goals, learning objectives. http://unesdoc. unesco.org/images/0024/002474/247444e.pdf. Accessed 16 Aug 2017

UNRISD (2017) http://www.unrisd.org/80256B3C005BF3C2/(httpPages)/8B13589A0273CF 3380257920003215AD?OpenDocument. Accessed 05 Aug 2017

UWE, Bristol http://www1.uwe.ac.uk/about/corporateinformation/sustainability/education/ourap proach.aspx. Accessed 2 Aug 2017

World Development Indicators (WDI) (2016) Highlights: featuring the sustainable development goals. Extracted from full version of WDI 2016 World Bank Group

Research Informed Sustainable Development Through Art and Design Pedagogic Practices

Fabrizio Cocchiarella, Valeria Vargas, Sally Titterington and David Haley

Abstract This paper explores a pedagogic case study, which embeds academic research activity into a masters level unit of study. Students were invited to work alongside the LiFE 'Living in Future Ecologies' research group at Manchester School of Art to collaboratively investigate themes for sustainable development within a city context. Pomona Island, a brownfield site on the boarders of Manchester, Salford and Trafford presented a context for complex issues of local government, and questions of international relevance on resilience and responsible urban planning. Through learning about the landscape and sensitive ecology of the island, students and researchers explored notions of context, climate, visions for future living, the opportunities and the responsibility of art and design practices in steering social reasoning within a neoliberal system. This paper presents a carefully considered enquiry-based framework, analysing academic questioning that has enabled the transformation of the ephemeral and immaterial into a methodology to address misguided political agendas. The paper articulates the different methods used to embed research practice in the learning environment. This type of project also fully illustrates innovative learning and teaching methods as ways in which art and design practices can uniquely engage with and stimulate thinking to influence and nurture change. Through presenting responses from a psychogeographical walk for Manchester European City of Science in July 2016, a conversational, transformative tool for learning was developed. Reflections on the project further evaluate the multi-disciplinary interpretations, already collated in a collaborative publication with the Pomona community and publisher Gaia Project.

F. Cocchiarella (✉) · V. Vargas (✉) · S. Titterington (✉)
Manchester Metropolitan University, Manchester, UK
e-mail: f.cocchiarella@mmu.ac.uk

V. Vargas
e-mail: v.vargas@mmu.ac.uk

S. Titterington
e-mail: s.titterington@mmu.ac.uk

D. Haley
Zhongyuan University of Technology, Zhengzhou Shi, China
e-mail: davidhaley@yahoo.com

© Springer International Publishing AG 2018
W. Leal Filho (ed.), *Implementing Sustainability in the Curriculum of Universities*,
World Sustainability Series, https://doi.org/10.1007/978-3-319-70281-0_13

Keywords Education · Sustainable development · Futures thinking
Art & design · Salon · Capable futures · Landscape

1   Introduction

The LiFE Professional Platforms unit ran from March to September 2016. It invited
Masters students at Manchester School of Art, across a spectrum of creative disciplines, to engage with the research group. This supportive sharing community,
investigated ways of shifting the way we think and create visions for the future
through ecological arts and design practices based on sustainable development
principles at a professional level. The reasoning for merging research culture with
postgraduate study was that students could learn important professional methods
through working with and alongside research professionals. The cultures of practice
have similarities so rather than mirror each other as tutor and student the interaction
was more as master and apprentice and as the apprentice becomes more skilled,
knowledgeable and flexible, they become more embedded within the 'professional
team.' The intention was to utilize the different experience levels and social lenses
into a way in which to re-evaluate, re-position and explore important mutual issues.
From a power perspective, this initiative permitted a more rihzomatic approach,
allowing the 'professional' researchers/teachers to learn from the learners what the
learners need to learn. In this way, a learning dialogue was formed that focused on
the topic of learning and the means by which all parties might learn and benefit
(Deleuze and Guattari 2014; Haley 2017b).

The LiFE (Living in Future Ecologies) research group was formed a year earlier
and is a cross-faculty collaborative group of researchers seeking to intervene in the
sustainability discourse to propose innovative responses to climate change and
design future scenarios in which to implement 'capable futures' (Haley 2008). The
group's main focus areas are around waste, biodiversity, urban food production,
future lifestyles and how these interrelate to create a circular economy. The group
also runs occasional events called 'Opiso' derived from the ancient Greek word,
which means behind or back, but refers not to the past but to the future. Early Greek
imagination envisaged the past and the present as in front of us, something that we
can see. The future was viewed as invisible, meaning that we are walking blind,
backwards into the future (Knox 1994).

These events became think tanks, sandpits and salons in which to showcase
futures thinking initiatives and to expand and explore specific future living issues.
M.A. students were invited to take part either individually or collectively to identify
an issue that related to or evolved from current LiFE group thinking. The students
were supported by researchers with professional experience and expertise in design,
art, space, architecture, anthropology, psychology, ecology, environmental policy
and education for sustainable development. Students acquired skills that enabled
them to tackle the theoretical and the physical, thereby materializing their thinking
through artefacts and practical applications in which to propose possible alternative

futures. By embedding research methods in the discourse between students and researchers, students became more confident and lateral thinkers. Ideas and theories carried more meaning through a shared collegiate urgency to respond and re-interpret opportunities. Students learnt how to position their thoughts and concepts more clearly as proposals for discussion with a knowledgeable community. This allowed students to explore ambitious creative applications for their work quickly and effectively. This, also gave members of staff the opportunity to 'play', improvise and invent beyond the curriculum, thereby joining students in the creative learning experience. In this way the LiFE research group was enabled to pursue its own agenda of seeking meaningful forms of inquiry and generate situations for transdisciplinarity to emerge (Nicolescu 2002).

As the students developed their questioning and thinking around LiFE issues the group was invited to respond to Pomona Island, a brownfield site between the Bridgewater and Manchester Ship canals, on the boarders of Manchester, Salford and Trafford. This presented an authentic context in which to deal with complex issues of local government, and questions of international relevance on resilience and responsible urban planning. Learning about the landscape through a series of talks and site visits the group collectively gained knowledge about the sensitive ecology of the Island. Through an enquiry-based approach to research and pedagogy, researchers and students synthesized information and context that emerged into new networks with local campaigners, artists, designers, film makers and ecologists who intervened in the planned development.

Pomona Island sits in a precarious situation, as the land is privately owned by one of the North of England's largest development companies, and it has strategic plans to develop city centre apartments rather than conserve sites of ecological importance. From an urban industrial residual space, deeply rooted in its historical and political context, Pomona has evolved new ecologics to become an important habitat for new species of plants and wildlife. The use of the site over many years has seen transitions from high industrial usage, as a main port connecting Manchester to the rest of the world, to a post-industrial wasteland. As the politics of place and the city changed it became an EU funded leisure facility. Then, as it again became an abandoned site, it became a paradise for nature and migrating birds who made Pomona their home.

Students and researchers responded to the proposed plans by developers with urgency. Art and Design practice here was seen as a way to intervene through imagining diverse futures for the island that could influence future development and present other alternatives to the insensitive systematic re-population of urban centres. Creative practice was used as a participatory, interdisciplinary and community-based process to investigate the appropriateness of potential solutions within a holistic context (Wahl 2016). The group recognized their potential impact on future development plans that society needs to re-invent future habits, new ways of living in urban environments, and provide a framework for facilitating the many stakeholders involved—humans from all sectors, and other-than-humans, coexisting in the urban landscape.

Through the synthesis of an enquiry-based process, analysing academic questioning enabled the transformation of the ephemeral and immaterial into methods of intervention into misguided political agendas. Proposals from the group questioned the foundations for 'a sustainable society' and the word sustainable itself was questioned as a confusing term that embeds a notion of maintaining the 'status quo' rather than inventing a framework to develop future capability for change (Haley 2017a). Recognizing the problem that current city development often exploits and degrades local ecosystems, responses interrogated the notion of 'capable futures' as a means of inventing ways in which to foster shared abundance for collaboration and regeneration (Haley 2008; Wahl 2016). Students and researchers created new visions for Pomona and presented these responses during a psychogeographical walk for Manchester European City of Science in July 2016, as a conversational tool for transformative learning. The multi-disciplinary interpretations were collated in a collaborative publication with the Pomona community and publisher Gaia Project to provide a creative form of project evaluation.

2 Collective Visioning

The LiFE research group gave students on the M.A. Professional Platform unit the opportunity to study and engage with research group members, through weekly seminars that presented LiFE research interests and tutorial sessions. These also gave students the opportunity to present and discuss issues and themes that they were investigating in their own practice. Through an enquiry-based pedagogic framework that supported the students autonomy as independent and collaborative researchers (Sierens et al. 2009), the unit also allowed students to become core members of regular research group meetings to help direct and discuss agendas being developed. This allowed the researchers to adopt roles as mentors. Those with specialist expertise and experience were able to nurture and develop the evolving curiosity of the students (Cocchiarella and Booth 2015).

At one of the research meetings, the group identified an opportunity to focus both their various enquiries around sustainability and their many disciplinary lenses on a more holistic project that would respond to Pomona Island as a real life situation. The collaborative nature of the project encouraged open dialogue through the dissemination of alternative visions for Pomona, which created the potential for cultural transformation.

By producing narratives that questioned the perceived 'unavoidable' future, conversation and collaboration helped to steer and create a move towards a more desirable co-created vision of a new regenerative culture. Re-positioning the responsibilities to maintain life within the biosphere rather than monetary economies, this shift in thinking allowed a more creative interpretation of how design for living needs to recognise and include the many stakeholders. With mutual respect, they may then negotiate a common ground between the developer, and community to prioritise co-existing in urban centres.

3 Learning Undisciplined Behaviour

In Higher Education, it seems that we have forgotten that sustainable development was devised as a response to climate change. Appropriated and applied to everything from financial viability to the *growth* of carbon industries, this ubiquitous phrase now, paradoxically, represents both the cure and the cause of our greatest concerns. The term sustainable development may indeed be a contributing factor in the wicked problem of climate change. Wicked problems often exacerbate the very situation they try to address. So educating people to sustain development provides us with a perfect paradox, perpetuated by ridged adherence to the disciplinary structures and strictures of higher education institutions and research councils. Indeed, most pleas for multi, inter and trans-disciplinarity are misunderstood, misguided or highjacked by continuing Cartesian cognition and fortifying the silos of disciplinary dogma.

Perhaps, 'undisciplinarity' is required in Higher Education for students to become ecologically resilient for adaptation? (Haley 2017a, b) Indeed, this may be the point at which art and design are finally valued, because traditional research methods, alone, inadequately address these issues. This project practically demonstrated that the creative arts can, potentially, offer the 'leverage points' (Meadows 1999) to provide the transition from order to disorder, thesis to antithesis, and structure to process that may then evolve as organisation, synthesis and pattern (form) for a critically robust curriculum for sustainable development, or 'Capable Futures' (Haley 2008).

Despite our compelling myths to the contrary, the world and most things in it are beyond our control, so we must learn to expect the unexpected, the uncertain, the indeterminate. And to prepare for such eventualities is to be ecologically resilient.

While 'engineered resilience' refers to how fast a system returns to equilibrium after a shock or stress, the term 'ecological resilience' is applied to how far a system may be perturbed before it collapses or becomes another state of being (Walker et al. 2006). The capacity to withstand disturbance is not just a question of how long the status quo can be maintained, but how we might emerge into a new world. Of course resilience is both pattern and process. And like ecology and art and design can be understood as emergent phenomena, not purposive, solution-led, problem-based methods or objects that can be fixed. So, perhaps, we need to reinvent art for the next revolution of life? And to do this, maybe we need to create disorder (diversity) from order (monoculture) and pursue the antithesis of civilisation, to synthesise our evolution to a new form of organisation, or undisciplined, 'regenerative culture'? Instead of thinking of these initiatives as separate or conflicting interests, it may be even more useful to consider resilience as dynamic revolution/resolution—resolving duality—a process emerging from one state to the next. Together, disciplinarity and undisciplinarity, then offers education for sustainable development the ability to evolve to meet the challenges of a rapidly changing world (Haley 2017a, b).

4 Education for Sustainable Development

This is the process of equipping students with the knowledge and understanding, skills and attributes needed to work and live in a way that safeguards environmental, social and economic wellbeing, both in the present and for future generations (QAA 2014). As in other processes of change there might be trends and shared experiences but at micro levels change varies according to the scenario and the people involved. At Manchester Metropolitan University there are policy frameworks that seek to support the integration of education for sustainable development in the university, probably due to variations in processes and culture between faculties or departments, there is a wide range of initiatives in this area. For instance, in the Art School, academics have worked towards embedding education for sustainable development into the curriculum through interdisciplinary and innovative approaches (e.g. Langdown and Vargas 2015; Haley et al. 2016). These include cross-faculty (i.e. design, art, architecture) and cross-university (i.e. engineering and fashion) collaborations, however, these are individually (i.e. staff and students) led initiatives so far. In this scenario, the LiFE research group is a pivotal enterprise for the integration of sustainable development, presenting opportunities for the local community, students and staff to work at the same level.

To follow, the paper will describe one of the key moments at the start of the student-staff dialogue for the development of a collective approach in this project.

After a few months of working together academics invited students to join the LiFE group. Before the one-hour meeting started, students and staff found their places in the meeting space. As it may be expected, students sat at one side of the table whilst staff sat at the other side. The first half of the meeting was, what could be described as 'business as usual', staff engaged in discussion and students silently listening. However, mid-way through the meeting, the Chair suggested different sitting arrangements, mixing academics and students. The dynamics in the group as a whole changed with students and staff interacting more actively with each other.

One interpretation of this event might be that power dynamics were disturbed to bring about a more equal approach to dialogue. Subsequently, the dialogue was open to the Pomona community seeking to foster active involvement of all members, which is one of the key principles for an institutional process of change towards sustainable development (Tilbury et al. 2005).

In this project, an action learning conversational approach started to emerge in which members of the collective brought up their questions and challenges to reflect on them, and to develop their practice through the cycle (Marsick and Maltbia 2009, p. 162):

- Framing of the challenge as a question.
- Unpacking meaning through sharing information about the context and prior action.
- Peer questioning (to which the problem holder does not immediately respond) to unlock mental models that make one blind to other points of view.
- Identifying assumptions that underlie current ways of framing the challenge.

- Reframing one's understanding of the situation.
- Making more informed decisions and taking informed action to address the challenge.

As this approach was not imposed on staff, students or local community, it might be possible to suggest that this was collectively developed instead of directed by an individual. However, future empirical research may be needed to develop conclusions.

5 A Creative Journey

The introduction to the subsequent book, 'Fruitful Futures: Imagining Pomona' included this statement about the Pomona Encounters walk for the Manchester European City of Science Festival:

> The Art of Fruitful Living was a walk inspired by the ecology of food, the potential for urban food production and Pomona, the Roman goddess of fruit. The walk took twenty one participants on a creative journey through the habitats of Manchester, Salford and Trafford to encounter a cornucopia of paradoxical tales about biodiversity and urban planning, carbon-free air miles, invasive species for healthy living, and old toolkits to design new nature. The walk acted as a method for testing and collecting ideas as part of Manchester European City of Science.

Although a trial walk had been conducted the previous week, the event itself took much longer than expected, as participants added extra material to their performances and contributions. Like the pilgrims of Chaucer's Canterbury Tales, these walkers gained energy from the stories, anecdotes, polemics and information they presented. The route of the walk itself generated another source of energy. The changing environments came as quite a surprise to many of those who thought that Manchester's urban landscape was mono-cultural. However, the link between the intense retail malls, over-designed civic spaces, monumental historic buildings, drab Modernist dwellings, mundane contemporary housing and secret wild pathways was the River Irwell. Glimpses and full vistas of the river, its confluence with the River Medlock and transformation into the Bridgewater and Manchester Ship Canals provided regular refreshment. As noted by retired, Senior Lecturer at Manchester School of Art, architect and geomancer, David Ellis, the Irwell contributed to the use of ley lines adopted by the Romans and the Church to site important buildings in Manchester (Ellis and Thomas 2001). The adoption of the river course both for this walk and in planning is, also, fundamental to Feng Shui, the ancient Chinese art/science/philosophy of harmonizing people and their environment. This form of Taoist ecology employs dragon lines, rather than the straight geometric lines of geomancy, and rivers play a very important part in this understanding of the interactions between natural and cultural phenomena. Both ley lines and dragon lines are known to align, attract and channel sources of energy. They may, also dissipate or dispel energy if not correctly related or attuned to.

However, the event permitted each walker, whatever their background, to bring their own research and interpretations to the walk, so in turn, all were experts, guides and followers. Similar to the 'V' formation or skein of migrating birds, each walker led to provide the slip stream for others, and then allowed others to lead the way physically and metaphorically. For the six-hour duration of the walk, this process provided great comradery and feelings of fraternity amongst those who participated, contributing knowledge in the context of a shared experience.

6 Walking and Talking as a Transformative Tool for Learning

Anderson (2004) suggests that talking whilst walking helps collaborative knowledge to emerge and that human identity is shaped through this practice by connecting the self with the outside (i.e. other humans, non-humans and the built environment). In addition, they suggest that 'places are not passive stages on which actions occur, rather they are the medium that impinge on, structure and facilitate these processes'(2004, p. 255). Haley had also deployed these methods in his programme of eco-urban, art-walks, 'A Walk on the Wild Side' (2004–8). He had learned the value of this process from psychologist, Professor Judith Sixsmith, who used walking therapy as a less confrontational way for her clients to provide relaxed interviews (i.e. interviewer and interviewee share the same horizon).

Pomona was a platform for further development of action learning conversations, which started in the meetings and sessions that were part of the students curriculum. Creating a collective through balancing power relations provided a bottom up approach where members shared responsibilities and ownership of the project. This approach fostered positive engagement for staff, students and local community members who presented their projects on the walk.

Walking provided space and time to reflect. Small groups formed at the beginning and soon individuals would be moving from group to group discovering others' perspectives. Walkers would also have time and space to engage with Pomona Island's non-human communities through structured activities such as the design interviews to different species developed by Crystal Chan, MFA Design Cultures student. Other contributions that were presented in the walk and later published in the Fruitful Futures collective book included tales, fables, poems, drawings, food, designs and analyses of social and political urban planning.

Students and staff performances during the walk were linked to the place at a conceptual level and through embodied experience, which aimed at fostering longer periods of reflection and multi-dimensional project developments (Lakoff and Johnson 1999). Walkers had the opportunity to explore their own identity, as well as that of the place, from varied perspectives engaging with the complexity of the interaction between human activity and nature. The temporal and spatial dimensions reinforced the context of the human identity exploration. People who knew Pomona island well from the three groups (i.e. staff, students and community) mainly provided the temporal dimension. The island, its inhabitants and the exchange of place related knowledge triggered by the walk, provided the spatial dimension.

Among the reasons that may suggest the success of the curriculum (i.e. unit and walk) and the potential for transformative learning through similar approaches include the following. Firstly, the engagement of students increased through the duration of the unit. Secondly, students engaged in this form of learning for the whole six-hour walk, which was probably the longest university session they had experienced. Thirdly, the interconnections that appeared through their conversations on the walk and through the whole unit of their course became more explicit in the writing process. In other words, the process of writing became a reflective continuation of the creative process.

7 Co-production

Through the creation of the book, editors, publisher Gaia Project and graphics collective Textbook Studio joined the editors (Cocchiarella, Haley and Vargas) to co-curate and publish a limited edition of 250 books that acts as a platform to disseminate the work produced by the collaborative efforts of the research group, students and Pomona community. The book is the synthesis of creative reasoning that acts as an interface to intervene in the sustainability discourse. Through the printed publication the group were permitted to document its visions and disseminate these messages to the developer, local community and international networks, hopefully inspiring a transition in thinking and eliciting a change in mindsets to address local and global issues.

The launch of the book took the form of a Salon event as part of Design Manchester in October 2016. The event showcased responses to Pomona in an exhibition to inspire and elicit dialogue, discussion and debate. The Salon format is a platform in which to network, share vision, specialism and non-specialism in response to the work presented as a way to initiate new networks and generate new collaborative initiatives.

The precise history and function of the Salon is varied, however in this particular context it is a place to bring people together from many different backgrounds and disciplines to exchange thoughts and ideas. Salons originated in Ancient Greece and over the centuries spread across Europe. Through the Literary Salons of Denmark and King Stanislav of Poland's 'Thursday Dinners' to the French 18th Century Salons, they were places to facilitate discussion and debate on: science, philosophy, radical theories and political matters. These meetings contributed to fundamental political and cultural change, inciting Enlightenment Philosophies and the French Revolution (Goodman 1994). There were also numerous salons held by women of various classes and some of the hostesses became very influential, which in turn played a seminal role in enabling women to be educated in politics, philosophy and play an active role in public life (Lougee 1976).

Whilst not being new, the Salon is an instrumental platform that can be forever modern, as it can be refreshed with new content and fed with a diverse demographic of people, subject matter and ideas. In this case, The Salon style of presenting the work was adopted to gather people of various communities, disciplines, ages and walks of life, to discuss, debate, exchange, entertain/be entertained and to think about LiFE.

The Salon also hosted a selection of work in response to the book by under-graduate students from B.A.(Hons) Three Dimensional Design, M.A. Landscape Architecture and Product Design students from ArtEz Institute for the Arts, Arnhem, Netherlands. Inspired by the M.A. Professional Platform project and facilitated by both researchers and M.A. students the 'Opiso City: Design Lab' project was a workshop based between Insitu (Architectural Salvage warehouse based in a Grade ll Victorian pub), Phoenix House (wood, metal and ceramic studios) and Pomona Island.

The Salon showcased an ambient backdrop for discussion and debate amongst a multitude of exhibits; designed and found objects, film projections, a performance of specially composed music and poetry inspired by Pomona. The mapping of ideas on rolls of wallpaper from the workshops, printed posters from the book, flowers and vegetation from Pomona adorned the walls and were suspended from the ceiling to make up what was a rapid curation with a 1970's Punk feel to it. This was a spon-taneous meeting of locals, Design Manchester festival patrons and all those involved in the content for the Salon and publication, which included students and staff, writers, artists, designers, educators, composers, poets, musicians, urban planners, zoologists, anthropologists, ecologists and filmmakers. It resembled a live, unrehearsed, unedited, unlimited, 3-Dimensional mixed version of Facebook, Instagram, Pinterest, Skype, and Face Time, that existed outside the digital arena, in real time.

It was both energizing and stimulating to experience such a large number of people connected together under one roof. Since so many of these people were presenting thoughts and ideas at the same time, it meant that there was a guaranteed audience with potential for shared interest and support for each other. Being part of a group show also reduces the pressure of showing work as a solo artist/presenter while at the same time an audience is given choice and variety of content.

To see the speed in which non-technological communication can operate is refreshing, there is no copying, trash can, digital delay, rendering, spell check or cut and paste, it's a form of unscripted, spontaneous and unguarded interaction with no information or electrical feeds required. It was pure 'Living in the Moment' (Jenkins and Deuze 2008).

Art galleries, theatre and opera present work in a similar way to the Salon, however the Salon is more informal and its dynamic positively encourages dialogue and engagement between what is on show, the presenter(s) and the observer(s), in equal standing.

Ultimately, the Salon format is very flexible, it can operate within or outside conventional spaces, with or without; public funding, commercial constraints, challenges and commitments. There is the potential to have regular Salons that can continue to grow. There can be shifts in focus and emphasis, bespoke gatherings can be tailored to the needs of those showing a particular type of work to reach a specific size and type of audience. It's a form of market research, marketing, promotion, public relations, communications, networking in a relaxed and informal way and the audience can mix, add and take what they will. Connecting, collecting and collating information and meeting people in the moment in a space(s) of interest, face to face contacts made, for future Salons, for meetings and dialogues.

Future Salons will maintain existing connections and expand new networks, the more support and the greater numbers of people brought into the Salon arena to share, discuss and present ideas then the stronger the community can become. The Salon is a format that can happily coexist and develop in tandem with social media to reach an ever-growing audience, both digitally and in real time. Participants may become a live stream of future Facebook pages, but the present and the now is in reality, face to face, no screens and digital edits can divide the social experience (Keller and Fay 2012).

People and content from varying backgrounds and disciplines can be represented, unite and exchange simultaneously within singular or multiple locations, both locally and internationally. From this, we can look at expanding and sharing our connections and knowledge, so that theories, visions and philosophies explore existing ecologies in which to develop, inspire and support Living in Future Ecologies.

8 Conclusion

As part of Design Manchester, the 'collective power' of the participants were showcased through the publication and Salon event. Scenarios for change were published and exhibited as mediators for conversation, connection and the development of new networks between public, planners, community and education.

Through creative visioning, students, researchers and community investigated the tools in which to implement new scenarios for Pomona. Through this collaboration the philosophical reasoning for responses was underpinned by specialist knowledge that was strengthened by professional experience. Responses although mainly conceptual were still plausible and possible. The creative interpretation of Pomona Island was holistic of its contexts, which lead to a deeper level of response and understanding of the importance of the works produced. Through co-design (Faude-Luke 2007), participants proposed visions of a new speculative future and how it could be achievable. These visions and responses re-interpreted the social, economic, political and ecological perspectives that were invested in the site. It was recognised that in order to implement many of the proposals there would need to be a move towards a bottom up, participatory democracy in which the community's voice is heard by the council, developer and planners. Furthermore, it was understood that in order to translate moral arguments of consensualism and evidence based realism (Batty 2001) into political structures for action, proposals needed to inspire a strategy to enable a 'power of change' to the current development strategy.

The Pomona Island Professional Platform unit explored the way in which art and design thinking, through creative discourse, can act as a conduit to innovate systems thinking and imbue future scoping in an educational context through 'pervasive creativity' (Bohm and Peat 2000). By asking students to critically analyse the social situation surrounding Pomona students learnt about local ecology, politics and social structures that shape the everyday. Through their participation in re-thinking and re-inventing a possible future, education facilitated a subversive force in which to intervene and initiate a change in mindset (Freire 1996).

Through futures thinking, art and design narratives generated by the unit in collaboration with the LiFE research group and local community, gave the opportunity to highlight social and political issues of how we choose to live in future cities, and provided the opportunity to make the public (localized audience) aware of capacities they already possess to elicit change and be more instrumental in developing socio-cultural capital (Head 2016). This holistic involvement of the many stakeholders of Pomona Island, through a systems approach (Westley et al. 2002) to strategic design thinking, allowed a more clearly defined egalitarian connection between the city (human) and landscape (nature), as a way to address forward-looking behaviour in response to a Panarchical system of regional resource and ecosystem management (Gunderson and Holling 2002).

Students and researchers working collaboratively on research focused agendas provided a platform for experimentation with ways of working, seeing, doing and thinking. Limitations of the project were turning these proposals into real changes beyond the conceptual and into real alternative realities for the Island. Unfortunately, the local councils and developer who owns the Island, have other agendas that follow a different rule of priorities. However, it's through this kind of project that we are allowed to explore notions of what is desirable in terms of re-inventing the city. As well as being a valuable creative and academic exercise through the project we can start to signpost changes in neoliberal ideals used to justify 'progress' which fundamentally taste less appetizing in the growing illumination of context and climate. The project gave a holistic vision from a community with many different connections to the place that runs far deeper than shallow financial motivations. The project although not physically changing the fate of Pomona has been an invaluable force in bringing communities of people together to re-think and re-iterate what is important in the changing social, political, cultural and ecological places we call home. Indeed, this concept is at the route of the word ecology or '*oikos*', that defined the design of an Ancient Greek home in its full familial, civic, environmental and cosmological contexts. And, as Gregory Bateson pointed out, ecology is 'the pattern that connects' (Bateson 2002).

Through conducting an imaginative transdisciplinary enquiry, investigations translated art & science methods to establish new measures in which to contextualize and respond to the multi-layered landscape. The inquiry imagined the context of Pomona as a wicked problem (Brown et al. 2015). Utilizing the many approaches and perspectives of participants permitted everybody to learn from each other's knowledge and experience, in turn contributing to the new visioning and development of strategies for Pomona. The open nature of the enquiry allowed for transdisciplinarity through the many perspectives of design, landscape, planning, governance, resilience and socio-environmental agendas. The many perspectives brought a holistic overview to the project which meant that creative visions became multi-faceted and multi-focused. Responses to the project engaged creativity to articulate socio-cultural experience and preference (Luck and Ewart 2012) allowing a variety of responses to an open critical enquiry.

References

Anderson J (2004) Talking whilst walking: a geographical archaeology of knowledge. Area 36(3): 254–261

Bateson G (2002) Mind and nature: a necessary unity. Hampton Press, London

Batty, S. (2001) The politics of sustainable development. In: Layard A, Davoudi S, Batty S (eds) Planning for a sustainable future. Spon Press, London, pp 19–32

Bohm D, David Peat F (2000) Science, order, and creativity, 2nd edn. Routledge, London, pp 229–271

Brown V, Harris J, Russell J (2015) Tackling wicked problems: through the transdisciplinary imagination. Earthscan, London, pp 16–30

Cocchiarella F, Booth P (2015) Students as producers: an 'X' disciplinary client-based approach to collaborative art, design and media pedagogy. Int J Art Des Educ, Wiley 34(3):326–335

Deleuze G, Guattari F (2014) A thousand plateaus. Bloomsbury Aacademic, London, p 1

Ellis D, Thomas M (2001) Gorton Monestary. In: McKennan G et al (eds) Manchester architecture papers. Manchester School of Architecture, Manchester UK

Fuad-Luke A (2007) Re-defining the purpose of (Sustainable) design: enter the design enablers, catalysts in co-design. In: Chapman J, Gant N (eds) Designers visionaries + other stories. Earthscan, London, pp 18–52

Freire P (1996) The pedagogy of the oppressed. Penguin Books, London, pp 11–22

Goodman D (1994) The republic of letters: a cultural history of the french enlightenment. Cornell University Press, Ithaca

Gunderson L, Holling C (2002) Panarchy: understanding transformations in human and natural systems (p 3–22). Island Press, London

Haley D (2008) The limits of sustainability: the art of ecology. In: Kagan S, Kirchberg V (eds). Sustainability: a new frontier for the arts and cultures. VAS-Verlag, Frankfurt, Germany

Haley D (2017a) Undisciplinarity and the paradox of education for sustainable development. In: Leal Filho W (ed) Handbook of Sustainable science and research. Series, Climate Change Management. Springer, Berlin (in press)

Haley D (2017b) Making our futures: accidental death of a planet. Paper to agents in the Anthropocene: trans/disciplinary practices in art and design education today. The Master of Education in Arts, Piet Zwart Institute, Rotterdam, (Unpublished)

Haley D, Vargas VR, Ferrulli P (2016) Weaving the filigree: paradoxes, opposites and diversity for participatory, emergent arts and design curricula on sustainable development. In Leal Filho W et al (ed) Handbook of theory and practice of sustainable development in higher education. Springer International Publishing, Berlin

Head A (2016) Here comes trouble: an inquiry into art, magic & madness as deviant knowledge, ZK/U Press, Berlin, pp 95

Jenkins H, Deuze M (2008) Convergence culture. Convergence: The International Journal of Research into New Media Technologies, Sage Publications London, Los Angeles, New Delhi and Singapore. 14(1):5

Keller E, Fay B (2012) The face-to-face book: why real relationships rule in a digital marketplace. Simon and Schuster, New york

Knox B (1994) Backing into the future: the classical tradition and its renewl. W.W. Norton and Company, Inc., New York

Lakoff G, Johnson M (1999) Philosophy in the flesh: the embodied mind and its challenge to western thought. Basic Books, New York

Langdown A, Vargas VR (2015) Integrating sustainable development within teaching fashion education. In: Leal Filho W et al (ed) Integrative approaches to sustainable development at university level. Springer International Publishing, Berlin, pp 539–550

Luck R, Ewart I (2012) Towards a living future of calm. In: Rogers P (ed) Articulating design thinking. Libri Publishing, England, pp 89–106

Lougee C (1976) Le paradis des femmes: women, salons and social stratification in seventeenth century France. Princeton University Press, Princeton

Marsick VJ, Maltbia TE (2009) The transformative potential of action learning conversations. In: Mezirow J, Taylor EW Associates (eds) Transformative learning in practice: insights from community, workplace and higher education. Jossey-Bass, San Fransico, pp 160–171

Meadows DH (1999) Leverage points: places to intervene in the system, sustainability institute. http://www.sustainer.org/?page_id=106. Retrieved 18 Nov 2016

Nicolescu B (2002) Manifesto of transdisciplinarity. State University of New York Press, New York

Quality Assurance Academy—QAA (2014) Education for sustainable development: guidance for UK higher education providers. Gloucester. http://www.qaa.ac.uk/en/Publications/Documents/Education-sustainable-development-Guidance-June-14.pdf. Retrieved from 1 Jun 2016

Sierens E, Vansteenkiste M, Goossens L, Soenens B, Dochy F (2009) The synergistic relationship of perceived autonomy support and structure in the prediction of self-regulated learning. Br J Educ Psychol 79(1):57–68

Tilbury D, Keogh A, Leighton A, Kent J (2005) A national review of environmental education and its contribution to sustainability in Australia: further and higher education. Australian Government Department of the Environment and Heritage and Australian Research Institute in Education for Sustainability (ARIES), Canberra

Wahl DC (2016) Designing regenerative cultures. Triarchy Press, Axminster, England

Walker BH et al (2006) A handful of heuristics and some propositions for understanding resilience in social-ecological systems. Ecol Soc 11(1):13. http://www.ecologyandsociety.org/vol11/iss1/art13/. Retrieved from 18 Nov 2016

Westley F, Carpenter S, Brock W, Holling C, Gunerson L (2002) Why systems of people and nature are not just social and ecological systems. In: Gunderson L, Holling C (eds) Panarchy: understanding transformations in human and natural systems. Island Press, London, pp 103–120

Author Biographies

Fabrizio Cocchiarella M.A.(RCA) FHEA FRSA Currently Programme Leader for B.A.(Hons) Three Dimensional Design at Manchester School of Art. Fabrizio's work as a lecturer, researcher and designer aims to inspire a new generation of critically aware creative professionals and contribute to shaping future scenarios for living. Fabrizio's specialism is routed in the production and manufacture of objects and installation projects utilizing the practice of Design as an interface in which to mediate and translate critical methodologies and commercial practices. Fabrizio is an Associate Fellow of the Higher Education Academy and the RSA (Royal Society for the Encouragement of the Arts, Manufacturing and Commerce).

Valeria Ruiz Vargas Currently Education for Sustainable Development (ESD) Co-ordinator at Manchester Metropolitan University, Valeria Vargas leads the Environmental Management System policy area on ESD (i.e. Teaching and Learning). She regularly develops collaborative multidisciplinary, pedagogical and research projects. Valeria has expertise in the practice of appreciative inquiry approaches based on creative arts to facilitate change in social and environmental practices for multidisciplinary settings. For over 10 years, Valeria Vargas has applied creative arts including music and sculpture to facilitate resilience of (human and non-human) communities.

Sally Titterington Associate Lecturer at Manchester Metropolitan University, M.A. unit 'Making Our Futures' in ecological arts and sustainable design, and a member of the research group LiFE. She works as a Designer and Media Artist producing art works from her own studio and collaborates with multidisciplinary studios; designing and producing products, media content and commissioned pieces for the creative, public and corporate industries. Since 2000, Sally is making an ongoing documentary and observational film 'Generous Acts' of the international furniture fair Salone del Mobile, Milan, Italy. She produces multimedia salons for international cultural exchange; creating soundscapes, film projections and collaborations with other artists from diverse creative fields.

Dr. David Haley Ph.D. HonFCIWEM RSA, creates ecological artworks and publishes internationally to promote 'capable futures' that address our transformative challenges. He is a Visiting Professor at Zhongyuan University of Technology, China. At Manchester Metropolitan University, UK, he was a Senior Research Fellow, Director of the Ecology In Practice research group and led the M.A. unit in Making Our Futures—ecological arts and sustainable design. David is, also, Vice Chair of the CIWEM Art & Environment Network, a member of UK MAB Urban Forum, Society for Ecological Restoration, Ramsar Culture Network Arts Group, and founding Trustee of Future's Venture Foundation.

A Critical Evaluation of the Representation of the QAA and HEA Guidance on ESD in Public Web Environments of UK Higher Education Institutions

Evelien S. Fiselier and James W.S. Longhurst

Abstract In June 2014 the Quality Assurance Agency for Higher Education (QAA) and the Higher Education Academy (HEA) published their education for sustainable development guidance for higher education. The guidance is a tool for supporting educators in embedding or including knowledge, understanding and awareness of sustainable development across the curriculum by identifying graduate learning outcomes and outlining approaches to teaching, learning and assessment. The purpose of the guidance is to help higher education institutions in training sustainability literate graduates who will contribute to an environmentally and ethically responsible society through the application of their skills, knowledge and experience. The guidance is now some 30 months old, which leads us to pose the question: to what extent has the UK higher education sector adopted and implemented the guidance in its curricula? A systematic web-based analysis has been performed of 139 higher education institutions' websites to identify the use of the guidance in the public web environments of UK higher education, especially regarding the design, delivery and review of curricula. To what extent do UK universities reference the role of ESD or the guidance in the specification of the graduate learning outcomes and the approaches to teaching, learning and assessment? In analysing the web environments for guidance related content we have also identified the presence of general information relating to estates sustainability and to general ESD concepts. In presenting the results a comparison is made between institutions, which contributed to the guidance and the rest of the sector. The analysis shows that 120 institutions provide information about estate sustainability, 82 general information on ESD, but only 16 institutions make public statements about their use of the QAA and HEA guidance for ESD. As such, this study provides the first comprehensive assessment of the presence of sustainability, and specifically ESD and the QAA and HEA guidance, in the online platforms of universities and colleges. This paper will share the results of the study with the UK

E.S. Fiselier (✉) · J.W.S. Longhurst
Faculty of Environment and Technology, University of the West of England,
Coldharbour Lane, Bristol BS161QY, UK
e-mail: Evelien2.Fiselier@live.uwe.ac.uk

© Springer International Publishing AG 2018
W. Leal Filho (ed.), *Implementing Sustainability in the Curriculum of Universities*,
World Sustainability Series, https://doi.org/10.1007/978-3-319-70281-0_14

223

higher education sector and in so doing hopes to encourage HEIs to engage with ESD by integrating it in their curricula.

Keywords Education for sustainable development · Higher education · QAA and HEA guidance · Web-based analysis

1 Introduction: Sustainability in a Global Competitive HE Sector

As competition in the higher education sector is increasing, it has become more and more important for universities and colleges to find out what differentiates them from the rest and to communicate their strengths, values, ambitions and the quality of the student experience to prospective students and staff (Chapleo et al. 2011). With a vast array of higher education institutions to choose from, there is a variety of factors that could determine the choice of study by a future student or place of employment for a future staff member. UK HEIs review these factors and adjust their recruitment strategies accordingly (Briggs and Wilson 2007, p. 59). The higher education institution's website has been described as 'the ultimate branding statement' that could strongly influence the student's decision to attend (Hannover research 2014). Lately, it has become essential for institutions to present themselves as being sustainable as a growing body of research suggests that prospective employees prefer working for sustainable organisations (NBS 2013; Global Tolerance 2015). People and Planet publish an annual university league table ranking UK universities by environmental and ethical performance (People and Planet n.d.). Accessibility of online information is an integral aspect of the rating system. Representation of information in the web environment on sustainability management of the estates and ESD, as such, will increase chances of ranking higher.

In addition, results of national surveys by NUS suggest that the majority of UK students '*believe that sustainability should be covered by their university*' and that it has become important to students that HEIs improve the sustainability of their operations and provide ESD across the institution (HEA 2014; Drayson et al. 2012).

2 Growing Recognition for ESD

Societies around the world are facing complex and interconnected issues i.e. climate change, resource depletion, poverty and environmental degradation and sustainable development is regarded as an appropriate approach to addressing these intricacies (Waas et al. 2010). All members of society are called upon to act according to sustainable standards. The role of the educational sector has been recognised as being pivotal to driving change, since it is deemed imperative to equip students who set out to become the future leaders of society with the skillset to deal with complexities and shape a sustainable world (Desha and Hargroves 2014).

Across the globe declarations and action plans have been developed to support the educational sector in its quest to sustainability (Desha and Hargroves 2014; Dale and Newman 2005). In the past three decades higher education institutions have also come to recognise the importance of education for sustainable development and are responding to the call for action. This is reflected in the growing number of institutions that are signing ESD declarations and statements of commitment to ESD, the efforts to reduce the impacts of the estate, and the rise of sustainability courses and degrees (Sterling et al. 2013; Lambrechts et al. 2013; Ferrer-Balas et al. 2010).

Even though the number of higher education institutions integrating ESD in the curriculum is increasing, change is occurring at a slow pace with the mainstream of the institutions greening the estate whilst sticking to the traditional education system (Yarime and Tanaka 2012; Su and Chang 2010; Tilbury 2011). The current educational system is, however, rooted in corporate globalisation, is ill designed for tackling the messy multidimensional complexities that society faces, and creates graduates who live their lives in an unsustainable fashion leading society along the same destructing path (Orr 1991; Everett 2008; Armstrong 2011; Martin and Jucker 2005).

In addition, there is a decreasing support for ESD from the UK government and higher education sector bodies. A report by the UK National Commission for UNESCO highlights the declining policy emphasis on sustainable development and education for sustainable development in England, Northern Ireland and Wales (UK National Commission for UNESCO 2013). Similarly, the higher education funding councils and the Higher Education Academy, through various sustainability initiatives, have been crucial for the development of sustainability in the UK HEI sector, but have reduced their support. The funding councils for England, Scotland and Wales supported institutional engagement in the sustainable development of the campus environment, HEFCE supported the NUS through the Students Green Fund to encourage students' unions to engage in sustainability activities and to address organisation sustainability impacts, and the Higher Education Academy encouraged HEIs to develop pedagogic, student engagement and curricula innovations in ESD through the Green Academy Programme and the ESD Advisory Group (HEFCE 2013; HEA n.d.).

The current resource constrained environment has, however, had a severe impact on the support offered to the HE sector for ESD or general sustainability initiatives by the funding councils and the HEA. Noticeably the next round of Green Academy has not been launched and despite publishing a sustainability framework in 2014 (HEFCE 2014), HEFCE appears to have withdrawn from an overt leadership role. The NUS and the Environmental Association of Universities and Colleges on the other hand continue to promote the importance of sustainability in both the estate management and teaching of HEIs.

Within this changing landscape of policy drivers and declining sector support for implementation how have UK HEIs responded? In considering a place of study or employment what will a prospective student or staff member now see represented in institutional web environments regarding sustainability and specifically their ESD commitments and ambitions?

As one test of this representation of ESD in HEI web environments this paper will explore the presence, extent or absence of reference to the guidance on ESD published by the Quality Assurance Agency for Higher Education (QAA) and the Higher Education Academy (HEA) (QAA and HEA 2014). The guidance is a tool for supporting educators in embedding or including knowledge, understanding and awareness of sustainable development across the curriculum by identifying graduate learning outcomes and outlining approaches to teaching, learning and assessment (Shephard and Dulgar 2015). The purpose of the guidance is to help higher education institutions in training sustainability literate graduates who will contribute to an environmentally and ethically responsible society through the application of their skills, knowledge and experience. By exploring the presence, extent or absence of reference to the guidance in the web environment, this study will be a first step in finding out to what extent sector wide interventions i.e. the guidance are effective in supporting and possibly driving the integration of ESD in curricula across universities and colleges, and indeed in finding out what is required to make this change happen.

3   Methods

139 UK higher education institutions' websites have been analysed over a four-week period to examine to what extent the UK higher education sector has adopted and implemented the QAA and HEA guidance for ESD. Quantitative methods have been used for the collection and analysis of data. The data presented here is merely a small part of a broader research for which a triangulation of methods was used. This was done to keep the size of the paper within the limits of the format.

The Recognised bodies list by Gov.uk includes 159 higher learning institutions that can award degrees and was taken as the basis for the web-based review. A quick assessment was done of websites of universities and colleges, which are specialised in a discipline that does not have a link with sustainability. Institutions were eliminated from the sample population when sustainability did not feature on their website, which led to 20 universities and colleges being excluded, i.e. Guildhall School of Music and Drama and the British School for Osteopathy, and a remaining sample size of 139 institutions. A comparison is made between institutions, which contributed to the guidance and the rest of the sector.

The research approach has been reviewed by UWE's Faculty Research Ethics Committee and approval to proceed granted under reference UWE REC REF No: FET.16.02.027 dated 21st April 2016.

Assessment criteria were created and tabulated to secure the consistency of the overall assessment. The table consists of three main categories;

1. QAA and HEA Guidance,
2. Education for Sustainable Development,
3. Estate sustainability

Each main category is subdivided into subcategories; the quantity, quality, type, and searchability of the information.[1] Each university is scored from 1 to 4 on the quantity, quality and searchability of information with 1 for very low quantity and quality, and very difficult to find information and 4 for very high quantity and quality, and very easy to find information. Furthermore, the ESD category has additional subcategories, which are the following: delivery of ESD in the curriculum, ESD graduate attributes, ESD pedagogy styles, ESD policy integration, and ESD awards.

Finally, for each university and college quotes have been gathered that are characteristic for the institution's view on sustainability, ESD or the guidance, and a quantitative content analysis has been performed, including a calculation of concept use per institution website. The concepts counted were: *sustainability, sustainable development, ESC, sustainability literacy, global citizenship, social justice, ethics and well-being, environmental stewardship, future thinking, holistic, interdisciplinary, critical thinking, systems thinking, self-reflection, inter- and intra-generational, adaptability, educational responsibility, societal challenges, future leaders,* and *whole institution approach.*

Qualitative content analysis focusses on the meaning of data, is systematic, flexible and reduces information, whilst quantitative content analysis holds the assumption that the most frequently mentioned words and sentences bare importance (Schreier 2014; Binsbergen 2013).

Data from the web-environment was organised into themes, and synthesised using quantitative content analysis. The advantages of document analysis are: high efficiency, easy obtainability of large volumes of data, ease of reaching a geographically diverse population and cost-effectiveness (Fielding et al. 2008; Bown Bowen 2009). An inter- and intra-comparison was done between the institutions, which provide information about the guidance on their website, and the total amount of institutions.

The web-review was done manually and solely relies on the judgment of the individual analyst, which implies a certain degree of subjectivity (Bauer and Scharl 2000). In order to test reproducibility of the scoring system a sample was resurveyed after the first analysis of the 139 institutions. The results are shown in Table 1. The sample population was established by taking every 10th HEI in the study population. In order to allow for the variation in scores ascribed a further four

---

[1]Searchability is reflected here as the level of accessibility of the information, which could range from easy to find to hidden information.

Table 1 Reproducibility test of web-review scoring

| Institution | First scoring | | | | | Second scoring | | | | | | Reproducibility |
| | ESD scoring | | Estate scoring | | Searchability | ESD scoring | | Estate scoring | | Searchability | |
	Quant.	Qual.	Quant.	Qual.		Quant.	Qual.	Quant.	Qual.		
10. Birkbeck university	1	1	/	/	1	/	/	/	/	/	±
20. University of Bristol	4	3	5	3	3	4	4	4	4	3	±
30. City University London	4	3	4	3	4	3	3	4	4	4	±
40. University of the East of London	/	/	/	/	/	/	/	2	2	1	−
50. Glyndwr University	1	1	2	2	3	1	1	3	2	1	+
60. Keele University	4	4	2	1	4	4	3	4	5	4	−
70. University of Lincoln	1	1	4	3	2	/	/	4	4	2	±
80. Loughborough University	/	/	4	2	4	/	/	4	3	4	+
90. Nottingham Trent University	4	4	4	3	1	4	4	4	3	1	+
100. Regent's University London	/	/	/	/	/	/	/	/	/	/	+
110. University of Southampton	4	3	2	3	2	4	4	2	2	3	±
120. University of Surrey	1	1	2	2	4	1	1	2	2	4	+

(continued)

Table 1 (continued)

Institution	First scoring					Second scoring					Reproducibility
	ESD scoring		Estate scoring		Searchability	ESD scoring		Estate scoring		Searchability	
	Quant.	Qual.	Quant.	Qual.		Quant.	Qual.	Quant.	Qual.		
130. University of the West of England	5	4	4	4	2	5	5	4	4	2	+
24. University of Cambridge	4	3	3	3	3	4	2	5	4	1	–
35. De Montfort University	2	2	5	4	4	4	4	2	2	3	–
65. Lancaster University	2	2	4	4	4	2	1	4	3	4	±
121. University of Sussex	1	1	2	1	1	1	1	2	2	1	+

Scoring legend
+ Equal scoring
± Partially similar scoring
– Different scoring

HEIs were added to give a robust sample for resurveying equivalent to 12% of the study population.

As shown in Table 1, the institutions have been given a symbol to illustrate the level of similarity between the first and second scoring. The plus symbol stands for an equal scoring to which institutions with a point-difference of 0 or 1 belong. The plus-minus symbol signifies slight differences between the scorings with a point-difference of 2 or 3. Note that this is the added difference and not a 2 point difference for one score (e.g. 2 total difference = 1 quantity ESD + 1 quantity estates). Finally, the minus symbol means significant differences in scoring. This group of institutions have a significant scoring difference of 2 or above for one element (e.g. 3 total difference = 2 quality estate + 1 quantity estate).

The outcomes show that there is largely an agreement between scorings, thus indicating internal consistency in the method.

4   Results

4.1   QAA and HEA Guidance for ESD

From the web-based review only identifies 16 out of 139 higher education institutions (12%) as providing information about the QAA and HEA guidance for ESD on their website. The 16 HEIs were assessed on the quantity, quality, type of information, and searchability of the information. Table 2 presents an overview of the findings.

Quantity

The quantity was scored on a scale from 1 to 3 with 1. one sentence, 2. one paragraph, and 3. more than one paragraph. The higher education institutions differed with six institutions scoring 1, seven scoring 2, and three institutions that scored 3.

Quality

The quality of the information was rated on a scale from 1 to 4. Universities or colleges, which only provided a link were scored 1, those providing background information about the guidance rated 2, those with a description of implementation scored 3, and the institutions that provided a detailed description of how the guidance was implemented scored 4. Five higher education institutions scored 1, four scored 2, six scored 3, and there was one that scored 4.

Searchability

The searchability rating was also done on a 1 to 4 scale with 1. very difficult to find to 4. very easy to find. For the majority of the websites, 9 out of 16, it was very difficult to find the information about the guidance. Three websites scored 2

Table 2 Web-review guidance categories

Number	Higher Education Instition name	Mission group	QAA and HEA Guidance						
			Guidance	Quantity info	Quality info	Type info	Searchability	Category
4	Anglia Ruskin University	MillionPlus Universities UK	Yes	2	2	As resource for staff to stimulate sustainability in curriculum, provision link	2	2
18	University of Bradford	Universities UK	Yes	1	1	Link to guidance is provided	1	1
19	University of Brighton	Universities UK University Alliance	Yes	1	1	Link to guidance is provided	1	1
24	University of Cambridge	Russell Group Universities UK	Yes	1	2	Brief description on the guidance provided	3	1
25	Canterbury Christ Church University	MillionPlus Universities UK	Yes	2	3	Sustainability at university supported by it, pilot, curriculum review based on 4 core themes	1	5
28	University of Chester	Universities UK	Yes	1	2	Adoption ESD definition, provision link	2	3
33	University for the Creative Arts	Guild HE	Yes	3	3	Adoption ESD definiton and 4 core themes, provision link	1	4

(continued)

Table 2 (continued)

Number	Higher Education Instition name	Mission group	QAA and HEA Guidance					
			Guidance	Quantity info	Quality info	Type info	Searchability	Category
42	University of Edinburgh	Russell Group Universities UK	Yes	2	3	Link provided, info on pilot, consultation, and intention	1	1
60	Keele University	Universities UK	Yes	2	2	Info on launch guidance, purpose, adoption definition and 4 core themes	3	4
62	King's College London	Russell Group Universities UK	Yes	2	1	Before publication	1	1
63	Kingston University	University Alliance Universities UK	Yes	1	1	Link to guidance is provided	1	1
71	University of Liverpool	Russell Group Universities UK	Yes	3	3	Annual report: publication, purpose, paper to request ESD policy and make recommendations implement QAA across uni	1	1
82	Manchester Metropolitan University	University Alliance Universities UK	Yes	2	3	Adoption definition ESD and 4 core themes, support curriculum development, provision link	2	5

(continued)

Table 2 (continued)

Number	Higher Education Instition name	Mission group	QAA and HEA Guidance					Searchability	Category
			Guidance	Quantity info	Quality info	Type info			
94	Plymouth University	University Alliance Universities UK	Yes	2	3	Adoption ESD definition, provision link		4	3
103	Royal Agricultural University	Guild HE	Yes	1	1	Link to guidance is provided		1	1
130	University of the West of England, Bristol	University Alliance Universities UK	Yes	3	4	In Quality Management and Enhancement Framework: graduate outcomes used for monitoring, developing programmes, annual module reports; Critical Evaluation Document; Link and whole document provided		3	6

implying relative difficulty to find information. On three other websites the information was relatively visible. Finally, one website scored 4 on searchability.

Type of information

Six categories emerged from the analysis, which are:

1. Informing
2. Stimulating
3. Partial adoption
4. Wider adoption
5. Partial implementation
6. Wider implementation

The 16 institutions were categorised according to the type of information that they provided about the guidance. Half of the group is categorised as '*Informing*', which are those institutions that provided the link of the guidance online with an occasional description of the consultation process and the purpose of the guidance.

One of the institutions used the guidance to stimulate academics to integrate sustainability in their course and was included in the second category. The two institutions that are in category 3 stated on their website to have adopted the ESD definition from the guidance. The fourth category comprises of a college and a university, which claim to have adopted the ESD definition and the four core themes. The fifth category includes those that report on the guidance supporting the design, delivery or review of the curriculum. Finally, one website contains information about implementation in multiple areas.

5 Education for Sustainable Development

The majority of the universities and colleges, 82 institutions (61%), discuss education for sustainable development on their website against 57 institutions (39%) that do not. All universities and colleges that mention the guidance also provide information on ESD integration.

Quantity and quality

The 57 institutions were scored from 1 to 4 on the quantity and quality of information. For quantity the scoring is as follows:

1. less or equal to one paragraph
2. more than 1 paragraph and less than 3 paragraphs
3. more than 3 paragraphs and less than 1 page
4. more than 1 page

For quality the institutions are scored on the variety of information provided, inclusive of information that is informing (e.g. definition ESD, background information on ESD), about how ESD is integrated, how staff and students are engaged

Table 3 Combined score for quantity and quality ESD information

Group	Combined score	Total institutions	Guidance institutions	% Guidance of total
1	2 or 3	28	0	0
2	4 or 5	17	1	6
3	6 or 7	15	5	36
4	8, 9 or 10	22	10	46

etcetera. Score 1 implies a low variety, whilst 4 stands for a high variety of information. Categories have been created for the combined score of quantity and quality, which is presented in Table 3.

Group 1 is the worst scoring group and group 4 scores the best. The total amount of institutions is divided in the following way: 28 are in group 1, 17 are in group 2, 15 are in group 3, and 22 are in group 4 of which 3 have scored exceptionally with a combined score of 10. As shown in Table 3, 22 out of 82 scored 4 of which 10 are guidance institutions and 11 are non-guidance institutions. For the guidance institutions that is 46% (10 out of 16) compared to a mere 18% for the non-guidance institutions (11 out of 66). Additionally, the majority of non-guidance institutions (28) had the lowest score on ESD information provision.

Delivery, graduate outcomes and pedagogy

The webpages have also been analysed to identify whether the institutions specify the delivery of education for sustainable development, graduate attributes, and teaching techniques. From the 82 institutions, 36 non-guidance and 10 guidance institutions reported on the delivery of ESD. The relative percentage per group was calculated. For the non-guidance institutions this is 55% (36 out of 66) and for the guidance institutions that is 63% (10 out of 16). The difference in percentage is below 10%.

Graduate outcomes are stated by a mere 17 universities and colleges of which 6 are guidance institutions and 11 non-guidance institutions. This is 38% of the guidance institutions versus 17% of non-guidance institutions signifying a 21% difference.

Finally, the pedagogy is discussed by 9 institutions of which 5 are guidance institutions and 4 non-guidance institutions. Again, there is a large percentage difference between 31% of guidance institutions and 6% of non-guidance institutions.

Policy documents

The websites were assessed for the extent to which it featured information about ESD embedded in policy documents. The results are presented in Table 4.

As evident from the table, the majority of the universities and colleges has integrated ESD in the Sustainability or Environment Policy, the Sustainability or Environment Strategy and the Sustainability or Environment Report. Less common is for institutions to mention ESD in the Vision Statement and the Sustainability or

Table 4 Level of embeddedness ESD in policy

Policy document	Total	Guidance
Vision statement	3	2
Strategic plan	13	5
Sustainability/Environment strategy	21	4
Learning and teaching strategy	15	2
Sustainability/Environment action plan	8	2
Sustainability/Environment policy	49	7
Annual report	0	0
Sustainability/Environment report	17	3
ESD documents	9	6
Other documents	9	4

Table 5 Compared categorisation of quantity and quality of estate sustainability information

Group	Combined score	Total institutions	Guidance institutions	% Guidance of total
1	2 or 3	36	6	17
2	4 or 5	40	2	5
3	6 or 7	30	4	13
4	8, 9 or 10	15	4	27

Environment Action Plan. The guidance institutions differ slightly, since they have mainly integrated ESD in the Sustainability or Environment Policy, an ESD document or the Strategic Plan.

5.1 Estate Sustainability

A large number of universities and colleges have provided information on estate sustainability on their website; 120 out of 139 (86%).

Quantity and quality

The quantity and quality of information has been scored for the 120 institutions. Both quantity and quality are rated from 1 to 4 and follow a similar method as for the other categories (Table 5).

One can deduce from the table that a large group of the 120 institutions provides information on estate sustainability on their website of relatively low quantity and quality and that it is less common for universities and colleges to provide high quantity and quality information. The guidance institutions are spread across the groups. There seems to be a slight majority that scores 1 and a slight minority that scores 2.

Searchability

The searchability rating was also done on a 1 to 4 scale. The majority of the websites (51) scored 1 on searchability. Sustainability had to be put in the university's search engine or in Google to find any information. For 26 institutions the information was relatively difficult to find. There were 20 institutions for which the information was relatively accessible. Finally, 36 institutions scored 4. For those institutions the sustainability webpage could be found by having to click once when starting on the homepage.

Concepts

For each of the 82 institutions' websites the presence and use of concepts from the QAA and HEA guidance for ESD has been calculated. Concepts of sustainability, SD, ESD and the four core themes of the guidance document were selected for analysis. The quantitative concept analysis for the 16 guidance institutions is shown in Table 6. The mode, mean, median and range have been calculated for guidance and non-guidance institutions and are in Table 7.

As becomes evident from the table, there are large differences between the guidance institutions in the concept use e.g. Manchester Metropolitan using *sustainability* only twice, whilst the University of the West of England using the word 139 times.

Table 7 shows that there is no big difference between the guidance and non-guidance, except for the use of the ESD and global citizenship terms.

Quotes

Table 8 gives an overview of a selection of quotes from institutional websites organised according to the three main categories of information provision; guidance, ESD and estate sustainability. Key themes were extracted from the quotes to be able to compare them. One can deduce from this that the first two groups do not differ considerably in their language use as both groups use concepts i.e. enabling graduates, sustainability in all subjects or practices, positive contributions, and balance of social, economic, environmental and cultural factors.

The guidance group differs in the use of the term *transformative educational experience*. A comparison with the estate sustainability group shows a slightly bigger difference in terminology use, as the emphasis lies on environmental improvement and managing the reduction of impact.

5.2   Involvement

From the seven universities that contributed to the QAA and HEA guidance on ESD only two have included information about the guidance on their website, namely Plymouth University and the University of the West of England, Bristol. All universities did, however, disclose information on their ESD achievements. The

Table 6 Concept use by the 16 guidance institutions

| Guidance institutions | Concepts | | | | | | | |
	Sustainability	Sustainable development	ESD	Global citizenship	Environmental stewardship	Social justice, ethics, well-being	Future thinking
Anglia Ruskin University	31	4	14	3	2	2	1
University of Bradford	9	10	14	0	0	1	0
University of Brighton	10	4	10	0	0	0	0
University of Cambridge	6	1	8	0	0	0	0
Canterbury Christ Church University	10	0	1	0	0	1	0
University of Chester	9	3	4	0	0	1	0
University for the Creative Arts	16	4	7	1	0	1	1
University of Edinburgh	12	0	11	0	1	0	0
Keele University	29	0	13	0	0	0	0
King's College London	17	0	3	0	0	0	0
Kingston University	30	0	7	0	0	0	1
University of Liverpool	27	4	6	1	0	0	0
Manchester Metropolitan University	2	4	18	20	2	2	1
Plymouth University	107	4	79	0	0	0	2
Royal Agricultural University	14	0	0	0	0	0	1
University of the West of England, Bristol	139	19	96	2	0	0	4

Table 7 Comparison concept use guidance and non-guidance institutions

	ESD		Global citizenship		Environmental stewardship		Social justice, ethics, well-being		Future thinking	
	G	Non-G	G	Non-G	G	Non-G	G	Non-G	G	Non-G
Mean	18.2	1	1.7	0.2	0.3	0.1	0.5	0.3	0.7	0.1
Mode	7 and 11	0	0	0	0	0	0	0	0	0
Median	9	0	0	0	0	0	0	0	0	0
Range	96	20	20	6	2	2	2	10	4	2

Table 8 Selection quotes about the guidance, ESD or estate sustainability

Quotes		
Guidance	Brighton	"This reflects the recognition that everything we do has long-term implications, and we need to enable our graduates to understand and balance the multiple and interacting dimensions—social, environmental, economic and cultural—of their actions. The vision is that this will help them to make positive contributions to our global society, now and for generations to come"
	Anglia Ruskin	"Sustainability is a cross-cutting theme that has a place in all university subjects, adding a dimension to learning rather than an extra topic to cover"
	Manchester Metropolitan Uni.	"At MMU we are working to incorporate aspects of ESD and Global Citizenship into our curricula and to translate these ideas into the outcomes of a transformative educational experience for students in all disciplines"
ESD	Uni. of Central Lancashire	"UCLan is committed to implementing environmental sustainability to benefit future generations, the local economy and community. Sustainability has a significant role within the University and we aim to include sustainable development in all of our practices"
	Uni. of Chichester	"Internationalism and global citizenship are key themes for the 21st century. These themes will be supported by the University through a focus on education for global citizenship and sustainable development that recognises the connections between the social, the cultural, the economic and the natural world"
	Uni. of Greenwich	"As a university we have a role in making further improvements in our own academic and operational sphere's of influence and also in enabling students and future graduates to have skills and knowledge to make positive sustainable change"

(continued)

Table 8 (continued)

Quotes		
Estate	Middlesex University	"At Middlesex University we are committed to continual environmental improvement through the systematic and responsible management of our environmental impacts"
	Uni. Of St. Andrews	"We have a mission to be recognised locally and internationally as a world-class institution that leads by example, actively implementing imaginative solutions and initiatives that achieve the aims of a more sustainable society"
	Uni. of Bath	"The University takes its environmental responsibilities seriously and is determined to reduce the impact of its activities"

scores for quantity and quality of ESD information varied considerably with the University of Gloucestershire, the University of Nottingham, Plymouth University, and the University of the West of England, Bristol scoring highest, Abertay University, and the University of Southampton scoring 3, and the University of South Wales scoring 1. Finally, all HEIs except for the University of Southampton mentioned the ESD delivery, graduate attributes and/or pedagogy styles on their website.

6 Discussion

In today's highly competitive global economy with the appearance of global ranking systems, the growing importance of human and knowledge capital for economic growth, and fewer resources, HEIs are in a competitive environment to recruit staff and students. Prospective students are in the position of carefully choosing the university or college that will bring them the most benefits (Hazelkorn 2015). In the last few years students have come to value institutions' efforts to manage the estates sustainably and to offer ESD within the curricula (HEA 2014). Similarly, future staff members prefer working for a sustainable organisation (Network for Business Sustainability 2013; Global Tolerance 2015). This increase in interest for sustainable institutions has challenged universities and colleges to pay more attention to it. In this same environment of growing emphasis on delivery of sustainability objectives, sector support from the UK government, HE funding councils and the Higher Education Academy is declining. Considering these events, how do they impact the extent to which HEIs manage their estates sustainably and offer ESD in curricula, and represent their achievements to prospective students and staff in the web environment?

The results of this study show that the majority of the 139 institutional websites that were reviewed—120 institutions—(86%) feature estate sustainability to some degree. The HEIs were divided into four groups according to the quantity and quality of information that they provided, as there was a large variety ranging from a university that merely mentioned employing carbon reduction targets, to an institution that gave a detailed account of rational for sustainability, targets, progress, case studies, awards etcetera for multiple estate areas.

More than half of the UK higher education sector (82 institutions, 62%) presents information on ESD on the institutional website. Again, universities and colleges have been clustered into four groups based on the quantity and quality of information presented on ESD. 46 out of 82 institutions reported on delivery of ESD on their website, 17 discussed ESD graduate attributes, but only 9 mentioned the use of ESD pedagogic styles.

An ever smaller group of HEIs referred to the QAA and HEA guidance for ESD (16 institutions, 12%). These have been divided into 6 categories based on the type of information that was provided on the institutional website, namely 1. Informing, 2. Stimulating, 3. Partial adoption, 4. Wider adoption, 5. Partial implementation, and 6. Wider implementation.

A small group of 14 HEIs have not included any information about the guidance, ESD or sustainability on their websites.

When comparing the 16 guidance institutions to the non-guidance institutions it becomes evident that the guidance institutions tend to report to a greater extent on their ESD ambitions and achievements. They also appear to present more information about ESD delivery, graduate attributes, and pedagogic styles. Remarkably, though, a slight majority of the 16 HEIs score lowest on information provision on estate sustainability. Moreover, the quantitative content analysis revealed that there are no substantial differences between the guidance and non-guidance institutions on concept use. Finally, quotes by HEIs on ESD are very similar regardless of whether an institution reported on the guidance or not.

The comparison between institutions that contributed to the guidance and those that did not, reveals that only two universities mentioned the guidance. Most of the involved institutions did, however, score high on quantity and quality ESD information provision.

7 Implications

Reflecting on the results, one could pose that for some UK HEIs sustainability and ESD is at the core of the institution and its values, commitments and ambitions, and it is important for them to differentiate the institution through presenting information about the guidance, ESD and estate sustainability. It is highly probable, however, that for the vast majority of universities and colleges sustainability and indeed ESD are one of many issues that need mention on the institution's webpage, but bares little significance. One needs to be careful, though, with drawing any conclusions about institutional practices regarding estate sustainability and ESD, since the focus of this study is on the representation in the web environment.

8   Limitations

A limitation of the research is that the selected method for assessing information presented on the web involves a certain degree of subjectivity. The reproducibility of the study has been tested through a rescoring exercise of which the outcomes show agreement between scorings, thus indicating internal consistency in the method.

Additionally, even though the research has provided an overview of the way in which UK higher education institutions represent their institution regarding sustainability, ESD and QAA and HEA guidance use on the online platform, it would benefit from a triangulation of methods to obtain a more in-depth view of the adoption and implementation of ESD by HEIs in the curricula, and the QAA and HEA guidance for ESD in particular.

Finally, this paper merely highlights key components of a broader research study. The data produced from a robust mixed multi-method research approach were too extensive to be shared in this format, which has limited this paper to the discussion of one method (Table 9).

Table 9   Categories of HEIs' online representation

Category	Name institution
QAA and HEA Guidance	*16 institutions*
1. Informing	Bradford, Brighton, Cambridge, Edinburgh, King's College London, Kingston, Liverpool, Royal Agricultural Uni.
2. Stimulating	Anglia Ruskin
3. Partial adoption	Chester, Plymouth
4. Wider adoption	Uni. for the Creative Arts, Keele
5. Partial implementation	Canterbury Christ Church, Manchester Metropolitan
6. Wider implementation	West of England
Education for sustainable development	*82 institution*
1. Low quantity and quality info	Bangor, Bath Spa, Birbeck London, Birmingham, Birmingham City, Bishop Grossteste, Buckinghamshire New, Cranfield, Cumbria, Derby, Dundee, Durham, Essex, Glyndŵr, Goldsmiths London, Harper Adams, Central Lancaster, Leeds, Lincoln, London Southbank, Newcastle, Queen Mary London, Royal Holloway London, South Wales, Stirling, Suffolk, Surrey, Uni. of the Arts London, West of Scotland

(continued)

Table 9 (continued)

Category	Name institution
2. Relatively low quantity and quality info	Bath, Brunel London, Cardiff, De Montfort, East Anglia, Glasgow Caledonian, Lancaster, Leicester, Liverpool, London Business School, The London School of Economics and Political Science, Loughborough, Royal College of Art, St. Andrews, Sussex, Wales Trinity Saint David, York
3. Relatively high quantity and quality info	Abertay, Aberystwyth, Bristol, Cambridge, Chester, Chichester, Coventry, Uni. for the Creative Arts, Hull, ifs Uni. College, Kingston, Southampton, Swansea, Westminster
4. High quantity and quality info	Aberdeen, Anglia Ruskin, Aston, Bedfordshire, Bradford, Brighton, Canterbury Christ Church, Cardiff Metropolitan, City London, Edinburgh, Exeter, Gloucestershire, Greenwich, Keele, King's College London, Manchester Metropolitan, Nottingham, Nottingham Trent, Plymouth, Royal Agricultural Uni., West of England, Worcester
Estate sustainability	*120 institutions*
1. Low quantity and quality info	Anglia Ruskin, Bangor, Uni. College Birmingham, Bishop Grosseteste, Bolton, Buckingham, Chester, Cranfield, Uni. for the Creative Arts, Cumbria, Derby, Essex, Gloucestershire, Goldsmiths London, Heriot-Watt, Keele, Kent, King's College London, Leeds Beckett, London, London South Bank, Middlesex, Newman Birmingham, Northampton, Queen's Belfast, Royal Agriculture Uni., Sheffield, Southampton Solent, St. Mark and St. John, Plymouth, St. Mary's Twickenham, Suffolk, Sussex, Arts London, West London, Winchester, Wolverhampton
2. Relatively low quantity and quality info	Abertay, Bath Spa, Bedfordshire, Birmingham, Arts Bournemouth, Bournemouth, Bradford, Buckinghamshire New, Cardiff, Chichester, Coventry, Dundee, Edge Hill, Edinburgh Napier, Glasgow, Glasgow Caledonian, Hertfordshire, Leicester, London Metropolitan, Uni. College London, Manchester Metropolitan, Newcastle, Portsmouth, Queen Margaret Edinburgh, Reading, Roehampton, Royal Holloway, Salford, South Wales, Southampton, Sterling, Sunderland, Surrey, Swansea, Ulster, Wales Trinity Saint David, Warwick, Westminster, York St. John

(continued)

Table 9 (continued)

Category	Name institution
3. Relatively high quantity and quality info	Ashton, Birmingham City, Brighton, Brunel London, Cambridge, City London, Durham, East Anglia, Exeter, Cardiff Metropolitan, Glyndŵr. Greenwich, Hull, Kingston, Central Lancashire, Leeds, Lincoln, Liverpool Hope, Liverpool John Moores, The London School of Economics and Political Science, Loughborough, Northumbria Newcastle, Nottingham Trent, Oxford, Oxford Brookes, Plymouth, Queen Mary London, Sheffield Hallam, St. Andrews, West of Scotland, Worcester, York
4. High quantity and quality info	Aberdeen, Aberystwyth, Bath, Bristol, Canterbury Christ Church, De Montfort, Edinburgh, Harper Adams, Lancaster, Liverpool, Nottingham, Staffordshire, Strathclyde, West of England
No sustainability information	*14 institutions*
1. No information on QAA and HEA guidance, ESD or estate sustainability	East London, Falmouth, Heythrop College London, Highlands and Islands, Leeds Trinity, Norwich Uni. of Arts, Open Uni., Regent's London, Robert Gordon Aberdeen, Teesside, Uni. College of Estate Management, Wales, Writtle College

9   Conclusions

Higher education institutions around the world are increasingly competing for students and staff in a global marketplace. It has become more and more important for universities and colleges to stand out from the rest by communicating their strengths, values, ambitions and the quality of the student experience to prospective students and staff in the web environment. The rise of sustainability has resulted in students demanding from their universities and colleges increasing efforts to manage the estates in a sustainable manner and to offer sustainability in curricula, and prospective staff members choosing employers that are operating in a sustainably sound way. This pressure may result in HEIs profiling sustainability and ESD in their web environments. Encouragement for sustainability and ESD initiatives and practices in the HE sector from the government, funding councils and the Higher Education Academy has been diminishing. This study has provided the first comprehensive assessment of the presence of sustainability, and specifically ESD, in the online platforms of universities and colleges.

Significantly, a large group of institutions mentions the management of estate sustainability on their websites, whilst a smaller group discusses ESD achievements and an even smaller group presents information about the institutional use of the QAA and HEA guidance for ESD. Mission group membership and contribution to the guidance both seem to have an effect on the degree of ESD information provision. Since this study focuses on representation opposed to institutional practice, one can merely postulate that for the majority of higher education providers sustainability is simply a means to an end and that for only a small number of universities and colleges can sustainability and ESD be considered to be approaching a position where it can be judged to be part of institutional policy, practice and lived experience.

References

Armstrong CM (2011) Implementing education for sustainable development: the potential use of time-honored pedagogical practice from the progressive era of education. J Sustain Educ 2:1–25

Bauer C, Scharl A (2000) Quantitative evaluation of Web site content and structure. Internet Res 10(1):31–44

Binsbergen J (2013) Quantitative content analysis. https://www.digitalmethods.net/MoM/QuantContentAnalysis. Last accessed 9 Jan 2016

Bowen GA (2009) Document analysis as a qualitative research method. Qual Res J 9(2):27–40

Briggs S, Wilson A (2007) Which university? A study of the influence of cost and information factors on Scottish undergraduate choice. J High Educ Policy and Manage 29(1):52–72

Chapleo C, Carrillo Durán MV, Castillo Díaz A (2011) Do UK universities communicate their brands effectively through their websites? J Mark High Educ 21(1):25–46

Dale A, Newman L (2005) Sustainable development, education and literacy. Int J Sustain High Educ 6(4):351–362

Desha C, Hargroves C (2014) Higher education and sustainable development: a model for curriculum renewal. Routledge, London, p 268p

Drayson R, Bone E, Agombar J, Kemp S (2012) Student attitudes towards and skills for sustainable development. York: Higher Education Academy. http://www.heacademy.ac.uk/assets/documents/esd/Student_attitudes_towards_and_skills_for_sustainable_development.pdf. Last accessed 8 Feb 2016

EAUC (n.d.) Four universities join elite Russell Group. Environmental Association for Universities and Colleges. http://www.eauc.org.uk/four_universities_join_elite_russell_group. Last accessed 7 Nov 2016

Everett J (2008) Sustainability in higher education: Implications for the disciplines. Theory and Res Educ 6(2):237–251

Ferrer-Balas D, Lozano R, Huisingh D, Buckland H, Ysern P, Zilahy G (2010) Going beyond the rhetoric: system-wide changes in universities for sustainable societies. J Clean Prod 18(7):607–610

Fielding N, Lee RM, Blank G (2008) The SAGE handbook of online research methods. SAGE publications, London, p 684p

Global Tolerance (2015) "The values revolution". Global tolerance. http://www.globaltolerance.com/wp-content/uploads/2015/01/GT-Values-Revolution-Report.pdf. Last accessed 18 Jan 2017

Hannover Research (2014) Trends in higher education marketing, recruitment, and technology. http://www.hanoverresearch.com/media/Trends-in-Higher-Education-Marketing-Recruitment-and-Technology-2.pdf. Last accessed 18 Jan 2017

Hazelkorn E (2015) Rankings and the reshaping of higher education: the battle for world-class excellence. Springer, London, p 270p

HEA (n.d.) "Green academy". Higher education academy. https://www.heacademy.ac.uk/work streams-research/themes/education-sustainable-development/green-academy. Last accessed 11 Jan 2017

HEA (2014) "Students want more on sustainable development on their higher education careers." Higher Education Academy. https://www.heacademy.ac.uk/about/news/students-want-more-sustainable-development-their-higher-education-careers. Last accessed 11 Jan 2017

HEFCE (2013) "Students' Green fund." Higher Education Founding Council for England. http://www.hefce.ac.uk/news/newsarchive/2013/Name,93911,en.html. Last accessed 19 Jan 2017

HEFCE (2014) "Sustainable development in higher education: HEFCE's role to date and a framework for its future actions." Higher Education Funding Council for England, http://www.hefce.ac.uk/pubs/year/2014/201430/. Last accessed 11 Jan 2017

Lambrechts W, Mulà I, Ceulemans K, Molderez I, Gaeremynck V (2013) The integration of competences for sustainable development in higher education: an analysis of bachelor programs in management. J Clean Prod 48:65–73

Martin S, Jucker R (2005) Educating earth-literate leaders. J Geogr High Educ 29(1):19–29

NBS (2013) "Three reasons job seekers prefer sustainable companies". Network for business sustainability. http://nbs.net/knowledge/three-reasons-job-seekers-prefer-sustainable-companies/. Last accessed 18 Jan 2017

Orr D (1991) "What is education for?" Context, 27, 53, 52–s58. People and Planet (n.d.) How sustainable is your university?

People and Planet. https://peopleandplanet.org/university-league. Last accessed 27 Jan 2017

QAA and HEA (2014) "Education for sustainable development: guidance for UK higher education providers." Quality Assurance Agency and Higher Education Academy, http://www.qaa.ac.uk/en/Publications/Documents/Education-sustainable-development-Guidance-June-14.pdf. Last accessed 8 Feb 2016

Schreier M (2014) Qualitative content analysis. The SAGE handbook of qualitative data analysis, Sage, London, pp 170–183

Shephard K, Dulgar P (2015) Why it matters how we frame 'Education' in education for sustainable development. Appl Environ Educ Commun 14(3):137–148

Sterling S, Maxey L, Luna H (2013) The sustainable university: progress and prospects. Routledge, London, p 334

Su HJ, Chang TC (2010) Sustainability of higher education institutions in Taiwan. Int J Sustain High Educ 11(2):163–172

Tilbury D (2011) Higher education for sustainability: a global overview of commitment and progress. High Educ World 4:18–28

UK National Commission for UNESCO (2013) "Education for Sustainable Development (ESD) in the UK—Current status, best practice and opportunities for the future", UK National Commission for UNESCO. http://www.unesco.org.uk/wp-content/uploads/2015/03/Brief-9-ESD-March-2013.pdf. Last accessed 11 Jan 2017

Waas T, Verbruggen A, Wright T (2010) University research for sustainable development: definition and characteristics explored. J Clean Prod 18(7):629–636

Yarime M, Tanaka Y (2012) The issues and methodologies in sustainability assessment tools for higher education institutions a review of recent trends and future challenges. J Educ Sustain Dev 6(1):63–77

Curriculum Review of ESD at CCCU: A Case Study in Health and Wellbeing

Adriana Consorte-McCrea, Chloe Griggs and Nicola Kemp

Abstract Canterbury Christ Church University (CCCU) has produced a curriculum review tool to help teaching staff and programme directors identify the four core components of Education for Sustainable Development (ESD), as defined by the Quality Assurance Agency (QAA) ESD guidance: Global Citizenship; Environmental Stewardship; Social Justice, Ethics and Well-being and Futures Thinking. The review tool was designed in order to realise the University's strategic aim '*to support curriculum innovation so that all students have the opportunity to engage with sustainability related issues relevant to their discipline and chosen field of work*' (CCCU Sustainability Framework 2015a), to be used by programme teams to identify sustainability content that is already present, and to support opportunities for curriculum innovation and development. This paper explores the process of creating the ESD review tool at CCCU and results of carrying out a pilot review on the Foundation Degree in Health and Social Care, within the faculty of Health and Wellbeing. By using the mapping tool, it became apparent that the concept of sustainability was present within the Foundation Degree, however it was not explicit. Its use motivated team members to question their own contribution to creating a more sustainable future. It has been noted that although climate change poses the most serious threat to global health in the 21st Century, for health professionals the threat seems removed and distant. Realization seems to challenge individuals to explore the potential posed by the theme. Results indicate the importance of allowing academic staff the time and space to think and talk, so that engagement with the topic is possible. They also identify a need to re-frame sustainability to evoke positive emotions, capitalising on the things that can be

A. Consorte-McCrea (✉)
Wildlife and People Research Group-ERG, Canterbury Christ Church University—CCCU, North Holmes Road, Canterbury CT1 1QU, UK
e-mail: adriana.consorte-mccrea@canterbury.ac.uk

C. Griggs
Foundation Degree in Health and Social Care, Canterbury Christ Church University, Canterbury, UK

N. Kemp
Early Childhood Directorate, Futures Initiative ESD Lead, CCCU, Canterbury, UK

© Springer International Publishing AG 2018
W. Leal Filho (ed.), *Implementing Sustainability in the Curriculum of Universities*, World Sustainability Series, https://doi.org/10.1007/978-3-319-70281-0_15

achieved, rather than creating a sense of enormity that results in disempowerment. By sharing the findings of a review of ESD curriculum mapping exercises; of the opportunities and hardships involved in developing CCCU's mapping tool; and the experience gained by piloting such tool, this paper will assist other universities that are interested in exploring the scope of ESD in their own curriculum.

Keywords Education for sustainable development · Curriculum review Sustainability in higher education · Health and wellbeing

1   Introduction: ESD at CCCU

Education for sustainable development is the process of equipping students with the knowledge and understanding, skills and attributes needed to work and live in a way that safeguards environmental, social and economic well-being, both in the present and for future generations (QAA 2014: 5).

The role that education can play in contributing towards Sustainable Development has long been recognised. Back in 1987, the Brundtland Report 'Our Common Future' called for a 'vast campaign of education'... (WCED 1987: 23). This message was reinforced in 1992 at the Earth Summit and in 2002 at the World Summit which restated the importance of education and learning to sustainable development goals. Then 2005–2014 was established as the United Nations Decade of Education for Sustainable Development.

In spite of this international policy support, universities have been slow to respond to the challenge. Jones et al. (2010) argue that there are a number of reasons for this delay. Firstly, there is concern about how a commitment to the principles of sustainability might affect academic freedom. In their recent report for the National Association of Scholars, Peterson and Wood (2015) refer to sustainability as Higher Education's new fundamentalism. Their claim that sustainability is somehow antithetical with academic liberty clearly depends upon a limited and limiting understanding of sustainability but this does not fail to make it a powerful and persuasive argument. Secondly, there is a perceived lack of relevance of sustainability beyond 'obvious' disciplines such as geography. As neither an academic discipline nor even a subject, sustainability does not fit easily into the current academic paradigm. Linked to this, the complex nature of organisational structures was another factor the researchers found to inhibit engagement with sustainability within Higher Education. The final barrier identified by Jones et al. (2010), and re-emphasised by Winter and Cotton (2012), concerned the lack of knowledge and understanding of sustainability by staff within universities.

These are complex challenges and ones which Universities across the world continue to grapple with (Scoffham & Kemp 2015). At CCCU, engagement with sustainability goes back some ten years and is based on the understanding that the key to addressing the barriers identified by Jones et al. (2010) is through staff development.

In 2011 the university made a successful bid to participate in the 'Green Academy' (an initiative run by the UK Higher Education Academy) and from this the Futures Initiative (FI) was born; a programme aimed at engaging staff from across the university through seed-funding small-scale curriculum innovation projects. Now in its sixth year it has supported around 100 projects. More significantly and drawing upon Deleuze and Guattari's (1999) rhizome metaphor, the projects have generated unimagined and unintended connections; the whole has become greater than the sum of its parts.

The FI "bottom-up", participatory approach to staff development in ESD is explained in detail by Stephen Scoffham, one of its founders (FI 2016, 6):

> Building the capacity and capabilities of colleagues holds out the promise of effective long term change. There is an underlying premise that subject knowledge is held by the academic staff who hold the key to embedding sustainability within the curriculum. This has meant that there can be no unified, top-down check list of whether a particular module or programme is 'sustainable.' Rather, through the Futures Initiative process, individual colleagues are encouraged to engage with ESD and bring their own academic disciplinary knowledge and understanding to the challenge. Building capacity takes time but then so does any lasting curriculum and attitudinal change.

The publication in 2014 of the HEA/QAA guidance on ESD in Higher Education designed to complement the Quality Code provided a vital endorsement for this curriculum development work. In 2015, CCCU developed its Sustainability Framework which includes an aim *'to support curriculum innovation so that all students have the opportunity to engage with sustainability related issues relevant to their discipline and chosen field of work'* (CCCU 2015a). In order to realise this aim, it was recognised that some kind of curriculum review would be helpful to surface existing curricular opportunities and to identify opportunities for future activity within academic programmes. It was to meet this need that the ESD curriculum review tool was developed.

This paper revisits CCCU's process of ESD curriculum review from its origin (identifying the need), through to the gathering information on similar reviews carried out by other universities and colleges, onto planning its own review platform, and finally looking at the first results of its application to a Foundation Degree course in the Faculty of Health and Wellbeing. By looking back and reflecting on this ongoing process, this paper allows for the identification of methods, aims and objectives, barriers and opportunities involved, which should be useful for other institutions aiming to explore a similar path. The present paper originates from a need to reflect on our own experience so far, providing considerations about the key achievements and limitations of the path we have trailed. This is pivotal to inform the development of our own sustainability review in the curriculum across faculties, to find the most effective ways within our structure to reach the whole institution.

2   ESD Curriculum Review: An Overview

The process of mapping the landscape of ESD in the formal curriculum at CCCU began with the vision stated in the University's Strategic Framework (2015–2020), where *'preparing individuals to contribute to a just and sustainable future'* constitutes part of its values. Concomitantly, its Education for Sustainable Development Policy recognized the need for a baseline audit of ESD, to help develop an understanding of the foundations that are already present and to identify gaps that would need addressing.

In 2015 the Sustainability team undertook a literature review of similar exercises that had been carried out by other HE and FE institutions (Dawe et al. 2003, 2005; Bunting et al. 2012; Eames 2012; Tierney and Tweddell 2012; Bunting et al. 2012; Bridgend College 2013; Hoover and Burford 2013; Ifs 2014; Kendal 2014). This review revealed a wide range of possibilities, commonalities, barriers and opportunities identified by teams from the various institutions. The associated ESD mapping exercises combined various quantitative and qualitative **methods**:

1. Using self-administered questionnaires, sometimes directed specifically at module leaders;
2. Carrying out conversations with teaching staff and with members of each faculty who were responsible for supporting curriculum development and enhancement;
3. Combining a qualitative assessment of course descriptions with a self-administered questionnaire;
4. Using a keyword search to survey module descriptions;
5. Selecting an existing ESD framework to function as common ground for the exercise, such as "The Futures Fit Framework" (Sterling 2011), the "Five Capitals model" (Dawes et al. 2005; Forum for the Future 2014), the HEA model (Dawes et al. 2005), the UNESCO's framework (Tilbury 2011), or the QAA Guidance for ESD in Higher Education (QAA 2014);
6. Using online surveys for academic staff and non-academic staff.

The overall **purpose** of the mapping exercises reviewed was to examine the provision of ESD across their institutions, varying in their scope and reach, assessing gaps and opportunities in ESD, and celebrating ESD coverage throughout their curriculum (Consorte-McCrea and Rands 2015). Largely, institutions were invested in moving away from a narrow sustainability focus on "green issues", towards a more holistic view that considers social, economic and environmental spheres (Kendal 2014). To address their purpose, the methods listed **aimed** to gather information about:

1. Institutional vison and strategy;
2. Staff familiarity with sustainability in university's strategic plan;
3. Sustainability skills of all academic and non-academic staff;
4. Relevance and understanding of ESD concepts;

5. Knowledge of global issues;
6. Student experience and Sustainable Development literacy;
7. Use of sustainability pedagogies;
8. The breadth and strength of ESD in the curriculum:

   a. Assessing the presence and the extent of coverage of sustainability related themes across modules and programmes;
   b. Assessing the opportunities for inclusion of sustainability related themes across modules, including where and how they could be introduced and assessed.

9. Student engagement with ESD; students' enrolment in sustainability related modules;
10. Personal choices in effecting change;
11. Partnership working;
12. Barriers to change;
13. Opportunities for change;
14. How individuals relate sustainability to employment, employers and business;
15. Familiarity with sustainability centres/support available at the institution;
16. Willingness to attend ESD staff development/training;
17. The need for support;
18. How to replicate and monitor future changes.

Expected **limitations** in staff and funding for ESD within HE institutions may guide restrictions in the focus of the mapping exercise, to avoid making it too wide reaching and time consuming. Ultimately, the strategy chosen and its reach have to align with the capacity available to support the exercise throughout, from planning to practice, analysis and dissemination of results, and follow up interventions in some cases. Ease of replicating the exercise to monitor future changes may also limit its scope. Other difficulties related to the ESD mapping process identified during this literature review were:

1. Lack of common ground: great diversity in the way in which sustainability is understood by individuals and in relation to different subject areas;
2. Narrow understanding of ESD amongst curriculum leaders;
3. Lack of clear records of student engagements with ESD;
4. Inconsistent visibility of ESD in module descriptions;
5. Lack of clarity regarding institutional objectives and the ways in which embedding ESD in the curriculum will further these;
6. Organisational hurdles, including time limitations and lack of staff.

Problems with a lack of shared understanding regarding sustainability at HE institutions have been reported at length by Djordjevic and Cotton (2011). Also regarding ESD **principles**, it became clear the importance of not only identifying the presence of ESD but of seeking a balanced view of its multifaceted components. Therefore, considerable **strengths** in the ESD curriculum reviews across institutions were:

- The establishment of a common ground;
- A focus on breadth and strength of ESF in the curriculum;
- Methodology that is feasible to replicate, to monitor future changes.

The review of practices in other institutions also revealed **opportunities** created by their mapping exercises. One of such is the opportunity for the identification of skills amongst staff, who may not be already engaged with ongoing sustainability initiatives, and of their openness to introducing ESD themes (Ifs 2014). As noted by leading voices in ESD pedagogy, good ESD is often equivalent to just good teaching practice (HEFCE 2008; Cotton and Winter 2010), allowing for reflection, exploration, participation and discussion of sustainability issues. ESD may involve creative, transformative and transdisciplinary approaches to learning, mixing '*knowledge, understanding, values and attitudes*' in an effort to develop sustainability literacy amongst students (Scoffham 2016: 296).

Mapping also creates the opportunity to locate knowledge gaps, outlining prospects for further work in order to plan training needs. Similarly, reviews have been useful to avoid duplication of efforts, helping identify where these should be most effective (Ifs 2014). In many cases, the ESD reviews created the necessary body of knowledge to help institutions understand the various ways in which sustainability is conceptualised and embedded within the curriculum (Hoover and Burford 2013). The planning of a consistent methodology can provide support for programmes to review their approach to ESD. Rather than limiting the scope of the exercise to auditing the presence/absence of ESD in the curriculum, the mapping may be used for promoting dialogue and engagement with sustainability issues, supporting developments in Faculties and Schools (University of Bristol 2013). Taking a long view, baseline mapping can be a tool in a long term process of progressive ESD integration in the curriculum. The resulting understanding of the ESD landscape within the institution provides planners with the opportunity to create and support action plans, which then can be monitored against the baseline (University of Bristol 2013). Throughout the mapping process academics may identify opportunities for further ESD development, and new ways to develop them (Ifs 2014). Maybe most importantly, it creates an opportunity to engage professionals to talk about what they are already doing in ESD, focusing on positive actions and offering support so that they may improve their holistic approach to sustainability.

These findings helped us refine our own aims and objectives, through identifying the areas that would be the most beneficial to explore, along with of our own sustainability strategies.

3   Curriculum Review at CCCU

At the core of FI's approach is a belief on the need to build capacity, providing '*a good grounding in sustainability*' amongst teachers and educators so that it may be related meaningfully to the curriculum, preparing students to deal with current issues in the real world (Scoffham 2016: 284).

The FI approach resonates with the guidance on ESD provided by the Higher Education Academy (QAA 2014) as it encourages innovative ways of teaching and learning, and recommends the use of experiential and interactive approaches involving interdisciplinarity.

After an initial period of focusing on cultivating knowledge, skills and experience of staff, which also resulted in the growth of a community of practice, FI participated in the development of a Framework for Sustainability (2015a). This was created to support the new University's Strategic Framework (2015–20) in which Sustainability features as a cross-cutting theme. Amongst the Framework for Sustainability's ESD targets:

> By 2020 all Schools and Faculties will have developed a response to ESD such that every student will have had the opportunity to learn about sustainability in the context of their chosen discipline and field of work.

Since the publication of the QAA guidance on ESD in June 2014, the ESD Working Group has focussed its attention on the development of a tool to support curriculum review and development using the ESD principles stated in the guidance (which draws on ESD principles established by UNESCO and the UN) as common ground. The ESD review tool emerged as an opportunity to promote alignment between the CCCU (2015b) Strategic Framework 2015–2020, the QAA (2014) guidance and Quality in HE directives, aimed at guiding the enhancement of student experience, recruitment and employability, to support change (Consorte-McCrea and Rands 2015).

Considering availability of resources for the implementation, follow up and overall support of the review process, a choice was made not to target primarily the ways in which sustainability issues are being taught (i.e. use of ESD pedagogies and educators as role models for putting sustainability principles into practice). Instead, our ESD review is focussed on assessing the present coverage of sustainability themes (as defined by the QAA ESD principles), the breadth and strength of ESD in the formal, informal and campus curricula. To do so, it relies on a combination of a top-down (deans of faculties) and bottom-up (programme and module leaders) approaches to publicise, support and carry out to use of the review tool.

Guiding our choice of focus is a belief that many modules and programmes already incorporate sustainability perspectives but there is little awareness or recognition of where this good practice exists. Equally there are opportunities to innovate and develop further in other curricular areas but again there is currently no coherent way to surface these.

Our ESD curriculum review is a facilitated process through which programme directors will have an opportunity to (Consorte-McCrea and Rands 2015):

- familiarise themselves with the CCCU sustainability framework
- understand the role of mapping in capacity building for ESD
- see their ESD initiatives recognised and make what they are already doing in ESD visible
- receive support towards constructing a holistic approach
- help develop a baseline, which can be used to monitor change

Therefore, the tool is designed to support curriculum innovation and development, not to 'account' for or 'rate' academic practice. Considering the nature of sustainability issues, our approach reiterates the belief that they are not to be an 'add on' to the curricula, but rather incorporated as part of a distinct approach to teaching (Winter and Cotton 2012). Instead of being an additional 'tick box' exercise, or one that highlights areas of poor practice, the review creates an opportunity to promote holistic provision. Here, all four areas of sustainability categorised in the QAA (2014: 5) guidance (Global Citizenship; Environmental Stewardship; Social Justice, Ethics and Well-being; and Futures Thinking) are used to inform learning; to raise awareness of sustainability related objectives, and of how to achieve them; and to provide a baseline of sustainability informed teaching and learning we can refer to in the future. Although the QAA categories provide a common framework that aids the ESD review process, it is important to consider that sustainability thinking goes beyond compartmentalisation and breaking down into meaningful categories. As suggested by Scoffham (in preparation) it is rather an '*integrating concept which cuts across subject boundaries*', more akin with '*a disposition or habit of mind which permeates the way we interpret experience*'.

To support module and programme directors using the curriculum review tool, the Futures Initiative team run staff development workshops and take part in faculty meetings to introduce the tool and discuss ways its use may be most beneficial. Talking directly with key individuals within the organisational structure (H.O.D.s, module coordinators) offers an insight into the sustainability culture within the organisational structure, and creates opportunities to discuss how best to develop sustainability thinking within their academic areas to meet CCCU's strategic goals.

4   Review Tool and Guidance

The review tool provides an initial scoring matrix, which allows programme directors and module leaders to consider and score the presence of each of four elements associated with sustainability (a) in their outfacing materials connected to programmes and modules (prospectus, web-pages, programme publicity and handbook), and (b) in each of the modules individually. Figure 1. shows the scoring matrix in an Excel format, which is easy to circulate and to use: the presence of each of these four elements can be scored from 0 (not applicable) to 3 (opportunities fully developed); three boxes are then provided, which allow brief comments to describe the form in which sustainability related contents are present (evidence), as well as the opportunities and barriers to further progress. At the end of the review a series of actions can be identified, which may include staff development activities, specific module developments, enhancement of programme publicity or handbook. Piloting the tool has allowed a deeper understanding of the diversity that exists at organisational level between different faculties and departments, which has to be taken into consideration. Initial feedback has indicated a need to make the rationale of the mapping exercise very clear to avoid it being seen as 'extra work and no added value'. The timing of the audit has also been discussed so that it may converge with other curricular reviewing mechanisms already in place (such as Annual Review or Periodic Programme Review, which at CCCU occurs every six years). The tool is accompanied by guidance on how to use it to highlight practice that is already in place as well as identifying gaps. The guidance also explains how the review aligns with the university's values and strategies.

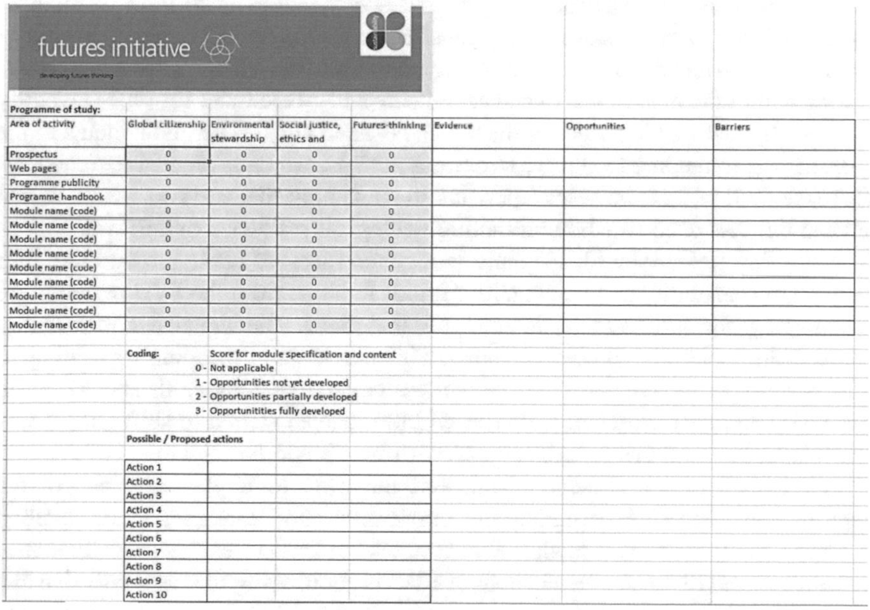

Fig. 1 The sustainability curriculum review tool

5 ESD Curriculum Review in the Faculty of Health and Wellbeing

The Faculty of Health and Wellbeing at CCCU pledged its commitment to ESD and recognised the importance of sustainability within all health and social care curricula. The Foundation Degree (FD) in Health and Social Care (HSC) was nominated as one of the programmes to pilot the mapping tool and lead what would be a faculty wide review of all programmes. The small multidisciplinary FD programme team (Adult, Child and Mental Health Nurses, an Occupation Therapist and Social Care worker) had little understanding of what sustainability meant and were unsure how it affected the programme. This was consistent with the findings of Richardson et al. (2015) who suggests that concepts of sustainability are weak within healthcare curricula and mirrored the experiences of Jones et al. (2010) discussed earlier. The team started this journey not knowing much at all about sustainability and how it related to education. Initially the team felt overwhelmed as the mapping tool looked complicated and time consuming. However, the initial scepticism was quickly replaced with a sense of optimism as the mapping process unearthed a whole host of sustainability related topics that were implicit within the programme.

The mapping tool was in fact straight forward to use and it quickly became apparent that the concept of sustainability was present within the FD, however it was not explicit. The validated programme specification made little mention of 'sustainability', in fact sustainability was cited as the longevity/continued benefit of the programme and the maintenance of capacity (Scheirer 2013). This is a common finding within the field of health and social care and Anaker and Elf (2014) identify two socially acceptable usages of the word 'sustainability'. The first pertains to the potential for something to survive over long periods, which is aligned to Scheirer's (2013) definition, and the second pertains to something that survives over a period of time but that does so while promoting ecological resilience or survival.

According to Dunphy (2013) there is a general lack of clarity around the links between sustainability and nursing. The initial findings from the ESD review were encouraging for the team as it emerged that many fundamental sustainability graduate attributes were present all along. This first stage of mapping was simply to take stock, gain a baseline from which we could move forward, and to create awareness amongst the team. Attitudes within the team started to shift once it was possible to see how this agenda aligned with health and social care.

The second phase of mapping allowed the team to implement some of the opportunities identified to enrich the curriculum. For example; Understanding Evidence for Practice is a module that enables students to engage with Evidence Based Practice (EBP). A core element of EBP is about using the most effective and efficient practice to achieve the best outcomes (Barker 2009), which also happens to be an aim of sustainability– making the very most out of the finite resources that we have in health care. With a slight change of language and a more explicit reference to sustainability the module was adapted easily to make coherent connections between existing programme content and sustainability. During this process it is however paramount that changes made are not purely semantic, but rather underline a new understanding of deeper connections between sustainability and the module's contents.

At this stage it was important to emphasise the importance of maintaining a focus on existing learning outcomes and not lose focus of programme aims. There is a need to balance all essential content without becoming "green-washed" (Bandura 2007) and as a result it became a reflecting re-framing exercise rather than creating additional content (Drayson et al. 2014). The emphasis was on our graduates and equipping them with the skills to be aware, responsible and empowered. Winter and Cotton (2012) warn that sustainability should not be viewed as 'another thing to add to an already overcrowded curriculum', they suggest a holistic view will allow the inter-connectedness of sustainability and health to emerge.

The potential benefits for students and the effects this new education could have on practice is an exciting prospect. Foundation Degree's are known for their transformative capacity (Yorke and Longden 2010) and the programme team have witnessed such changes in the past. Students begin the programme with low self-esteem, low confidence and full of self-doubt but are transformed into confident individuals, who have a voice at work and feel empowered to challenge clinical practice (Griggs 2013). This illustrates the potential power that education could have in creating attitude and behaviour changes. The FD HSC has been able to make some short-term changes but in reality these are quick wins and carry risks. The risk for any established curriculum is that sustainability feels like an add-on or an after-thought. The true power of the ESD will be seen when the programme is revalidated and sustainability is firmly woven through the fabric of the entire programme.

Aside from the benefits that have come from the mapping and curriculum development, there have been a number of personal gains. This is consistent with the discussion offered by Gower (2013), who suggests teams have started to question their own contribution to creating a more sustainable future, from consumer habits, to travel, to caring for the environment. Talking to colleagues, family and friends about connections to nature, where our food comes from and 'doing our bit' is becoming a more frequent occurrence and this is influencing others to starting thinking, processing and allowing those uncomfortable subjects (that we would rather forget) to be present in the foreground. As a team we have begun to appreciate the position of influence that we hold and with this comes a sense of duty and responsibility.

5.1 Further Steps

The mapping tool has been further developed by the Faculty of Health and Wellbeing who are taking the opportunity to be innovators and leaders in implementing strategies for embedding SD across its programme and pathways. The aims are not only to support the University to meet its obligation but also to create sustainably literate graduates who can be leaders in practice with extra value to their qualification.

One way in which this has been done is through the development of Sustainable Development Graduate Outcomes for students in the Faculty. These have been developed using the Quality Assurance Agency's (QAA, 2014) guidance on education for sustainable development (ESD), the Kent Joint Strategic Needs Assessment (Kent JSNA 2013) and the Sustainable Development Unit's (SDU) strategic plan for 2015–2020 (SDU, 2015). NHS & Public Health drivers have given the opportunity for the Faculty to respond more specifically and to develop a bespoke set of graduate outcomes, which can be applied across all programmes in the faculty. This is a unique approach and has put CCCU at the forefront in the UK. Following the pilot curriculum review of 2015–2016, a decision has now been taken within the Faculty to introduce the sustainability curriculum review tool from level 4 upwards and to employ the graduate outcomes in revalidation of all programmes. Its aim is to inform the development of the new programmes and those up for revalidation and be used across whole programmes and embed within individual modules and placements.

6 Conclusions

With the growing challenges of developing a sustainable future, ESD is a catalyst for innovative ways to address them through teaching and learning '*to deal with change, complexity, controversy and uncertainty*' (Tilbury 2011; Nolan 2012: 65). These transposable skills apply to all disciplines and can be connected to personal and professional development. Considering that one of the major barriers to embedding sustainability in the curriculum relates to a lack of belief in its relevance (Winter and Cotton 2012), the case study in the Faculty of Health and Wellbeing suggests ways in which ESD curriculum review may promote and support positive change.

As the case study exemplifies, not all sustainability related content may be explicit in the literature of programmes and modules, and could possibly elude a sustainability word search. Meanwhile, further links between subject matter and sustainability may only need to be surfaced to start shifting attitudes within programmes, as alignments between ESD and their own agendas become clear.

Overall, sustainability curriculum review exercises can be instrumental in creating opportunities for the identification of areas that could benefit from developments in the formal, informal and campus curriculum, and in professional development concerning ESD related issues. Furthermore, a review may facilitate an alignment between curriculum and the strategic objectives of the institution; increase the visibility and awareness of sustainability related policies, initiatives and support network amongst the academic and professional body (Consorte-McCrea and Rands 2015). However, as Sterling (2001) points out, the resulting map of ESD in the curriculum must not be seeing as a blueprint, but rather as a point in a continuum of change towards a sustainability minded design.

There are limitations in our primary focus on assessing the breadth and strength of ESD in the curriculum, which has not been explored within the scope of this paper. Although a sustainability informed curriculum is paramount, ultimately what matters is the way in which the acquired understanding is used, which depend on each person's beliefs and value systems (Parkin 2010, Scoffham personal communication). An attempt to capture changes in students' attitudes and beliefs may be carried out by means of before-and-after surveys, which are outside of the scope of our investigation. Moreover, making changes to the contents of a module to include sustainability themes risk being treated as an easier option, as it may be more difficult for individuals and certain disciplines to promote '*the changes in pedagogies that sustainability seems to require*' (Cotton and Winter 2010: 42). A high level of integration of ESD into the HE curriculum ultimately requires a shift in the way teaching and learning are seeing by the practitioner, including an alignment with sustainability's core principles (Cotton and Winter 2010; Tilbury 2011).

Other programmes within different faculties have embraced the use of the ESD Curriculum Review tool, and further more are planning to use it. We may say that the overall guiding intention is that knowledge and understanding gained from the mapping exercise will inform the creative thinking of future actions to advance the sustainability agenda, working towards our vision of a sustainable university. However, decentralizing the mapping process by handing it over to Deans of Faculties can result in loosing track of the overall picture. We have yet to find effective ways to capture the findings, to monitor and share them, so that we may support change where it is needed.

As a concluding thought, we believe that '*sustainability education prepares people to cope with, manage and shape social, economic and ecological conditions characterised by change, uncertainty, risk and complexity*' as suggested by Sterling (2011: 9), Therefore even small changes in teaching and learning as a result of an ESD review can cause a positive impact on student experience that may encourage choices in their personal and professional lives towards a sustainable future (Winter and Cotton 2012).

References

AnAaker A, Elf M (2014) Sustainability in nursing: a concept analysis. Scand J Caring Sci 28(2):381–389. doi:10.1111/scs.12121

Bandura A (2007) Impeding ecological sustainability through selective moral disengagement. Int J Innov Sustain Dev 2(1):8–35. doi:10.1504/IJISD.2007.016056

Barker J (2009) Evidence-based practice for nurses. Sage Publications Ltd, London

Bridgend College (2013) Health, Safety and Environmental (HSE) management system. Sustainability survey. HSESS01. Version 01. 14.03.2013. Bridgend College, UK, p 6

Bunting G, Davidson J, Osborne P (2012) Sustainability skills survey: staff questionnaire. University of Wales, University of Wales Trinity Saint David, Swansea Metropolitan University, UK, p 20

CCCU (2015a) Sustainability framework. Canterbury Christ Church University, Kent, UK

CCCU (2015b) Strategic plan 2011–2015. Canterbury Christ Church University, Kent, UK

Consorte-McCrea A, Rands P (2015) Mapping the future: the search for alignment between curriculum and the university's education for sustainable futures objectives. In: Lynne Wyness (ed) Education for sustainable development: towards the sustainable university. PedRIO paper 9 (8):27–30. Plymouth University. ISSN:PedRIO paper ISSN 2052-5818

Cotton D, Winter J (2010) It's not just bits of paper and light bulbs: a review of sustainability pedagogies and their potential for use in higher education. In: Jones P, Selby D, Sterling S (eds) Sustainability education: perspectives and practice across higher education. Earthscan, London, pp 39–54 (Chap 3)

Dawe G, Grant R, Taylor R (2003) Kingston University: sustainability in the curriculum [PDF]. www.kingston.ac.uk/sustainability/includes/docs/final%20report.pdf. Accessed 21 Jul 2014

Dawe G, Jucker R, Martin S (2005) Embedding ESD into HE: final report for the Higher Education Academy. The Higher Education Academy http://www.heacademy.ac.uk/assets/documents/sustainability/sustdevinHEfinalreport.pdf. Accessed 16 Jul 14

Deleuze G, Guattari F (1999) A thousand plateaus: capitalism and schizophrenia. Athlone, London, UK

Djordjevic A, Cotton DRE (2011) Communicating the sustainability message in higher education institutions. Int J Sustain High Educ 12(4):381–394. doi:10.1108/14676371111168296 (Emerald Group Publishing Limited)

Drayson R, Bone E, Agombar J, Kemp S (2014) Students attitudes towards and skills for sustainable development. Available from http://www.iau-hesd.net/sites/default/files/documents/student_attitudes_towards_and_skills_for_sustainable_development-2014.pdf. Accessed 19 Jan 17

Dunphy JL (2013) Enhancing the Australian healthcare sector's responsiveness to environmental sustainability issues: suggestions from Australian healthcare professionals. Aust Health Rev 37(2):158–165

Eames K (2012) Curriculum audit of sustainability content for 2011–12 academic year. Kingston University London, London, UK

FI (2016) Five years of the Futures Initiative (2011–2016): building capacity for an uncertain world. Futures initiative, sustainability development. Canterbury Christ Church University, Kent, UK, p 39

Forum for the Future (2014) The five capitals model—a framework for sustainability. http://www.forumforthefuture.org/sites/default/files/project/downloads/five-capitals-model.pdf. Accessed 16 Jul 14

Gower G (2013) Sustainable development and allied health professionals. Int J Ther Rehabil 20(8):403–409

Griggs C (2013) The impact of a foundation degree: graduate perspectives. Br J Healthc Manag 19(12):590–595

HEFCE (2008) HEFCE strategic review of sustainable development in higher education in England. Report to HEFCE by the Policy Studies Institute, PA Consulting Group and the Centre for Research in Education and the Environment, University of Bath, p 151

Hoover E, Burford (2013) Curriculum audit full report: sustainability/sustainable development. University of Brighton, UK

Ifs (2014) Sustainability report: the story so far… e-version available online at www.ifslearning.ac.uk/sustainability ifs University College, UK. Accessed 20 Jan 17

Jones P, Selby D, Sterling S (eds) (2010) Sustainability education: perspectives and practice across higher education. Earthscan, London

Kendal J (2014) Mapping at southampton. Personal communication

Kent JSNA (2013) Joint Strategic Needs Assessment and Joint Health and Wellbeing Strategy development process in Kent. Kent Public Health Department. http://www.kpho.org.uk/__data/assets/pdf_file/0011/46388/JSNA-JHWS-development.pdf. Accessed 07 Feb 2017

Nolan K (2012) Shaping the education of tomorrow: 2012 Report on the UN Decade of education for sustainable development, Abridged. UNESCO, p 89. ISBN:978-92-3-001076-8

Parkin S (2010) The positive deviant: sustainability leadership in a perverse world. Routledge, UK

Peterson R, Wood P (2015) Sustainability: higher education's new fundamentalism. Report by National Association of Scholars available at https://www.nas.org/images/documents/NAS-Sustainability-Digital.pdf. Accessed 01 April 15

QAA (2014) Education for sustainable development: guidance for UK higher education providers. The quality assurance agency for higher education, Gloucester, UK. http://www.qaa.ac.uk/en/Publications/Documents/Education-sustainable-development-Guidance-June-14.pdf. Accessed 07 Feb 17

Richardson J, Grose J, O'Connor A, Bradbury M, Kelsey J, Doman M (2015) Nursing students' attitudes towards sustainability and health care. Nurs Stand 29(42):36–41

Scheirer MA (2013) Linking sustainability research to intervention types. Am J Public Health 103 (4):e73–e80

Scoffham S (2016) Grass roots and green shoots: building ESD capacity at a UK University. Challenges in higher education for sustainability. Springer International Publishing, Switzerland, pp 283–297

Scoffham S, Kemp N (2015) It's contagious! Developing sustainability perspectives in academic life at a UK University. In: Integrating sustainability thinking in science and engineering curricula. Springer International Publishing, Switzerland, pp 89–102

SDU (2015) Sustainable Development Strategy. http://www.sduhealth.org.uk/policy-strategy/engagement-resources.aspx.

Sterling S (2001) Sustainable education: re-visioning learning and change. Schumacher briefings no 6. Schumacher UK, CREATE Environment Centre, Seaton Road, Bristol, BS1 6XN, England. p 96

Sterling S (2011) The future fit framework: an introductory guide to teaching and learning for sustainability in HE. The Higher Education Academy. https://www.heacademy.ac.uk/system/files/future_fit_270412_1435.pdf. Accessed 24 Jan 17

Tierney A, Tweddell H (2012) University of Bristol ESD mapping tool. Bristol, UK

Tilbury D (2011) Education for sustainable development: an expert review of processes and learning. UNESCO, p 132. http://unesdoc.unesco.org/images/0019/001914/191442e.pdf. Accessed 07 Feb 2017

University of Bristol (2013) Education committee. Education for Sustainable Development Strategy 2013–2016. Bristol, UK, p 16

WCED (1987) Report of the World Commission on Environment and Development: our common future. Oxford University Press, Oxford. http://www.un-documents.net/our-common-future.pdf . Accessed 07 Feb 2017

Winter D, Cotton J (2012) Chapter 3: sustainability pedagogies: what do they offer for educational development? In: Cotton D, Sterling S, Neal V, Winter J (eds) Staff and Educational Development Association (SEDA) Special 31: Putting the S into ED—Education for Sustainable Development in Educational Development. SEDA, Woburn House 20–24 Tavistock Square, London, pp 23–26

Yorke M, Longden B (2010) Learning, juggling and achieving students' experiences of part-time foundation degrees. Foundation Degree Forward (fdf) Publication, Staffordshire, UK

Author Biographies

Dr. Adriana Consorte-McCrea, works in Education for Sustainable Development since 2012, at CCCU. As part of the Futures Initiative team she provides support to academic staff and students on projects that propose to embed sustainability related issues into the curriculum. She has a Ph.D. in Ecology and is also a lecturer in ecology and conservation in the School of Human and Life Sciences, CCCU, since 2004. As a conservation biologist her research focuses on how people relate to biodiversity and to the conservation of native wildlife, including the role of zoos in forming attitudes. She chairs the Wildlife and People Research Group, as part of the Ecology Research Group, at CCCU.

Chloe Griggs, has a background as a Registered General Nurse working with the elderly in community settings across the South East of Kent. Chloe moved into Higher Education in 2007 and her role at Canterbury Christ Church University is Programme Director of the Foundation Degree in Health and Social Care. Chloe specialises in the development of the Nursing support workforce including the Assistant Practitioner role and the Nursing Associate. Chloe has a keen interest in the environment is currently completing a Ph.D. titled 'Psychological mechanisms to cope with the realities of healthcare associated anthropogenic global warming: an exploratory sequential study of nurses within the United Kingdom'.

Dr. Nicola Kemp, has an academic background in rural and environmental geography and worked in the Rural Regeneration team at Kent County Council for ten years before moving to CCCU. She has a particular interest in children's experiences of the natural environment and has developed the 'Connecting Children and Nature' Network for Kent which aims to bring together staff, students and a range of local organisations to explore and develop this agenda through research, knowledge exchange and curriculum development. She co-leads the 'Ecopedagogy' research theme group and her research interests include home education, alternative curricular and Education for Sustainable Futures.

A Unifying, Boundary-Crossing Approach to Developing Climate Literacy

Ann Hindley and Tony Wall

Abstract Empirical evidence suggests that educational approaches to climate change remain limited, fragmented, and locked into disciplinary boundaries. The aim of this paper is to discuss the application of an innovative unifying, boundary-crossing approach to developing climate literacy. Methodologically, the study combined a literature review with an action research based approach related to delivering a Climate Change Project conducted in a mid-sized university in England. Findings suggest the approach created a unifying vision for action, and did so across multiple boundaries, including disciplinary (e.g. psychology, engineering, business), professional services (e.g. academic, library, information technology), and identity (e.g. staff, student, employee). The project generated a number of outcomes including extensive faculty level climate change resources, plans for innovative mobile applications to engage people in climate literacy, and new infrastructural arrangements to continue the development of practice and research in climate change. This paper outlines empirical insights in order to inform the design, development, and continuity of other unifying, boundary-crossing approaches to climate literacy.

Keywords Boundary crossing · Climate literacy · Business and management education

1 Introduction

In May 2016, UNESCO's International Institute for Educational Planning (IIEP) convened to debate the role of higher education and the Sustainable Development Goals (SDGs) (UNESCO 2016). The debate recognised the vital role higher

A. Hindley · T. Wall (✉)
International Thriving at Work Research Group, University of Chester, Parkgate Road, Chester CH1 4BJ, UK
e-mail: t.wall@chester.ac.uk

A. Hindley
e-mail: ann.hindley@chester.acuk

© Springer International Publishing AG 2018
W. Leal Filho (ed.), *Implementing Sustainability in the Curriculum of Universities*,
World Sustainability Series, https://doi.org/10.1007/978-3-319-70281-0_16

education has in achieving the SDGs. The IIEP not only argued that higher education is important in inculcating the values, beliefs, and knowledge which are crucial to the achievement of the SGDs, it argued that higher education is also an interdependent part of all educational and training systems, including secondary, vocational, and non-formal education systems. These silos within higher education institutions and across educational systems were both seen as problematic for being able to share practices and understandings of the SDGs to develop a coherent set of actions. It was also acknowledged that although integral to the SDGs, higher education still 'had a way to go' to fulfil this status, and needed to be better informed and mobilised to engage in the SDG agenda (ibid).

It seems scholars in business and management faculties echo this message, with a stark criticism of the failings of business and management teaching and learning approaches in terms of ethics and sustainability (e.g. Wall 2016; Wall and Perrin 2015; Wall and Jarvis 2015). Miller and Xu's (2016) recent empirical work, for example, points to the positive relationship between an MBA education (a flagship of business and management education) and self-serving behaviours at the expense of wider organisational outcomes. As part of this, scholars criticise a lack of a clear theoretical framework for understanding pertaining to responsibility and sustainability (Nonet et al. 2016), and criticise diverse pluralism and related 'academic provincialism' in presenting alternative perspectives in the business and management classroom (de los Reyes et al. 2016). Further, they criticise a fragmented, silo and 'bolted on' approach to education for sustainable development (Painter-Morland et al. 2016).

Within the broader practice literature of ethics and sustainability, there is an equally challenging problem, reminiscent of the IIEP debate: a dearth of literature in relation to the specific area of teaching and learning approaches for climate change, or climate literacy. A search of EBSCO Business Source Elite (all databases) using the terms 'climate literacy', 'climate change literacy', 'climate change education' or 'curriculum' and 'climate change', identified five full-text academic peer reviewed journal articles out of total 1446 using the term 'climate change'. In journals recognised as leading in business and management (ranked as being 3 and 4 star in the 'Management Development and Education' domain, CABS 2015), this search resulted in zero returns. The scholarship in this area, therefore, remains limited.

This paper discusses the application of an alternative approach to climate literacy which encourages a unifying purpose across multiple subject and practice boundaries. It presents and discusses an action research informed case study of a Climate Change Project conducted in a mid-sized university in England, in order to provide empirical insights which can inform the design, development, and continuity of other unifying, boundary-crossing approaches to climate literacy. This paper is structured as follows. First, it discusses the current approaches to climate literacy and highlights some of the ongoing characteristics and limitations of a fragmented landscape of climate literacy scholarship and apparent practice. Second, it outlines the methodological approach and methods used. Third, it discusses the findings of adopting this approach, including indicators of how dimensions of culture appeared to have developed (or not). Finally, it outlines and discusses empirical insights into conducting the unifying, boundary-crossing approach to developing climate literacy in higher education, and concludes with some recommendations.

2 Boundaries and Fragmentation

In 2009, UNESCO released a dialogue document which argued that anthropogenic climate change should be addressed as part of Education for Sustainable Development (ESD) (Pavlova 2013). Within this context, 'climate literacy' can be conceptualized as an understanding of the climate system, an ability to communicate climate change information in a meaningful way, and make informed responsible decisions on actions that may impact the climate (Bofferding and Kloser 2015; Veron et al. 2016). It reflects three domains of 'declarative knowledge'—'system knowledge' (such as understanding how CO_2 can increase global temperatures); 'action knowledge' (understanding the actions and behaviors that influence a situation); and 'effectiveness knowledge' (making decisions between options by weighing the potential positive or negative impact of behavior) (Bofferding and Kloser 2015). However, there are a number of challenges in delivering this agenda.

Scholars have noted tensions between the content-driven educational approaches which aim to produce 'correct' behavior, and the critical thinking approaches which aim to develop the capacity to address uncertainty (Blum et al. 2013). This is partly related to the different conceptions of climate change education, which straddle disciplinary boundaries. Anyanwu et al. (2015), for example, suggest that climate change education straddles the biological, social and physical science disciplines, amounting to an 'interdisciplinary enterprise'. Despite the focus in the science disciplines and physical geography (Kagawa and Selby 2015), there are calls for climate change education to be integrated into all disciplines. As Pavlova (2013, p. 735) considers, "ethical development is a core business of education", and calls for focusing on 'the *why*' we are teaching climate change education, not 'the *how*'. Evidence suggests, however, that barriers are created in trying to locate climate change education within specific disciplinary curricula (Martin and Mahaffy 2011).

Evidence also suggests that a broader, societal issue is that young people appear to be either ambivalent or uncertain about environmental problems (Ojala 2012). As part of this situation, 'black and white' thinking can result in distancing, de-emphasising or denying climate change issues, and the use of negative emotions or helplessness can be used as strategies for avoiding responsibility (ibid). As future policymakers and key civic leaders young people appear to be pessimistic about global problems, and their lifestyles suggest they lack sufficient outdoor experience to understand and prevent environmental damage (O'Malley 2015). Therefore, interconnectedness between the natural environment and 'human community' is a necessary aspect for an educational experience that can help address global problems (Pavlova 2013).

However, misconceptions on climate change are not solely the domain of students, and some scholars have called for professional development programs to equip teachers with contemporary understanding of climate change concepts (Anyanwu et al. 2015). Misconceptions include the conflation of climate change with depletion of the ozone layer (Bofferding and Kloser 2015; Liu et al. 2015); incorrectly relating climate change or global warming to air pollution (Liu et al. 2015; Veron et al. 2016); confusing weather with climate (Liu et al. 2015); and misunderstanding the relationship between littering and climate change (Bofferding and Kloser 2015).

Approaches which promote active learning appear to be effective in tackling these issues (Porter et al. 2012). These include inquiry-based and experiential learning approaches, rather than lecturing and the memorization of scientific facts (Korsager and Slotta 2015). Examples include a 1-day educational intervention in a botanical garden (Sellmann and Bogner 2013), experiential learning in an energy efficiency program (Purnell 2004), and online visualization and discussion resources (Martin and Mahaffy 2011). Collaborative approaches which cross professional boundaries also offer effective educational opportunities, such as collaborative climate change education networks for non-science teachers and students (Pruneau et al. 2006), peer collaboration among international students (Korsager and Slotta 2015), or inter-organisational collaborations including universities, schools, and museums (Melrose 2010; Veron et al. 2016). The following case study joins these themes together, adopting aspects of experiential learning, multi-professional working, and climate change knowledge.

3 Methodology

In the context of the above issues related to the fragmented and highly-bounded nature of climate literacy, an action research study was conceptualised in a mid-sized university in England. This approach, whilst not claiming representation across all higher educational institutions across cultures, aimed *to generate empirical insights into the practice of instituting a unifying, boundary-crossing approach*. It was important to the team to ground this approach with an authentic context of actually attempting to institute change in an organisation, *in-action* (Wall 2010; Stokes and Wall 2014). Within this broader research purpose, the team agreed the high level, driving research questions which then guided phases of the research and the choice of methods. The broad questions and methods used were:

1. *What action will we take?* The team chose to start their action with a large-scale curriculum benchmarking exercise across the university, to identify good or promising practices, which could then be used to inform the decisions about what action should be taken next. This involved an online search for terms in programmes and module descriptors (see below).

2. *What were our experiences of taking action?* The team chose to undertake cycles of action and critical reflection, which were recorded by personal and team logs and reflections in team meetings. Given the constraints of this article, we are only presenting the empirical insights which have been deemed the most significant by the authors.

4 Findings

4.1 Stage 1: Benchmarking (What Action We Will Take?)

The first key finding was that the term "climate change" appeared in only 5 of circa 500 Programme Specifications (see Table 1). The team noted that not only was there very *limited* reference to climate change, there was also *very limited* reference to climate change in Conservation Biology (only in the educational aims section) and International Development Studies (only in the programme structure and features sections). Within Natural Hazard Management there was reference within the programme structure and features section and the module structure. Only the Programme Specifications for Geography show a *high level* of reference to the term "climate change". These key insights are summarised in Table 1. Overall, although the intention was to identify good or promising practices across the institution, the team concluded that there was very limited reference to climate change or literacy at the programme level, and 'the climate' was only positioned as an important issue in very limit pockets in the institution.

The second key finding was that 43 module descriptors used the term "climate change". These appeared mainly in the subject area of geography (17 times), biology (6 times), and health and social care (6 times). Of the 43 module descriptors identified, 21 were removed (7 did not actually contain the term "climate change"; 1 lacked relevance; 3 were 1-day continuing professional development courses; 5 were not accessible to check; and 5 were delivered by partner organisations). The remaining 22 modules were examined to identify the areas of the descriptor where climate change were mentioned (see Table 2).

Overall, the term "climate change" appeared in the Module Content section of 12 modules and in the Key References of 10 modules. However, it only appeared 4 times in the Aims or the Learning Outcomes sections. Only 5 modules contained the term "climate change' in two or more sections of the module. From this perspective, it seems that climate change would generally appear in lectures or seminars by chance, rather than design. This provided a strong prompt for the design of a different approach, which will now be discussed.

Table 1 Programme specifications including the "climate change" term—results indicated by yes/no response (Undergraduate validated programmes, 2016/17)

Programme specification	International tourism management	Conservation biology	International development studies	Natural hazard management	Geography
Faculty	Business and management	Medicine, dentistry and life sciences	Social sciences	Social sciences	Social sciences
Subject benchmarking group	Hospitality, leisure, sport and tourism	Ecology and environmental biology	Geography	Geography	Geography
Educational aims of programme	No	**Yes**	No	No	No
Programme outcomes—knowledge and understanding	No	No	No	No	**Yes**
Programme outcomes—cognitive	No	No	No	No	**Yes**
Programme outcomes—practical and professional skills	No	No	No	No	**Yes**
Programme outcomes—communication skills	No	No	No	No	**Yes**
Programme structure and features	No	No	**Yes**	**Yes**	**Yes**
Module structure	No	No	No	**Yes**	**Yes**
Subject benchmark statements	No	No	No	No	**Yes**
Learning, teaching and assessment methods	No	No	No	No	No
Careers and employability	No	No	No	No	No
Level of reference to term "climate change"	⟶				

Table 2 Identified modules searched for "climate change" term—results indicated by yes/no response

Module Descriptor (search term "climate change")	Introduction to physical geography and geology	Hazard processes and human vulnerability	Challenges of development	Sustainable futures	The (Un)sustainability challenge	Leadership for sustainability	Politics of sustainability	Journalism of fear	Geo-hazards and conservation	Canada: environment and resources	Communicable diseases in context
Faculty	Social sciences	Social sciences	Social sciences	Social sciences	Social sciences	Social sciences	Social sciences	Arts and media	Social sciences	Social sciences	Health and social care
Module title	No	No	No	No	No	No	No	No	No	No	No
Module content	No	No	No	No	No	No	No	No	Yes	Yes	No
Aims	No	No	No	No	No	No	No	No	No	No	No
Learning outcomes	No	No	No	No	No	No	No	No	No	No	Yes
Key references	Yes	Yes	Yes	Yes	Yes	Yes	Yes	Yes	No	No	No
Level of reference to "climate change"											

Module descriptor (search term "climate change")	Global health in context	The environmental determinants of health	Education for sustainable development	Ex-situ conservation	Biodiversity conservation and sustainability	Environment and animals: theology and ethics	Communicable diseases	Biogeography: ecosystem response to change	People, hazards and resources	Tourism futures	Climate change: the recent record and future prospects
Faculty	Health and social care	Health and social care	Education and children's services	Science and engineering	Medicine, dentistry and life sciences	Humanities	Health and social care	Social sciences	Social sciences	Business and management	Social sciences
Module title	No	No	No	No	No	No	No	No	No	No	Yes
Module content	Yes	Yes	Yes	Yes	Yes	Yes	No	Yes	Yes	Yes	Yes
Aims	No	No	No	No	No	No	Yes	Yes	No	Yes	Yes
Learning outcomes	No	No	No	No	No	No	Yes	No	Yes	No	Yes

(continued)

Table 2 (continued)

Module descriptor (search term "climate change")	Global health in context	The environmental determinants of health	Education for sustainable development	Ex-situ conservation	Biodiversity conservation and sustainability	Environment and animals: theology and ethics	Communicable diseases	Biogeography: ecosystem response to change	People, hazards and resources	Tourism futures	Climate change: the recent record and future prospects
Faculty	Health and social care	Health and social care	Education and children's services	Science and engineering	Medicine, dentistry and life sciences	Humanities	Health and social care	Social sciences	Social sciences	Business and management	Social sciences
Key references	No	No	No	No	No	No	No	No	No	Yes	Yes
Level of reference to "climate change"											

4.2 Stage 2: Action/Reflection Cycles (What Were Our Experiences?)

The project was conceptualised as a collaborative workplace project requiring six Research Assistants, and was offered to students within the university-wide employability work based learning module. This is a Level 5 (second year) module, which featured as a core module for the majority of the university's undergraduate programme, and was designed as a self-directed experience driven by reflective learning principles over a 5-week period. The project had both a formal workplace task, plus a personally negotiated dimension, which aimed to inculcate a lifelong learning approach to education.

In terms of their formal workplace task, the goal was to collectively generate online resources and Endnote lists of "climate change" related resources for each of the different faculties. The wider aim here was to encourage engagement with the subject of climate change and to use the empirical insights generated by the project to inform 'climate literacy' activity in the faculty of business and management, but as a consequence of the collaborative approach, activities across the institution. The induction of the Research Assistants involved various skills training sessions delivered by professionals across the university, positioning the Research Assistants as part of a much larger collection of professionals. The skills training included EndNote training and practical skills in project planning and time management, team building and leading, reflective learning and diary keeping. To provide a better understanding of roles, functions and available assistance within the institution, the Research Assistants were introduced to the library specialists for each faculty, the institution's sustainability unit, and the careers and employability staff.

In terms of the negotiated aspect of their experience, the Research Assistants were asked to consider drawing on their previous experience to design a climate change themed project that would involve them working collectively, but which would result in an educational resource of interest to Level 4 students (students in their first year of study). They quickly selected recycling as the theme for their educational resource, as they had identified this to be a particularly relevant and recurring issue for Level 4 students, the institution, the local council and the local community. By their own admission, the students did not know how to recycle as their parents had always completed that task, and some of them had experienced a local council fine for not recycling in accordance with local guidelines. It was also recognised that not recycling appropriately, e.g. street littering, negatively impacts various local stakeholder relations.

To inform their decisions, the Research Assistants reflected on engagement within academic courses and leisure time activities to determine what does and what does not engage a Level 4 student. Early on they recognized the strong attachment students have with their mobile phones, and specifically the usefulness of the institution's own mobile app to provide personalised timetables of where they needed to be, when, and with who. They recognised the importance of apps to entertain during short periods of waiting, such as queuing at bus stops, restaurants and nightclubs, but specifically that the institution's mobile App was a 'lifeline' during studies.

Using this knowledge the Research Assistants generated ideas to utilise the institution's mobile App as a gateway to a new *recycling game App*, which was, in itself a way to directly connect students with climate literacy, and do so in a way which was paperless. The idea of the App was to gamify aspects of climate literacy, taking inspiration from a popular basketball shooting game: students would flick different types of rubbish into the various council recycling bins, and would be rewarded when they had flicked the right type of item into the right type of recycling bin. The idea was that difficulty levels increased with completion of levels and that easily digested messages about the climate and climate change would be presented throughout the game. Their view was that engagement would be promoted through the progression of levels and a competitive element linked to various prizes and awards. They designed an integrated scoreboard to facilitate competition between individuals, programs, departments and faculties. This reward element was seen as essential for the Research Assistants, as they suggested that '*students tend not do things for nothing*'. The recycling game app is now currently under development.

In assessing the outcomes of the project, the project team considered the extent to which the project had changed aspects of culture within the university, so adopted an integrative framework to prompt reflection (Giorgi et al. 2015). For the team, an important reflection was how the project had facilitated a unifying vision for action which was then positively enacted by a diverse group of people across multiple boundaries, including students (e.g. psychology, engineering, business), professional services (e.g. academic, library, information technology), and identity (e.g. staff, student, employee). The project also generated an extensive range of faculty level climate change resources, which had encapsulated and explicated a new range of values and stories about ESD, and which had positioned ESD ideas within a workplace and educational space for staff and students.

Equally as important, the project also mobilised new streams of action in the form of the development of an innovative mobile App to engage others in climate literacy and new infrastructural arrangements, in the form of a Climate Change Special Interest Group, to continue the development of practice and research in climate change. A summary of indicators suggesting developments in cultural dimensions are summarised in the second column of Table 3. That said, the team also recognised some of the difficulties in engaging some staff within the climate change project, such as conceptualising climate change as an extra curricula activity rather than relevant or important to the core of a subject area. A summary of the tensions and issues experienced by the project team are highlighted in the third column of Table 3.

Table 3 Indicators of development, tensions, and issues in relation to Climate Literacy

Cultural dimension (based on Giorgi et al. 2015)	Indicators of development	Indicators of tension or issue
Values	• Helping others to learn about climate and climate change • Merging of climate/climate change and technology applications • Personal growth and confidence • New special interest group	• Focus on subject-specific content-teaching • Staff indifference towards climate
Stories	• Collective presentation (story) at the university's staff conference to demonstrate its collaborative nature • Building self-confidence by working with others • Working together, with different students from different disciplines/ courses, different professional groups • Case study adopted to encourage further internal and external placement offers	• No time for extra curricula activity
Frames (or focus)	• Project as workplace • Workplace learning as lifelong learning • Workplace project as climate literacy • Networks and connections	• Subject content is key • Climate as subject (but not valued) • Working in silos (not interdisciplinary) • Climate change is an issue for others—other departments (geography), experts • Climate change education is the domain of faculties of education • Climate change education should be delivered by scientists
Categories (socially constructed)	• New resources specifically for climate, climate change, climate change education, climate literacy	• Focus on subject content versus non-subject content
Toolkits (sets of the above, practices, etc)	• Sets of faculty-level resources for learning about climate and climate change • Phone app to embed lessons about climate and climate change • New information sharing between sustainability unit and staff	• Lack of resources, funding or investment in local contexts

5   Discussion

There are three main areas of discussion to highlight in this paper. The first relates to the nature of the curricula which served as an initial pedagogic container for the Climate Change Project, but which also seemed to influence how it manifested. The project was constructed within a module which had three important and interdependent curricula features: (1) pan-university/cross-disciplinary (2) negotiated, and (3) work-based (Wall 2010, 2013). In this case, as there was a single infrastructure to deliver and resource the module, it meant the agent initiating the project was enabled to source a group of interested students from different disciplines. In addition, the negotiated nature of the curricula enabled the learners to co-construct their own learning with their employer. This allowed for a personally relevant but strategically located learning focus and approach (ibid). And finally, as the module was configured within a work-based learning frame, the vehicle for learning was conceptualised as action, reflective practice, and evidence-based evaluation, or in broad terms, an approach to lifelong learning (Boud and Solomon 2001; Rowe et al. 2016; Wall 2010).

Framed and conceptualised in this way, such an approach offers an insight into how higher education can interconnect with other educational forms and systems, that is, by infusing climate change knowledge, skills and attitudes into a wider cluster of transferable learning relevant for life-long circumstances (de los Reyes et al. 2016). This directly speaks to the concerns expressed at UNESCO's IIEP, raised earlier in this paper. Specifically, this pan-university, negotiated work based learning approach not only positions climate literacy across disciplinary and professional boundaries, but also across formal/informal boundaries, and as part of a lifelong learning agenda (Wall 2013). Such an approach recognises the emerging recognition of the complexity and tensions of living and working with sustainability in mind (Longo et al. 2017). However, rather than a space of fragmentation, this approach provided a space for unifying action, towards the shared goals for the students, tutors, professional groups realised through this project. This is reminiscent of the vision of a systemic approach to embedding sustainability articulated by Painter-Morland et al. (2016), where climate literacy is interwoven throughout the curriculum.

Although reminiscent of a unifying vision, the final area of discussion relates to the extent to which the approach has reached into other curricula in university. There is also a perspective which might suggest that the pan-university approach to negotiated work based learning is, itself, a 'bolt on', and therefore contains and limits the more widespread integration into or transformation of other curricula. Indeed, such curricula are often housed in a particular organisational unit given the distinctly different philosophical approach to education, such as the location of knowledge generation outside of university classrooms (Billett 2014). Similarly, in the context of this project, in attempting to engage academics in the Climate Change Project, stories emerged of climate change not forming 'part of' the standard curricula in subject areas, or that climate literacy should be retained within particular disciplinary domains. In this way, it echoed the culturally fragmented 'academic provincialism' discussed by de los Reyes et al. (2016) and Painter-Morland et al. (2016). Though these stories existed, the faculty level resources appeared to provide counter-stories and counter-frames in a way which was culturally positioned and relevant.

6 Future Directions

Tracking how the Climate Change Project develops from the negotiated, work based learning curriculum space, and the extent to which it influences other cultural dimensions across the institution will provide new insights into how climate literacy can be initiated and integrated in higher education institutions. Importantly, as Akrivou and Bradbury-Huang (2015) have pointed out, inculcating deep knowledge, beliefs and values with respect to sustainability and ethics requires an equally deep transformation in the wider organisational structures which deliver the educational experiences. As the Climate Change Project was positioned in a pan-university space, not necessarily aligned on a content level to any particular faculty, it is not yet clear how this may play out across the university.

Such a location may have facilitative or inhibitive effects, and such dualistic outcomes is a consequence of the complex and fragmented space of higher education (Barnett 2000; Wall 2016). Such a dynamic could provide insight into how climate literacy might be further integrated into higher education curricula (Cotton et al. 2013). At the same time, it is important to recognise that it may not be enough to change the organisational structures alone. Rather, wider macro structures at governmental and policy levels need to promote knowledge and value relevant to sustainability and ethics (Akrivou and Bradbury-Huang 2015). This returns us to the issues raised at UNESCO's IIEP and ultimately means that pedagogical practices in and outside of the classroom, such as the Climate Change Project, need to be accompanied by longer-term policy work needs to amplify the importance of the SDGs.

7 Recommendations

This case provides a number of considerations that other higher education change agents may adopt in initiating or implementing climate literacy in their own organisation. These will of course be located culturally and historically, and the particular context will provide boundaries as to what might be permissible or not. Some key considerations are:

- Consider curricula options which cross boundaries, such as work based learning, internship, or professional development modules, to introduce climate literacy initiatives. Consider setting up a project that people can easily join and be part of. This can provide multiple incentives for staff and students.
- Consider developing resources which can adopt localised identities such as departments or faculties, and which reflect the different ways in which climate literacy can manifest. Also make these resources available across departments to make it easy to see alternative approaches.
- Consider involving a range of people from a range of disciplines, including students and professional groupings. Professional service staff from library, learning support, sustainability units and facilities, tend to have experience of working across boundaries, so can help support projects.
- Consider seeking a champion or sponsor with significant power within the organisation. In the case of pan-university projects, it is useful to seek the support of a Vice Chancellor or a Pro-Vice Chancellor. Dean level support is useful, but may not be able to influence other Deans.

8 Conclusions

Calls for higher education to do more to contribute to SDGs are increasing amidst a backdrop of fragmented approaches and understandings, and a dearth of scholarship in relation to climate literacy. This paper discussed the application of a Climate Change Project which generated a unifying vision for action, and did so across multiple boundaries. Action research into the delivery of the project gave some indicators of positive cultural change, but also indicators of tensions and issues which indicated minimal or no cultural change. Relating these themes back to the extant literature highlighted that the themes were experienced more broadly than the specific case study. However, the constraint of this paper is that it is exploratory in a particular context, that is, a mid-sized university in England. The findings may not translate into other higher education systems with different histories and cultures, and more empirical work needs to be done to explore different contexts. This will enable researchers and practitioners to build knowledge about the relative applicability and different manifestations of unifying, boundary crossing approaches to climate literacy.

References

Akrivou K, Bradbury-Huang H (2015) Educating integrated catalysts: transforming business schools toward ethics and sustainability. Acad Manag Learn Edu 14(2):222–240
Anyanwu R, Le Grange L, Beets P (2015) Climate change science: the literacy of geography teachers in the Western Cape Province South Africa. S Afr J Edu 35(3):1–9

Barnett R (2000) Realizing the university in an age of supercomplexity. Open University and Society for Research into Higher Education, Buckingham

Billett S (2014) Learning in the circumstances of practice. Int J Lifelong Edu 33(5):674–693

Blum N, Nazir J, Breiting S, Chuan Goh K, Pedretti E (2013) Balancing the tensions and meeting the conceptual challenges of education for sustainable development and climate change. Env Edu Res 19(2):206–217

Bofferding L, Kloser M (2015) Middle and high school students' conceptions of climate change mitigation and adaptation strategies. Env Edu Res 21(2):275–294

Boud DJ, Solomon N (2001) Repositioning universities and work. In: Boud D, Solomon N (eds) Work-based learning: a newer higher education?. Society for Research into Higher Education and Open University Press, Milton Keynes, pp 18–33

CABS (2015) Chartered Association of Business Schools academic journal guide. Chartered Association of Business Schools, London

Cotton D, Winter J, Bailey I (2013) Researching the hidden curriculum: intentional and unintended messages. J Geog H Edu 37(2):192–203

de los Reyes G, Kim TW, Weaver G (2016) Teaching ethics in business schools: a conversation on disciplinary differences, academic provincialism, and the case for integrated pedagogy. Acad Manage Learn Edu http://amle.aom.org/content/early/2016/04/29/amle.2014.0402.abstract

Giorgi S, Lockwood C, Glynn MA (2015) The many faces of culture: making sense of 30 years of research on culture in organization studies. Acad Manage Ann 9(1):1–54

Kagawa F, Selby D (2015) The bland leading the bland: landscapes and milestones on the journey towards a post—2015 climate change agenda and how development education can reframe the agenda. Pol Prac Dev Edu Rev 21:31–62

Korsager M, Slotta JD (2015) International peer collaboration to learn about global climate changes. Int J Env Sci Edu 10(5):717–736

Liu S, Roehrig G, Bhattacharya D, Varma K (2015) In-service teachers' attitudes, knowledge and classroom teaching of global climate change. Sci Edu 24(1):12–22

Longo C, Shankar A, Nuttall P (2017) "It's not easy living a sustainable lifestyle": how greater knowledge leads to dilemmas. Tensions and Paralysis, J Bus Eth (online)

Martin B, Mahaffy P (2011) Using climate change to create rich contexts for physics and chemistry education. Alb Sci Edu J 42(1):19–27

Melrose P (2010) Climate Solvers. Teach Sci J Aus Sci TA 56(1):45–49

Miller D, Xu X (2016) A fleeting glory: Self-serving behavior among celebrated MBA CEOs. J Manage Inq 25(3):286–300

Nonet G, Kassel K, Meijs LJ (2016) Understanding responsible management: emerging themes and variations from European business school programs. J Bus Eth 139:717–736

Ojala M (2012) Regulating worry, promoting hope: how do children, adolescents, and young adults cope with climate change? Int J Env Sci Edu 7(4):537–561

O'Malley S (2015) The relationship between children's perceptions of the natural environment and solving environmental problems. Pol Prac Dev Edu Rev 21:87–104

Pavlova M (2013) Teaching and learning for sustainable development: esd research in technology education. Int J Tech Des Edu 23(3):733–748

Painter-Morland M, Sabet E, Molthan-Hill P, Goworek H, de Leeuw S (2016) Beyond the curriculum: integrating sustainability into business schools. J Bus Ethic 139(4):737–754

Porter D, Weaver AJ, Raptis H (2012) Assessing students' learning about fundamental concepts of climate change under two different conditions. Env Edu Res 18(5)

Pruneau D, Doyon A, Langis J (2006) When teachers adopt environmental behaviors in the aim of protecting the climate. J Env Edu 37(3):3–12

Purnell K (2004) Creating Sustainable Futures through experiential learning in an energy efficiency in schools program. Geog Edu 17:40–51

Rowe L, Perrin D, Wall T (2016) The chartered manager degree apprenticeship: trials and tribulations. H Edu Skills Work Based Learn 6(4):357–369

Sellmann D, Bogner FX (2013) Climate change education: quantitatively assessing the impact of a botanical garden as an informal learning environment. Env Edu Res 19(4):415–429

Stokes P, Wall T (2014) Research methods. Palgrave, London
UNESCO (2016) Three challenges for higher education and the SDGs, Paris, UNESCO: http://www.iiep.unesco.org/en/three-challenges-higher-education-and-sdgs-3556
Veron DE, Marbach-Ad G, Wolfson J, Ozbay G (2016) Assessing climate literacy content in higher education science courses: distribution, challenges, and needs. J Coll Sci Teach 45 (6):43–49
Wall T (2010) University models for work based learning validation. In: Roodhouse S, Mumford J (eds) Understanding Work Based Learning. Gower, Farnham, pp 41–54
Wall T (2013) Diversity through negotiation. In: Bridger K, Reid I, Shaw J (eds) Inclusive higher education. Libri Publishing, Middlesex, pp 87–98
Wall T (2016) Žižekian ideas in critical reflection: the tricks and traps of mobilising radical management insight. J Work App Manage 8(1):5–16
Wall T, Jarvis M (2015) Business schools as educational provocateurs of productivity via interrelated landscapes of practice. Chartered Association of Business Schools, London
Wall T, Perrin D (2015) Slavoj Žižek: A Žižekian Gaze at Education. Springer, London

Author Biographies

Dr. Ann Hindley is Senior Lecturer in the Department of Tourism and Marketing Management, and researcher at the International Thriving at Work Research Group, University of Chester, United Kingdom. She is Principal Investigator on the Climate Change Project and Convenor of the Climate Change Special Interest Group.

Professor Tony Wall is Director of the International Thriving at Work Research Group at the University of Chester, UK, and international visiting scholar at research centres in the UK, Australia and the US. He leads a number of research projects and champions sustainability in various international professional bodies.

Monitoring Progress Towards Implementing Sustainability and Representing the UN Sustainable Development Goals (SDGs) in the Curriculum at UWE Bristol

Georgina Gough and James Longhurst

Abstract This paper will discuss the methods used to identify the baseline status of sustainability in the curriculum of UWE Bristol programmes and to begin to assess the contribution which the university is making towards the objectives outlined in the United Nations Sustainable Development Goals (SDGs). Further, it will consider how this knowledge will inform future curriculum development and research activities within the institution and future tracking thereof. The experience of UWE Bristol in engaging with the SDGs is likely to be of use to other institutions (in higher education or elsewhere) who are similarly aiming to contribute positively to sustainable development. UWE has undertaken a baseline assessment of its engagement with issues highlighted in the SDGs. Alignment between the SDGs and the primary disciplinary focus of each of the university's faculties has already been identified, as has alignment between the core themes of the institution's Sustainability Plan and the SDGs. The next phase of this process involves an examination of UWE's portfolio of programmes of study, its public and community engagement and its research activities with respect to the extent to which they contribute to the achievement of the Goals. The university has begun a programme of mapping the curriculum against the SDGs which will enable this offer to be enhanced and made more comprehensive in terms of the scope and level of visibility of sustainability issues embedded in programme design and delivery. The disciplines of midwifery, environmental science and public health have already been comprehensively assessed whilst accounting and finance, geography, and computer science are in the final stages of assessment. Other UWE disciplines will be assessed over the 2016/17 academic year.

Keywords Sustainable development · Higher education · Education for sustainable development · Curriculum development · Responsible education

G. Gough (✉) · J. Longhurst
University of the West of England, Coldharbour Lane, Bristol BS16 1QY, UK
e-mail: Georgina.Gough@uwe.ac.uk

© Springer International Publishing AG 2018
W. Leal Filho (ed.), *Implementing Sustainability in the Curriculum of Universities*,
World Sustainability Series, https://doi.org/10.1007/978-3-319-70281-0_17

1   Introduction

On September 25th 2015 at the General Assembly of the United Nations, 193 nations adopted 17 Sustainable Development Goals (SDGs) to end poverty, protect the planet, and ensure prosperity for all as part of a new sustainable development agenda (United Nations 2016). The Sustainable Development Goals (SDGs) provide an agenda for sustainable development that will stimulate collective action to address areas of critical importance for humanity and the planet.

UWE, Bristol welcomes the clarity and universality of the SDGs and recognises the urgency attached to their implementation. Subsequently, the university has begun a process of deep reflection on how it contributes to the achievement of the SDGs. There are many frameworks and tools available to organisations for identifying their level of engagement with sustainability issues. The Global Reporting Initiative (GRI) et al. (2015a) advise that baseline responsibilities for business include "recognition of the responsibility of all companies—regardless of their size, sector or where they operate—to comply with all relevant legislation, uphold internationally recognized minimum standards and to respect universal rights" (p. 10). UWE Bristol has been working for over a decade to continue to improve its environmental performance and has successfully achieved institution-wide ISO14001 accreditation for the past three years and won numerous awards for its environmental management in recent years. The university has further been successful in achieving certification, accreditation and awards for its commitment to gender and disability equality, to fair trade, to health and well-being and to responsible sourcing. The university has made additional commitments to it employees, including, for example by offering flexible working and through inclusive and consultative processes and decision-making, all promoted within the SDGs.

Education is key to many of the Sustainable Development Goals and their subordinate targets and as a higher education institution (HEI) education is core to UWE's purpose. The university has a long history of engagement with education for sustainable development (ESD) which aims to enable students to develop the knowledge and understanding, skills and attributes needed to work and live in a way that safeguards environmental, social and economic wellbeing, both in the present and for future generations" (Quality Assurance Agency (QAA)/Higher Education Academy (HEA) 2014, p. 5). The engagement of UWE's educational provision (both formal and informal) with the SDGs will be critical in determining the extent to which the university is able to meaningfully contribute to the 2030 agenda. This paper outlines UWE's work to date in considering the university's current and future alignment to the SDGs. The processes which we have engaged with to undertake this review and the interventions which we have put in place to promote understanding of and action for sustainable development, could be replicated by other higher education institutions (HEIs) and indeed, by organisations in other sectors. Thus, it is proposed that this paper could be of use in promoting further and deeper engagement with sustainable development by a range of other contexts.

2 Organisations, Higher Education Institutions (HEIs) and Contributions Towards Meeting the SDGs

Much work has been done in relation to sustainability in the higher education sector. In Europe, Australia and North America particularly, the environmental sustainability of university campuses is now a common concern. The development of education for sustainable development within higher education has similarly gained momentum over the past decade or so, supported in the UK by sector-facing publications by the Quality Assurance Agency (QAA), Higher Education Academy (HEA) and Higher Education Funding Council for England (HEFCE) (2014). These documents call for 'sustainable development to be central to higher education' (HEFCE 2014, p. 2) and for 'educators working with students to foster their knowledge, understanding and skill in the area of sustainable development' (QAA-HEA 2014, p. 4). The Sustainable Development Goals (SDGs) offer a new framework for the HE sector to engage with in considering its contribution to sustainable development.

The SDG Compass provides a five step process for organisations to follow to ensure that they make a positive contribution to the global sustainable development agenda. These steps are to: understand the SDGs, define priorities, set goals, integrate sustainability into core business and to report and communicate on performance in relation to sustainable development (GRI et al. 2015a). This process could equally be utilised by an HEI as a guide to assessing its contribution to sustainable development. The Compass states that the baseline responsibilities for business include 'recognition of the responsibility of all companies—regardless of their size, sector or where they operate—to comply with all relevant legislation, uphold internationally recognized minimum standards and to respect universal rights' (GRI et al. 2015a, p. 10). Such recognition sets a useful context for meaningful engagement with the SDGs.

Understanding the SDGs is the first step to aligning an organisation's activities to their achievement. The first of these is to understand the SDGs. This process may take some time, particularly if an organisation has not previously engaged with sustainability or aligned agendas, such as business ethics or corporate social responsibility (CSR). In the context of higher education, there is much scope not only to ensure that staff and students understand the Goals, but to discuss and debate them, to develop competencies for achieving them and to create new knowledge for solutions to the challenges which they reflect. The GRI et al. (2015a) state that organisations should next ensure that the SDGs are reflected in business priorities and goals, are integrated across all functions and inform reporting and communication.

Defining priorities can involve a process of mapping the value chain to identify impact areas. It is interesting to consider how this process might be applied to the context of an HEI (outside of a procurement context). What are the components of the value chain and where are the opportunities both to increase positive impact and to minimise negative impact? A more familiar process to HE colleagues is that of the use of indicators. The SDG Compass advocates the selection of SDG-relevant indicators and collection of data to inform the identification of priorities (both in policy and operational terms). HE (in the UK at least) already reports against a number of national sector-wide indicators, particularly in the context of students experience and employability outcomes. Similarly, agreed standards and criteria exist for general environmental (and increasingly sustainability) management, including those which form the ISO 14001 environmental management standard. Some attempts have been made to create education, even higher education, specific sustainability frameworks. These include the Association of University Directors of Estates (AUDE) Green Scorecard, the National Union of Students (NUS) Responsible Futures accreditation, the People and Planet Green League and the Association for the Advancement of Sustainability in Higher Education (AASHE) Sustainability Tracking, Assessment & Rating System™ (STARS). Each of these initiatives offers a range of criteria which an HEI might use to determine the extent and effectiveness of their sustainability engagement.

Further indicators might be identified from with the IRIS metrics, developed by the Global Impact Investment Network (GIIN) to measure the social, environmental and financial performance of an investment. However, these metrics also offer potential as criteria for considering sustainability of organisations, including HEIs. Examples of these metrics include those designed to measure the social and environmental impact objectives pursued by an organization, the sector(s) which an organisation seeks to influence and those with more direct educational relevance such as widening participation and access to education, teacher qualifications, staff student ratio and facilities for teaching and learning.

The SDG Compass offers a large set of business indicators (largely derived from the Global Reporting Initiative G4 Guidelines) which can be used to monitor sustainability engagement, commitment and practice. Some of these relate to generic business practice, such as policies regarding flexible working and the extent to which these are utilised by both male and female employees. Inclusive decision making, reporting processes for consultation between stakeholders and the highest governance body on economic, environmental and social topics and average hours of training per year per employee by gender, and by employee category also fall into this category. Other potential indicators have a more explicit sustainability relevance, such as the extent of impact mitigation of environmental impacts of products and services, measures taken to develop and enhance the highest governance body's collective knowledge of economic, environmental and social topics and the number, type and impact of sustainability initiatives designed to raise awareness, share knowledge and impact behaviour change, and results achieved. All of these could readily be applied to a higher education context.

Within the goal setting stage, organisations are encouraged to define the scope of their goals and select KPIs. Further, there is advice to define a baseline and set a level of ambition. This then enables announce of commitment to SDGs. Once commitment has been made and operational objectives and plans established, organisations can look to integrate the SDGs across their operations. This can be achieved by anchoring sustainability goals within the business, embedding sustainability across all functions and engaging in partnerships. In the context of an HEI, this would mean consideration of the application of sustainability principles to, for example; estates management; financial management; human resource management; all other areas of operations and professional services; educational provision; student experience; research and knowledge creation; and, public and business engagement. HEIs have significant responsibility in relation to influencing other, be they individuals, organisations or government, staff or students or members of the public. To fully embrace the sustainable development agenda, and thus contribute wholly to the achievement of the SDGs, an institutional should seek alignment between a strong institutional commitment to sustainability and operationalisation of this commitment across all functions of the institution. This could be argued as one way to demonstrate sustainable development as being central to higher education.

The final stage of action detailed by the SDG compass is that of effective reporting and communicating on SDG performance. Prior to the development of the SDG Compass, a number of other tools, processes and frameworks existed which enable organisations, and in some cases HEIs specifically, to assess the extent to which sustainability is embedded within their processes and operations and to offer guidance, assure quality and facilitate comparability of sustainability reporting. The Global Reporting Initiative (GRI) was established to standardize sustainability performance reporting across the globe and, particularly in more recent versions, to promote continuous improvement and to encourage transparency. The corporate social responsibility movement, latterly just corporate responsibility, has persisted as a forum in which the obligations and actions of businesses, institutions and organisations in relation to people and planet are demonstrated, discredited, disputed and debated. Other aforementioned initiatives also encourage comprehensive and transparent reporting and communication of sustainability performance, including ISO 14001 and the NUS Responsible Futures initiative.

3 Baseline Assessment of the Alignment of the Policies and Actions of UWE Bristol with the SDGs

Using the stages outlined in the SDG Compass, UWE has begun a process of trying to understand its current contribution to meetings the objectives of the SDGs. UWE has been engaged in education for sustainability and in sustainable estates management for many years and has always sought continuous improvement in these activities. Success in supporting existing sustainability champions and in facilitating some curriculum and behavioural change has been reinforced by the inclusion of sustainability into core institutional policies, strategies and practice. However, the SDGs offer a new, globally accepted framework for sustainability and the university seeks to utilise the Goals to further enhance its action for sustainable development. The SDG Compass is one tool which can be used to explore current sustainability practice and to identify key opportunities for further action. The remainder of this paper articulates the results of a process of document, process and outcome analysis undertaken by the authors in relation to education for sustainable development at UWE, following the stages of business engagement advocated by the SDG Compass.

For the moment, the stage of understanding the SDGS will be set aside. In relation to defining our priorities, we have considered both the institutional Strategy 2020 and the university's Sustainability Plan (2013–2020). Within Strategy 2020 there is much evidence of institutional commitment to issues represented in the SDGs, not least of all in the overall ambition for UWE to be known for 'our inclusive and global outlook and approach' (UWE n.d., p. 4). Three out of the four key priorities of Strategy 2020 explicitly mention sustainability, including in the context of developing graduates who are 'well equipped to make a positive contribution to society and their chosen field of work or further study; and primed to play their part in developing a sustainable global society and knowledge economy' (UWE n.d., p. 5). The university's Sustainability Plan (2013–2020) (UWE n.d.) further articulates a visions for UWE to generate a net positive sustainability footprint and makes commitments to sustainability performance in relation to enhancing the staff and student experience, education for sustainable development and resource effectiveness. It also considers the institution's wider corporate social responsibilities.

These documents are also places where the university sets out its sustainability goals, which include enhancing the health, sustainability and prosperity of the University, Bristol and our wider region [from Strategy 2020 (UWE n.d.)]; defining the mechanisms by which sustainability is embedded within the University's day to day operation and effective decision making [Sustainability Plan (UWE n.d.)]; and, developing a sustainable estate infrastructure (Sustainability Plan (*ibid.*). The university's graduate attribute framework sets out a further sustainability goal; to ensure that our graduates are, among other things, globally responsible and future-facing.

Some of the institution's goals are derived from externally generated frameworks or audit criteria, such as those which enable UWE to meet the expectations of ISO 14001. Our institutional risk register identifies, among other things, key sustainability impacts of UWE's activity and provides a framework for the identification of controls which ensure that we manage our impacts to return a net positive footprint where possible. In this area of institutional management, and others, UWE has chosen sustainability goals for itself which are not mandated by HE sector or organisational obligations (in the legal sense) but which it feels reflect its ambition to deliver high quality education which benefits not only its graduates, but wider society in the broadest sense. It is now on a journey of integrating this ambition fully into its operations.

One of the indicators identified by the GRI et al. (2015b) as illustrative of activity which supports the SDG agenda, specifically in relation to inclusive and equitable quality education and lifelong learning opportunities for all, is the number, type and impact of sustainability initiatives designed to raise awareness, share knowledge and impact behavior change, and results achieved. The university's own self-assessment of relative high performance across these areas and functions has been verified by the success of the university in achieving both ISO 14001 accreditation and NUS Responsible Futures accreditation. UWE recently achieved the highest ever score for an HEI in its NUS Responsible Futures accreditation audit. The university has also been successful in winning awards for its sustainability and education for sustainable development work. This demonstrates the level of ESD and broader sustainability activity which is already being undertaken at UWE. Engagement with the SDGs will enable this work to explicitly contribute to the global sustainability agenda and will enable staff and students to be aware and feel proud of their personal contribution to meaningful action.

In the integration stage, the SDG Compass refers to anchoring the SDGs in all business goals and embedding sustainability across all functions. As part of UWE's baseline assessment of its engagement with issues highlighted in the SDGs, alignment between the Goals and the primary disciplinary focus of each of the university's faculties has already been identified, as has alignment between the core themes of the institution's Sustainability Plan and the SDGs (Longhurst 2016). The Sustainability Plan ensures that sustainability goals are set for all parts of the business and the institutional ISO 14001 certification provides assurance that international standards of environmental management across all functions are being met. Further certifications and accreditations also reflect the university's high levels of practice in relation to social sustainability, for example Fair Trade, equality and diversity, disability, mental health and workplace wellbeing. Discussion of reviews of curriculum and research activity are detailed below.

To promote engagement across all functions, presentations and other staff development activities on sustainability and the SDGS specifically are delivered at key times, such as staff induction, but also throughout the year and to key groups, such as new staff, trainee academics, programme leaders, heads of department and associate deans with responsibility for teaching and learning. A dedicated sustainability team support sustainable practice across the university but responsibility is also distributed across all roles and responsibilities. Terms of reference of boards, groups and committees are also under review with such distributed responsibility in mind. To further support engagement by academics, templates for annual monitoring and reporting and for curriculum review, all now include a requirement to consider education for sustainable development (ESD) and dedicated staff and resources support staff in meeting this requirement. A requirement for programme teams to ensure that their graduates demonstrate the UWE graduate attributes further encourages engagement with issues aligned to the SDGs and is core the significant impact which the university can have in helping to achieve the SDGs, via its graduates.

The SDG Compass advocates engaging in partnership as an effective way to achieve integration of the Goals. Partnership is a key strategic priority for UWE and the university has reflected on ways in which existing partnerships support the SDGs. Working with others for sustainability is facilitated via, for example: a knowledge exchange for sustainability education (KESE) which comprises academics from all departments as well as professional service staff; a staff sustainability network comprised of staff from all parts of the university; the Responsible Futures steering group comprised of senior managers, academics and members of the Students' Union; the programme team for a MSc Sustainable Development in Practice which is drawn from all faculties of the university; and, the Sustainability Board which includes representatives of the senior management team, academics, professional services and the Students' Union. The university's work with its students' union helps to ensure that students are offered a wide variety of formal and extra curricular opportunities to engage with sustainable development. The university is also actively engaged with external partners to support sustainability, not least of all via the Bristol Green Capital Partnership, a community interest company which exists to bring organisations in Bristol together to promote sustainability in the city. UWE is a founding and active member of the Partnership with a seat on the Board of Directors and has committed to continue to support sustainability in the city in whatever ways it can.

In final reference to the SDG Compass stages, UWE has been considering its baseline position with regard to reporting and communicating. The challenge is not that the university does not report or communicate its sustainability. Progress towards achievement of UWE's sustainability objectives and targets is monitored and reported quarterly. UWE then produces annual reports on the Sustainability Plan, drawn from quarterly update reports, as well as reports on individual sustainability themes and for a number of internal and external purposes, including annual ISO 14001 certification and biannual Principles of Responsible Management Education (PRME) sharing information on practice (SIP) reports. Rather, the focus for future reporting will be consideration of how various reporting requirements can be met with a single, or reduced number, of individual reports.

Finally, UWE is keen that communication about sustainability is not a one way process. It actively engages in sharing and listening, both internally and externally. Being conscious of the views of others and developments in other fields, fora and sectors are critical as part of a quest for continuous improvement. Ensuring that sustainability work draws inspiration from and engages with the widest possible set of ideas and people is key to this two way communication.

4   SDGs and the Curriculum

In addition to considering current sustainability engagement across its range of activities, UWE has also been undertaking an examination of its programmes of study with respect to the extent to which they meet the expectations of the Goals. This work is a continuation of its education for sustainable development (ESD) work which has existed for many years. The university has conducted surveys of its curriculum with regard to opportunities for students to engage with sustainability issues for a number of years and actively promotes the further incorporation of sustainability into the curriculum. It has now reached the point where 100% of all programmes include an ESD offer to students. However, mapping the curriculum against the SDGs will enable this offer to be enhanced and made more comprehensive in terms of the scope and level of visibility of sustainability issues embedded in programme design and delivery. This process is a key mechanism by which the university is helping staff and students to understand the SDGs. Each discipline is supported to undertake the mapping exercise in a way which works for the discipline and its community members. Templates and personal support are available to staff undertaking the mapping. Staff and students within discipline areas are. encouraged to consider ways in which their discipline links to each SDG and its associated targets. They might further consider links between associated professions and each SDG. Reflection on the extent to which the university's programmes of study offer students the chance to meaningfully engage with these issues is also encouraged. Midwifery is one discipline which has nearly completed this exercise. The final draft output can be seen in Fig. 1 below and identifies issues relevant to the discipline and issues covered in UWE's midwifery programme. The process and output matrix are being used within the discipline as tools for considering short- and long-term curriculum review (both from a content and delivery perspective). Environmental science and public health have also already been comprehensively assessed whilst accounting and finance, geography, and computer science are in the final stages of assessment. All other UWE disciplines will be assessed over the 2016/17 academic year. The same will then be done in relation to public and community engagement and research. Whilst the

Fig. 1 Output of SDG mapping of the discipline of midwifery at UWE

outputs from these exercises are useful for engaging staff, students and partners of all types in discussion about the SDGs, the benefits from this extend more widely into the local, national and international community, and thus the outcomes of undertaking the mapping are also significant. The process requires deep thinking about the discipline, about teaching and learning, about professional practice, about higher education and about responsibility.

5 Future SDG Engagement and Curriculum Development

The processes which UWE has undertaken so far in promoting sustainability engagement across the university, to establish a baseline of sustainability performance and to assess contribution to the SGDs have enabled it to develop a comprehensive overview of the existing ways in which sustainability has a place and informs activity at UWE. These processes have revealed a strong moral commitment within the university to the values represented in the SDGs and a desire to take more conscious action to contribute and to influence others to contribute. UWE is now looking at ways in which can articulate and put into practice a more ambitious set of goals. These are being set out in a refreshed Sustainability Plan. Further, our curriculum monitoring and review processes are being amended to require more considered attention to the place of the SDGs and ESD in all programmes with the goal of deepening and making more explicit, these connections. A key motivation for UWE in undertaking the review work outlined above is to affect culture change within the organisation in support of the SDGs but also to enhance the staff and student experience and to ensure that the university continues to make progress towards its ambition of net positive impact on both the environment and society.

6 Limitations and Conclusion

The work summarised here has been undertaken over a period of twenty years (when the full journey of engaging in discussions at UWE about environmental management of the campus and inclusion of sustainability in the curriculum is taken into account). However, there is still some way to go to ensure that the institution's staff, students and activities are truly aligned to the goals of sustainable development. The most recent phase of action, facilitating discussions about the university's contribution to the sustainable development goals, represents the university's desire to continuously improve its ESD performance and to actively engage with global agendas. This paper has only outlined the processes and thinking which UWE is now undertaking to broaden and deepen its sustainability engagement but will hopefully be of use to others in considering their own contribution to sustainable development.

UWE is working hard to understand the ways in which its current policies and practice align, or do not align, with the ambitions represented in the Sustainable Development Goals. Over the next few years it will continue to engage in dialogue and partnerships both internally and externally to ensure that we contribute to knowledge creation and transfer and demonstrate behaviours which enable our staff and students to be more positive agents of change for sustainable development.

References

Global Reporting Initiative (GRI), The UN Global Compact, The World Business Council for Sustainable Development (WBCSD) (2015a) SDG compass—the guide for business action on the SDGs, Available online at http://sdgcompass.org/wp-content/uploads/2015/12/019104_SDG_Compass_Guide_2015.pdf. Accessed Jan 2017

Global Reporting Initiative (GRI), The UN Global Compact, The World Business Council for Sustainable Development (WBCSD) (2015b) SDG compass—inventory of business indicators. Available online at http://sdgcompass.org/business-indicators/?filter_sdg_goal=&ref_filter sdg_target=&filter_biz_theme=&filter_indic_type=&filter_indic_source=&filter_date=&custom_tool_search=share+knowledge. Accessed Jan 2017

Global Impact Investing Network (GIIN) (2017) IRIS Metrics. https://iris.thegiin.org/metrics. Accessed Jan 2017

Higher Education Funding Council for England (HEFCE) (2014) Sustainable development in higher education: HEFCE's role to date and a framework for its future actions. Higher Education Funding Council for England, Bristol

Longhurst J (2016) Meeting the UN sustainable development goals: the contribution of the University of the West of England, Bristol. Available at http://www1.uwe.ac.uk/about/corporateinformation/sustainability/policiesplansandtargets.aspx. Accessed Jan 2017

Quality Assurance Agency (QAA)/Higher Education Academy (HEA) (2014) Education for sustainable development: guidance for UK higher education providers. Quality Assurance Agency, Gloucester

United Nations (2016) Sustainable development goals. Available online at http://www.un.org/sustainabledevelopment/sustainable-development-goals/. Accessed Jan 2017

University of the West of England (UWE) (n.d.) Strategy 2020. Available online at http://www1.uwe.ac.uk/about/corporateinformation/strategy.aspx. Accessed Jan 2017

University of the West of England (UWE) (n.d.) Positive footprint: phase 2. UWE Bristol Sustainability Plan 2013–2020. Available at http://www1.uwe.ac.uk/about/corporateinformation/strategy/strategydocuments.aspx. Accessed Jan 2017

Teaching Accounting Society and the Environment: Enlightenment as a Route to Accountability and Sustainability

Jack Christian

Abstract This paper describes a postgraduate course taught for several years and outlines how this aimed to enlighten and empower students and increase their understanding of accounting and its purpose. First some key concepts are discussed: the purpose of education, pedagogy and accountability. Then the course is described in detail, finishing with some reflections. Enlightened students expect more from accounts; they see them as society or multi-stakeholder orientated and potential indicators of a need for change, rather than as a historical narrative claiming to substantiate the efficient use of resources. In the spirit of interpretive accounting research as discussed by Ahrens et al. (Critical Perspectives on Accounting 19: 840–866, 2008) this paper is offered in the hope it helps other educators develop their understanding of how we can enlighten, and empower, our students whilst making them more aware of their role in building a sustainable future.

Keywords Enlightenment · Empowerment · Pedagogy · Accounting and accountability

1 Introduction

Lovelock (2006) speaks of the end of life on our planet if global warming continues unabated, fired by humanity's incessant demand for economic growth and its ensuing carbon emissions. Jones (2014) notes the looming sixth mass extinction of species in our planets history. Everywhere clean water is disappearing as rivers no longer reach the sea (The Independent 2006) whilst the seas themselves grow increasingly polluted (UNESCO 2015). The call of the Brundtland report (United Nations World Commission on Environment and Development 1987) for

J. Christian (✉)
Department of Accounting, Finance and Economics, Manchester Metropolitan University Business School, All Saints Campus, Oxford Road, Manchester M15 6BH, UK
e-mail: j.christian@mmu.ac.uk

© Springer International Publishing AG 2018
W. Leal Filho (ed.), *Implementing Sustainability in the Curriculum of Universities*,
World Sustainability Series, https://doi.org/10.1007/978-3-319-70281-0_18

'development that meets the needs of the present without compromising the ability of future generations to meet their own needs' appears to go unheeded as governments and businesses—and accountants—set their sights on economic targets, paying little if any attention to social and environmental impacts (Deegan 2007; Gray 2010; Solomon et al. 2013; Spence 2007).

This paper explores what accounting education can do to improve the situation. It begins by querying the nature of education suggesting the role of education is to inculcate a sense of uncertainty. An uncertainty though that is positive, feeding an eager curiosity that explore life's landscape and always looking for a better future. However for many this journey will be fraught with anguish. They may turn their thoughts away, looking for comfort in the status quo. How then does education equip students with the courage to take on life's uncertainties? This is the realm of pedagogy.

First students must broach uncertainty as the teacher reveals a multitude of potential futures. Then the teacher emphasises the positivity in this scenario, the freedom from false consciousness and the blank canvas free and available for the student to fill in. This is enlightenment.

Then the student joins friends and contemporaries. As they all stare into the future they are encouraged to share past stories and visions for the future. Together they begin to colour the canvas, shape paths into the future. They are empowered.

This journey to enlightenment and empowerment follows a path forged by two somewhat dissimilar groups, existentialist philosophers (Warnock 1970; Wilson 2001) and followers of the Frankfurt school of critical theory (Alvesson and Skoldberg 2000; Dillard 2007). As the students learn choose their own futures they are emancipated. But how can this 'halt the rush to an unsustainable future'?"

Frankly this paper cannot answer that question with certainty. However unencumbered by false consciousness and empowered to design a future of their own making, the students are in a position to challenge the status quo should they wish. At this point the students are introduced to the concept of responsibility and with it the notion of accountability. Dillard (2007) and Gray et al. (2014) describe accountability as a relationship between two parties where one party agrees to account for some benefit bestowed upon it by the other party and the latter takes responsibility for holding them to account. Dillard builds an ethic of accountability on this foundation and with it places a duty on accountants in their wider public interest role [see also Collison et al. (2007)] to hold organizations responsible for their use of social and environmental goods.

It is not for the teacher to tell the students how to behave or what to do, the teacher's role is to open new vistas for exploration. In their paper *Social and environmental reporting in the UK: a pedagogic evaluation* Thomson and Bebbington (2005) refer to Paulo Freire and his concepts of banking and dialogic education. In the former the teacher tells the students what they should know; in the latter the teacher opens a dialogue and teacher and students explore that which is to become known. Thomson and Bebbington suggest that all too often accounting education is banking education; education that tells students what to think and what to do. An education that does not listen where there is no exchange of understanding and, as a result, little or no enlightenment or empowerment. Ultimately Thomson and Bebbington call for a more dialogic accounting education.

Following their lead and the work of Dillard (2007) students are invited to consider and discuss the impacts of accounting on society and on the environment and what these mean in terms of the profession's wider public interest role and accountability in general. For most students this opens a much wider perspective on accounting allowing them a richer choice of futures in the profession. Further this works to the benefit of society and the environment by increasing accountability for the use of natural and community resources.

This paper describes a postgraduate course developed over several years at Manchester Metropolitan University Business School and outlines how this aimed to enlighten and empower students and increase their understanding of accounting and its purpose. Enlightened and empowered students expect more from accounts; they see them as society or multi-stakeholder orientated and potential indicators of a need for change, rather than as a historical narrative claiming to substantiate the (efficient?) use of resources by an incumbent hegemony.

It is offered in the spirit of interpretive accounting research as discussed by Ahrens et al. (2008) and Parker and Northcott (2016). That is as a contribution to our understanding of how we as practitioners—in this case as accountancy educators—can empower our students and make them more aware of their role in building a sustainable future. First some key concepts are discussed: the purpose of education, pedagogy and accountability. Then the course is described in detail with commentary on the lessons learned its four year history. It finishes with some reflections on the strengths and weaknesses of this paper and then on the course itself.

2 Key Concepts

'...education is the most vital of all resources...education which fails to clarify our central convictions is mere training or indulgence. For it is our central convictions that are in disorder, and, as long as the present anti-metaphysical temper persists, the disorder will grow worse. Education far from ranking as man's greatest resource will then be an agent of destruction.' (Schumacher 1973, pp. 64, 83).

Increasingly, it seems, higher education is being linked to employability. This brings with it good news and bad news. The good news is that sustainability awareness is seen as a necessary adjunct to employability which raises the former's profile, the bad news is that education is often confused with training in the minds of students. In the brief literature review that follows the paper offers some views on the purpose of education, pedagogy as appropriate to social and environmental accounting, and the nature of accountability—the reason why we account in the first place.

2.1   The Purpose of Education

The paper differentiates between training and education as follows. Training is concerned with how to do it (whatever 'it' happens to be) whilst education is more concerned with why we do it at all. Thus in accounting training students are informed how to prepare financial statements, how to carry out an audit, how to budget and prepare management accounts, how to use various business tools such as break-even analysis, even how to assess risk. But in many instances they are not told why they do these things or ask why they should. That is taken for granted, it is based on precedent or perhaps dictated by the professional accounting bodies.

Accounting education however should of course be asking why we do these things, what do we hope to achieve and do we achieve it. [Or perhaps why we don't do these things? See for example Puxty (1993) or Scapens (2006) on the influence of power]. This is the skill of critical analysis so highly valued by academics.

Of course the problem with asking 'why' is that it can go on forever and this problem leads this paper to that branch of philosophy known as existentialism. Existentialist philosophers from Kierkegaard to Nietzsche, from Heidegger to Sartre pursued the question of "why" and ultimately concluded we cannot know. As Heidegger argues we do not have access to Truth, only partial and perhaps inauthentic truths (Blattner 2006; Polt 1999). Warnock (1970) takes this analysis further. In reviewing the work of the great existentialists she argues their goal was emancipation, but with the realization that emancipation could only come when we accepted the limits of our knowledge.

The paper is then drawn to critical theory. In discussing their Reflexive Methodology and their quadri-hermeneutic (multiple interpretations) technique Alvesson and Skoldberg (2000) refer to the Frankfurt school and critical theory, portraying the latter as open-ended and in search of emancipation. They point to writers such as Adorno, Horkheimer and Habermas seeking freedom from 'psychic prisons' and 'false consciousness' with little patience for 'bookkeeper mentality' (p. 131). Their goal is summed up in three words—enlightenment, empowerment and emancipation.

The relevance of this for educators in general, and accounting educators in particular, is that we must accept uncertainty and we must make our students aware of this uncertainty. This opens the door to emancipation as they will now understand the need to ask why, the need to explore and the eternal need to look for a better way. In other words they are enlightened. The next step then, the next job of education, is empowerment.

2.2   Pedagogy

In their paper *Social and environmental reporting in the UK: a pedagogic evaluation* Thomson and Bebbington (2005) refer to Paulo Freire and his concepts of banking and dialogic education; they compare environmental reporting to banking education. Banking education is education that tells people what to think and do, it does not listen and takes no account of what people actually do think and want to do. Too often it seems accounting education is banking education; there is no exchange of understanding and as a result little or no empowerment. Educators they argue must learn to listen to their students and share the future with them as they shape their ideas.

Further they must to leave time and make opportunity for their students to develop their understanding. They should not fill their minds with their own thoughts, they must allow them to build their own understanding of the world. Another Brazilian, indeed another Paolo, Paolo Coelho wrote a novel describing a young boy's search for treasure—The Alchemist (Coelho 2006)—which was, in effect, a description of a journey of learning. Whilst 'an Englishman' in the story studied maps, texts, formulae and symbols the boy studied the desert, i.e., the real world, and ultimately found the treasure they were both searching for. The lesson here was that learning is not simply a rational process, people experience the world holistically, it comes with intuitions, emotions and feelings. Concepts, that is theoretical representations of the world (Llewelyn 2003), are meaningless without some form of attachment to the real world. McGilchrist (2009) argues at some length that humanity is in danger of losing touch with the real world—the world of society and environment—as we focus more and more on the maps rather than the landscape itself.

This paper is based on the argument that words like society, environment and sustainability are concepts and as such they have to be filled with some sort of understanding. Simply standing in front of a class describing one's own and a few other peoples' perceptions of them is unlikely to interest or hold the attention of students. The words need to be linked with experiences of them in some way. This can only be done by student and teacher listening to each other and sharing their experiences and perhaps enhancing these experiences with some actual contact with the environment—natural and human. It is in the sharing that students are empowered as they come to realise their contribution is worthwhile and valuable.

2.3 Accountability

Collison et al. (2007) critique both accounting academics and the accounting profession in their approach to social and environmental, or sustainability, accounting. They accuse the former of holding to a hidden curriculum with 'value-impregnated ontological and epistemological underpinnings' and an 'ideology of shareholder capitalism' (pp. 330, 331) whilst they challenge the profession to make a 'genuine re-avowal of the primacy of the wider public interest over the commercial interests of its members or its clients' (p. 337).

Dillard (2007) takes up the expectation that the profession should act in the public interest [see also McPhail and Walters (2009)] to justify what he terms the social accounting project. In building his justification he also calls on critical theory, referring to the social accounting project as a critical undertaking looking to foster the realization of human potential. Ultimately he constructs an ethic of accountability whereby organizational management is granted fiduciary responsibility (rights) over society's economic resources on the premise that they agree 'to be held accountable by those who grant these rights, and those who grant the rights accept the responsibility for holding the recipients accountable for the related outcomes' (Dillard 2007, p. 43).

He goes on to explain that an ethic of accountability is an ongoing conversation between the affected parties and process of determining appropriate actions and outcomes takes place within in an ongoing community. Further he outlines a role for accounting academics in facilitating and engaging in the dialogue among members of the community and promoting 'scholarly activities leading to enlightenment, educational innovation leading to empowerment, and community action/interaction facilitating emancipation' (p. 46).

So far, it would appear, so good. However as Gray et al. (2014, p. 53) point out there are very few examples 'where there is any sort of congruence between an organization's defined responsibility and its discharged accountability'. Worse, according to Cooper and Johnston (2012), the term accountability has been taken over by the rich and powerful and now serves to immunize them from criticism. They can claim to be accountable when in fact they manipulate outcomes or performance indicators to serve their own purposes, all the while claiming to give voice to the less powerful.

It therefore appears that there is a need for accounting academia to find ways to promote enlightenment and empowerment. This paper continues by outlining a course developed with those specific aims in mind.

3 The Course

3.1 Background

The Accounting Society and Environment course (ASE) was a 20 credit unit and as such it was one of six units presented to postgraduate students undertaking business or accounting programmes. As well as taking these six units all the students were required to submit a dissertation to obtain their qualification. At various times, and in different programmes, the unit has been either mandatory or an optional unit. The student numbers ranged from 8 to 24 over the unit's four year history.

What now follows is a description of the course in its most recent manifestation. Effectively the course is ongoing action research undergoing constant revision in the light of different student intakes and their reactions to the course material. Commentary is offered on the different lessons to reflect how students have reacted to the course material. No firm conclusions are drawn on the "right way" to teach the subject because the author believes the concepts involved are highly contextual with each student bringing their own interpretation of them. Indeed a large part of the course is aimed at helping the students understand that.

3.2 The Lessons and Assessment

The course consisted of 2×2 hour seminars every week for ten weeks. In essence each seminar consisted of a lecture and a class discussion. There were two activities that did not fit this mould—a scenario planning game and an (un)sustainability walk —which were fitted into the course at an 'appropriate' moment. Appropriate in this instance was a function of the weather (for the walk) and student appetite/interest. These activities are described as Weeks 8 and 9 respectively but they may actually take place anywhere between weeks 4 and 9.

Week 1

The first thing to emphasise is the course is a shared journey. From the start of Week 1 students are asked to work together whilst considering pertinent issues. After a brief introduction to the course the students are asked to work in pairs to discuss their knowledge and experience of accounting. What does an accountant do? The students then feedback to the class and this leads to whole class discussion. Invariably at this stage feedback represents mainstream views centred round the financial reporting framework and decision making. Most of the students have usually just finished an undergraduate degree in accounting and/or finance so this is perhaps no surprise. Their thoughts have already been put in a box.

The second seminar of Week 1 begins a process of enlightenment. The students are introduced to academic views on accounting which include critical theorist, feminist and postmodern views. Now some begin to venture their own hitherto suppressed opinions about the role of business and the focus on profit at all cost. It is perhaps true to say that these thoughts are usually opined by overseas students first, who often relate tales of damaged communities and broken promises. Some however stay aloof, they are listening but do not wish to deconstruct their concepts of business and accounting. They are not disinterested, this is revealed by the occasional comment, they are listening and wondering.

Both seminars are focused on the role of the accountant. The first is a standard discussion around financial and management accounting, treasury and internal audit etc. Most of the students were comfortable with this though elements were new to some business students. The second seminar is more novel and enlightening. In four years of running the course there wasn't a student who did not express surprise in the range of academic views on accounting (Dillard 1991; Hines 1992; Scapens 2006; Spence 2009; Hopwood 2009). In so far as the academic views mirrored professional activities, discussion was limited, subdued even, however when accounting was portrayed as a technology with controlling, even dominant, tendencies eyebrows were raised and discussion became more animated. This level of interest increased still further when the students are introduced to the concepts of social and environmental accounting and the public interest role of the professional (Collison et al. 2007; Dillard 2007).

Week 2

Week 2 focuses on ontology and epistemology with ideas from crude realism to pure idealism being presented. The students are then asked how they perceive the world and what can truly be known. At this point an astonishingly large number of students reveal religious backgrounds, some quite devout. There have been several lay preachers on the course. Alongside Christians there have been Moslems, Buddhists and one Daoist. The majority of students are however from secular backgrounds but many still take an interest as the class discusses the discourses that shape our thinking, e.g., how in western Europe the Ancient Greeks and Christianity have shaped our thinking. From here the class considers how science developed and how knowledge came to be defined in rational terms from empirical evidence. They discuss whether that is the only knowledge. The second seminar focuses on the history of science and the way theories develop (Chalmers 1999; Leahey 2004). The students are asked what they mean when they refer to a theory. It is only fair to note that not every student is interested in this discussion or participates with great enthusiasm; some listen on the edge of the discussion perhaps revealing a liking for a good argument with the odd critical comment, others simply do not see any relevance to their accounting studies.

Week 3

Week 3 closes in on accounting theories, or rather theories on why we account, how we account and what we account for. By now the students are more receptive to new ideas (and they are all still attending—they are intelligent, curious people). In particular they are beginning to recognise the role of power in shaping accounting with its focus on economics and capital employed; and its apathy towards social and environmental issues.

The first seminar draws on Ryan et al. (2002) and Deegan and Unerman (2006) introducing accounting research specifically and the theories that have perhaps dominated social and environmental accounting research in recent history; legitimacy theory, stakeholder theory and institutional theory. The second seminar draws on work by Thomson (2007) to explore the space occupied by social and environmental accounting research and also introduces the first of the two course assignments. From Thomson the students see the width of social and environmental research as he presents an analysis of several hundred papers highlighting the theoretical frameworks and research methods, the locations and entities involved, and the purposes of the research. In the assignment students are asked to work as individuals to find five recent academic papers covering a social or environmental topic of interest to them personally and to write a literature review around these papers.

By the end of Week 3 it is probably true to say the students have come to understand that accounting is a much broader topic than they had originally thought. This is not to say they have become social and environmental accountants. Most still want to be financial accountants or business consultants or financial advisers because that suits their temperament, their skills, their view of the world; but they do it knowing there are other ways of seeing accounting and its role.

Enlightened, and empowered by the skills they have garnered in their undergraduate years, they are more emancipated in that they can choose their career less restricted by false consciousness.

Week 4

The first seminar is concerned with social and environmental reporting and the second with corporate social responsibility (CSR). It starts with the Brundtland Report (United Nations World Commission on Environment and Development 1987) definition of sustainable development and the Triple Bottom Line concept (Elkington 1997) and goes on to introduce standards such as the Global Reporting Initiative and the Accountability Stakeholder Engagement Standard. The value of standard Reports is discussed and the merits and demerits of voluntary versus mandatory reporting. In the second seminar the focus is on ISO 26000, the international standard on social responsibility. In this seminar discussion focuses more on concepts, i.e., what is the purpose of business and what issues should a responsible business concern itself with.

Week 5

This week is specifically about Environmental Management Accounting and the lecture covers human impact on the environment before leading onto a review of relevant legislation and strategic and operational tools to ensure compliance and perhaps encourage best practice. Topics, for example, include various environmental management systems and ideas such as life cycle analysis and biomass recording. Discussion also includes ways to internalize external costs and the pros and cons of trading schemes versus taxation schemes.

This discussion leads into the afternoon session which covers full cost accounting and tools such as SIGMA and the Sustainability Assessment Model. The possibility or otherwise of attaching monetary values (as distinct from costs) to social and environmental phenomena such as community libraries and meeting places, areas of natural beauty or living ecosystems is also discussed.

Week 6

Week 6 moves onto ethics with a whole new discussion centred round right and wrong. Different perspectives on what constitutes right and wrong—deontological and teleological—are mooted and whether the means should ever justify the ends. False consciousness is revisited as students find different cultures have different ideas on, for example, property rights, and these ideas are derived from different histories. This underwrites a need for stakeholder engagement to explore the social impact of business and the notion of accountability now raises its head. Students are invited to discuss the accountants' public interest role again and to what extent the profession should perhaps stand apart from business and insist on accountability as well as working with business to produce accounts.

The first seminar draws on McPhail and Walters (2009) to cover ethical theories and why accountants should behave ethically. It also draws on MacIntosh and Quattrone (2010) to introduce Enron and its collapse. Ensuing discussion covers ethical codes, governance and the efficacy of management control systems.

The second lecture covers stakeholder engagement. Who business should engage with, and how, when making business decisions? Various stakeholder engagement standards are considered such as Accountability SES 1000.

Week 7

By now students are asking about alternative business models as the neo-liberal competitive business model has been subject to much scrutiny and critique over the last few weeks. This week co-operative and not-for-profit business models are introduced. Subject matter includes the history of the co-operative movement, the different types of co-operative organizations and the global reach of the movement. Students are often astonished to find that in G7 countries member owners of mutual enterprises outnumber individual shareholders or that the Spanish co-operative, the Mondragon group employs more than 70,000 people in 21 countries around the world (Mayo 2015; Williams 2007). Charitable organizations are also discussed and again students are surprised at the size of some charities and the complexities of charity accounting. For the majority of them this the first time they have been made aware of the extent of this 'other economy' that gets relatively little publicity in a curriculum dominated by mainstream practice.

Week 8

The scenario planning game

The scenario planning game brings senses of reality and responsibility to the course. The students are asked to think about the future and their role in it. Together they colour in a sketch of a possible future and this brings home to them how actions today will impact on tomorrow.

This game was developed by Landcare Research in New Zealand. The students are placed in small groups. They are then asked to imagine they are living on a small island state sometime in the future, say around 2050. They are told they will shortly be given one of four possible scenarios that describe the world they will be living in. First though they are each given a card describing technological or lifestyle changes that have effected everyday life over the last 20 years and asked to consider where we will be in another 20 years. They discuss these changes in pairs.

They are then given a card looking at drivers for change. Again they discuss these in pairs. Next they receive the scenarios. These are related and each of which is explained—essentially there are two axes.

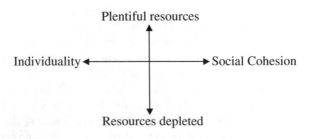

Each group member is given a role to play in their group scenario. After individual reflection they discuss these roles in their group. Finally wild cards are handed out which 'shock' the scenarios and ask the students to reflect again on how this might affect them in their roles.

Throughout the students are asked to make notes of their thoughts and at each stage of the game they are asked what the consequences might be for society and the environment. The game takes about 2 hours and the students respond well with lots of discussion and commentary.

Week 9

This week is about the (un)sustainability walk. This takes approximately 2 hours and the weather usually determines exactly when it takes place. The class goes out in the morning and discusses the walk in the afternoon. What was seen, was it unpleasant, could things be different in cities of the future? The discussion provides a useful lead into the course assignment.

The (un)sustainability walk

The walk is through East Manchester where students can physically observe the social and environmental impacts of business. The aim is to touch on the real, as compared to the simply rational, impact of business and help the students fill concepts such as sustainability, social impact and environmental impact with feelings and meaning, i.e., make them real too.

As the students walk they note the various transport systems and visit some suspect areas yet to be regenerated. They also look over (and hear and smell) a waste recycling operation.

This all leads to discussion about the social changes that have occurred over the last two centuries and indeed the last few decades. They are asked what they think the cities of the future will look like and how will they source resources such as food and energy. Two or three rivers and canals are passed and these contain tiny glimpses of nature—birds, trees and plants. The students are asked if this is all the nature there will be in the future.

The students were also asked to consider what might be described as the forgotten places and forgotten people; people who were sleeping in doorways, selling themselves on the canal towpath and running drugs from isolated communities; people for whom sustainability and the environment have little or no immediate meaning in their lives. How could these people be re-integrated into the social mainstream?

The really good thing about learning outside the classroom is that it makes what is taught real. Students are not simply seeing photographs of something happening 'somewhere else'; or listening to some kind of simplified model, wordy definition or formula, remote from everyday experience; they are shown them how it is "out there" in all its complex, messy glory. Sometimes students think learning is just about putting things in your head, remembering them and reproducing them in an assignment or examination. Outside though they see the value of analysis and the importance of enquiry, reflection and problem solving; those academic skills suddenly become relevant.

It is in the real world where we gain aesthetic awareness, form attitudes, and develop skills and understanding. This lesson draws sustainability and ethical themes together and to make concepts real.

Week 10
The final week of the course is reserved for group presentations which form part of the second assignment. This assignment requires the students, working in groups, to review the annual reports of a company of their choice and discuss the way it accounts for its social, ethical and environmental policies and impacts. The students are also asked to suggest ways in which the company might improve its accounts, e.g., make them more meaningful or relevant.

Course assessment

The course assessment reinforces the key elements of the course. The first assessment by asking the students to read and critique relevant literature with heavy weighting, i.e., allocation of marks, attached to critique. The assignment is set at the end of Week 3 and the aim of the course at this stage is enlightenment. The second assignment is a group assignment as this is seen as empowering. Again there is a focus on critique, this time on reporting praxis, which represents the first step in holding the reporting organization to account; further though, this time the students are asked to make recommendations, to initiate the conversation that would hold the reporting organization to account, that would demand accountability.

4   Concluding Reflections

This paper

This paper has sought to share ideas with fellow educators. It is pluralist in its outlook in that it makes no attempt to offer a single best way forward. It aims simply to join the discussion of how best to prepare today's business, and particularly accounting, students for the complex challenges they will face in their future careers. For some this may be seen as a weakness or limitation of the paper, others may see strength in it pluralistic approach.

Some may also want to see objective evidence that the course does make students think more deeply about sustainability. Personally I am happy to judge the impact of the course in my day to day dealings with the students I teach. Are they interested, do they participate in class discussions for example? On that basis I believe the course does encourage students to think more deeply. However other researchers might reasonably have expected some sort of analysis of the student's assignments or perhaps a questionnaire. Again the lack of such may be seen as a weakness or limitation of the paper.

The course

The primary objectives of the course are first enlightenment and empowerment then second understanding the nature of accountability. Notably though it also hits some of the more mainstream objectives of higher education. Critical analysis, reflection, communication and team working are all key academic and employability skills. These skills are all enhanced by the very nature of the course. Also today's employers are reputedly keen that their employees are sustainability and ethically aware, again this course hits these targets.

It also aims to produce more rounded accountants aware of their public interest duties to a wider range of stakeholders (Bebbington and Larrinaga 2014) and the raison d'etre behind modern day calls for integrated reporting (Adams 2015).

Finally it is also worth noting that whilst the course has run it has invariably attracted dissertations from students. Students have frequently decided, whilst attending the course, to change their dissertation topic to cover some form of investigation into social and/or environmental reporting. Approaches vary from mainstream empirical investigations comparing some definition of financial success to some definition of corporate social responsibility or sustainability reporting, to exploratory research into the impacts of business in various countries. The course stimulates new thinking which is arguably what education is all about.

References

Adams SCA (2015) The international integrated reporting initiative: a call to action. Crit Perspect Account 27:23–28

Ahrens T, Becker A, Burns SJ, Chapman CS, Granlund M, Habersam M, ... Scheytt T (2008) The future of interpretive accounting research—a polyphonic debate. Crit Perspect Account 19:840–866

Alvesson M, Skoldberg K (2000) Reflexive methodology: new vistas for qualitative research. SAGE Publications, London

Bebbington J, Larrinaga C (2014) Accounting and sustainable development: an exploration. Acc Organ Soc 39:395–413

Blattner W (2006) Heidegger's being and time: a reader's guide. Continuum International, London

Chalmers EF (1999) What is this thing called Science. Open University Press, Maidenhead

Coelho P (2006) The Alchemist. HarperCollins, London

Collison D, Ferguson J, Stevenson L (2007) Sustainability accounting and education. In: Unerman J, Bebbington J, O'Dwyer B (eds) Sustainability accounting and accountability. Routledge, London, pp 327–344

Cooper C, Johnston J (2012) Vulgate accountability: insights from the field of football. Account Audit Accountabil J 25:602–634

Deegan C (2007) Organizational legitimacy as a motive for sustainability reporting. In: Unerman J, Bebbington J, O'Dwyer B (eds) Sustainability accounting and accountability. Routledge, Abingdon, pp 127–149

Deegan C, Unerman J (2006) Financial accounting theory. McGraw-Hill Education, Maidenhead

Dillard JF (1991) Accounting as a critical Social Science. Account Audit Accountabil 4:8–28

Dillard JF (2007) Legitimating the social accounting project: an ethic of accountability. In: Unerman J, Bebbington J, O'Dwyer B (eds) Sustainability accounting and accountability. Routledge, Abingdon, pp 37–54

Elkington J (1997) Cannibals with forks: the triple bottom line of 21st Century business. Capstone, Oxford

Gray R (2010) Is accounting for sustainability actually accounting for sustainability…and how would we know? An exploration of narratives of organisations and the planet. Account Organ Soc 35:45–62

Gray R, Adams C, Owen D (2014) Accountability, social responsibility and sustainability: accounting for society and the environment. Harlow, Pearson Education

Hines RD (1992) Accounting: filling the negative space. Acc Organ Soc 17:313–341

Hopwood AG (2009) Accounting and the environment. Acc Organ Soc 34:433–439

Jones M (2014) Accounting for biodiversity. Routledge, Abingdon

Leahey TH (2004) A history of psychology: main currents in psychological thought. Pearson Education, Upper Saddle River, NJ

Llewelyn S (2003) What counts as "theory" in qualitative management and accounting research? Introducing five levels of theory. Account Audit Accountabil J 16:662–708

Lovelock J (2006) The revenge of Gaia. Penguin, London

MacIntosh N, Quattrone P (2010) Managing accounting and control systems: an organizational and sociological approach. Wiley, Chichester

Mayo E (2015) The co-operative advantage: innovation, co-operation and why sharing business ownership is good for Britain. Co-operative UK, Manchester

McGilchrist I (2009) The master and his emissary: the divided brain and the making of the Western World. Yale University, New Haven and London

McPhail K, Walters D (2009) Accounting and business ethics: an introduction. Routledge, London

Parker LD, Northcott D (2016) Qualitative generalising in accounting research: concepts and strategies. Account Audit Accountabil J 29(6):1100–1131

Polt R (1999) Heidegger: an introduction. UCL Press, London

Puxty AG (1993) The social and organizational context of management accounting. Academic Press, London

Ryan B, Theobald M, Scapens RW (2002) Research method and methodology in finance and accounting. Thomson Learning, London

Scapens RW (2006) Understanding management accounting practices: a personal journey. Br Account Rev 38:1–30

Schumacher EF (1973) Small is beautiful. Abacus, London

Solomon JF, Solomon A, Joseph NL, Norton SD (2013) Impression management, myth creation and fabrication in private social and environmental reporting: insights from Erving Goffman. Acc Organ Soc 38:195–213

Spence C (2007) Social and environmental reporting and hegemonic discourse. Account Audit Accountabil 20:855–882

Spence C (2009) Social accounting's emancipatory potential: a Gramscian critique. Crit Perspect Account 20:205–227

The Independent (2006) Rivers: a drying shame. http://www.independent.co.uk/environment/rivers-a-drying-shame-469598.html. Accessed 10 Aug 2015

Thomson I (2007) Mapping the terrain of sustainability accounting. In: Unerman J, Bebbington J, O'Dwyer B (eds) Sustainability accounting and accountability. Routledge, Abingdon, pp 19–36

Thomson I, Bebbington J (2005) Social and environmental reporting in the UK: a pedagogic evaluation. Crit Perspect Account 16:507–533

UNESCO (2015) Facts and figures on marine pollution. http://www.unesco.org/new/en/natural-sciences/ioc-oceans/priority-areas/rio-20-ocean/blueprint-for-the-future-we-want/marine-pollution/facts-and-figures-on-marine-pollution/. Accessed 10 Aug 2015

United Nations World Commission on Environment and Development (1987) Our common future. Oxford University Press, Oxford

Warnock M (1970) Existentialism. Oxford University Press, Oxford

Williams RC (2007) The co-operative movement: globalization from below. Ashgate Publishing, Aldershot

Wilson C (2001) The outsider. Orion Books, London

Professional, Methodical and Didactical Facets of ESD in a Masters Course Curriculum—A Case Study from Germany

Markus Will, Claudia Neumann and Jana Brauweiler

Abstract Universities are expected to contribute to transforming societies towards sustainable development by proving transdisciplinary knowledge and to educate and prepare students for their future roles as decision-makers, entrepreneurs, and academics. The purpose of this chapter is to describe the integration of sustainability aspects into a master course program at a German university, to share experiences and to inform and motivate other educators to also "walk the talk". The article provides insight into the approach taken to integrate sustainability aspects into the didactical concept such as the instrument of learning diaries or the implementation of problem-based learning modules. Furthermore the article introduces the specific way of quality assurance and evaluation which is used to structure the continual improvement of the master courses. It is shown on one hand, that the combination of all of these elements is necessary to ensure ESD in a holistic way. On the other hand this demands high standards of the lecturers, which is why a didactic consulting partnership was established to accompany this process with various offers for self-improvement, which are finally presented. Highlights of the papers are: a set of competencies of a sustainability manager and an overview to related modules in the study program, a description on pedagogical and didactical instruments (learning diaries, PBL related instruments such as case study work and theory-praxis transfers, colloquia with industry partners), insights into the approach of quality assurance and continual improvement (i.e. evaluation, several formats of didactical consultations).

M. Will (✉) · J. Brauweiler
University of Applied Sciences Zittau/Görlitz, Theodor-Körner Allee 16,
02763 Zittau, Germany
e-mail: m.will@hszg.de

J. Brauweiler
e-mail: j.brauweiler@hszg.de

C. Neumann
International Institute IHI Zittau, Technical University Dresden, Markt 23,
02763 Zittau, Germany
e-mail: claudia.neumann@tu-dresden.de

© Springer International Publishing AG 2018
W. Leal Filho (ed.), *Implementing Sustainability in the Curriculum of Universities*,
World Sustainability Series, https://doi.org/10.1007/978-3-319-70281-0_19

Keywords Education for sustainable development · Problem-based learning
Learning diaries · Continual improvement of study programs

1 Introduction and Background

In the context of the public debates about education for sustainability, the 21st century's skills and the role of universities are often discussed (Lozano et al. 2015; Davim and Leal Filho 2016; Leal Filho 2011; Wals 2012). Universities are expected to contribute to transforming societies towards sustainable development. They are supposed to offer transdisciplinary knowledge and to educate and prepare students for their future roles as decision-makers, entrepreneurs, and academics (f.i. Disterheft et al. 2013). Due to the commitment to implement ESD in teaching, research and operation, universities may contribute to the five-year World Program of Action "Education for Sustainable Development" of the United Nations (2015–2019) and to the achievement of the Sustainable Development Goals of the United Nations. However, the research and education they have provided also has contributed in many ways to the strengthening of unsustainable path-dependencies (Goshal 2005; Sterling and Scott 2008; Wals 2008). Despite the variety of declarations and university policies, sustainability in and at universities is often early stage and remains to be a tremendous challenge (Lozano et al. 2013; Mulder et al. 2012; Waas et al. 2010). The ambiguity of the terms sustainability, sustainable development, and ESD, their normative notions, the equivocal validity of assessments lead to various challenges and obstacles, which are discussed in further detail elsewhere (Stough et al. 2017; Sylvestre et al. 2013; Urbanski and Leal Filho 2014; Jickling 2016; Ávila et al. 2017; Leal Filho et al. 2017; Jickling and Wals 2012). Obviously this is also an institutional change management issue (Kapitulčinová et al. 2017; Will et al. 2015). The value of being engaged in ESD is that due to this work, improvements towards sustainable development in companies may be stimulated.

The goal of this paper is to describe the approach taken at the University of Applied Sciences Zittau/Goerlitz, Germany to an interested audience. No results of empirical research are presented, instead, the application of different didactical instruments in a problem-based-learning context, such as learning diaries, theory-praxis-transfer modules. The paper also discusses the way how continual improvement is sought beyond the curriculum design stage.

The paper may offer the chance to compare it with other approaches to ESD and to provide impulses for curriculum development in general. The paper therefore contributes to sharing good practice and provides insight how education for sustainable development can be tackled in the context of a relatively small University of Applied Science.

The article pays attention to an example of how to integrate sustainability aspects into a Master Course Program. Based on the concept of sustainability competencies (Chapter "Enabling Faith-Inspired Education on the Sustainable Development Goals Through e-Learning"), the implementation of sustainability aspects in the curriculum (Chapter "Sustainable Architecture Theory in Education: How Architecture Students Engage and Process Knowledge of Sustainable Architecture") and the didactical concept as well as the evaluation and improvement philosophy of the study course (Chapter "Education for Sustainability in Higher Education Housing Courses: Agents for Change or Technicians? Researching Outcomes for a Sustainability Curriculum") is outlined.

2 Competence Profiles for Sustainability Managers

Competence-based approaches have been widely adopted in higher education for designing curricula and defining learning objectives (Van den Bergh et al. 2006). However, the concept of competencies is also criticized for several reasons, in particular with regard to the economization of education and other more conceptual burdens as well as organizational problems with regard to holistic implementation (Lambrechts and Van Petegem 2016; Lansu et al. 2013). There is also the suspicion that all the 'ado about competencies' is just window dressing in order to meet the checklist-criteria of accreditation agencies. In the end the competency approach may offer a way to evolve towards problem-based education. Conventional pedagogic paradigms have focused on the contents and subjects which should be taught (EQF 2017). The competencies-based approach instead is more output-oriented, i.e. focusing on what should be learned beyond theoretical knowledge. The desired set of competencies usually includes knowledge, attitudes, and values as well as managing abilities, problem-solving strategies, application of analytical and profound expertise (Stoof et al. 2002). Competencies-based education should enhance the employability in organizations that have to deal with sustainability issues. Thus, competencies can be understood as a set of skills and functions, a graduate student should fulfil in order to perform in a defined work setting. A good starting point, therefore, seems to be the job description of a sustainability manager. This is a professional, who works in the sustainability or EHS department of a company, taking the responsibility for the implementation and effective operation of standardized management systems (ISO 14001, ISO 9001, ISO 45001, ISO 50001, ISO 27001 etc.). This is nowadays status quo in a majority of enterprises and most of them seek to integrate different management systems in order to take advantage of synergies. These so called integrative management systems are usually defined as a set of interconnected processes sharing information, structures, and responsible employees in order to fulfil a number of requirements and to satisfy stakeholders (Karapetrovic and Willborn 1998; Bernado et al. 2017). Hence, integration is regarded as the process of aligning strategy, methodology and auditing at the level of management systems and business processes (Domingues et al. 2015; Nunhes et al. 2017; Bernado et al. 2017). These integrated management systems are of high relevance for businesses and societies. They are expected to be a key factor for business sustainability and environmental protection as they help to identify unsustainable business practices and support mitigation of environmental impacts related to resource and energy consumption. However, sustainability needs a lot more effort than business-as-usual management capabilities. Companies often operate in a rather unsustainable manner (Will 2017), causing pollution and wastes, using non-renewable and scarce resources, off-shoring "dirty" manufacturing processes to developing countries etc. Therefore, the sustainability manager also is an innovator and change agent, more than an administrator, who organizes a change to more sustainable strategies, products, and processes, i.e. a more sustainable value creation (Laszlo and Zhexembayeva 2011; Atkisson 2010; Hesselbarth and Schaltegger 2014). With the help of the competencies compiled in Table 1, the obligations of a sustainability manager in a company could be fulfilled. The master course's competence map and learning objectives have been derived based on this set of competences.

The challenge during the design of the curriculum was, however, to integrate the development of the competencies effectively in the study program. A 'combined-integration approach' was taken (Lambrechts et al. 2011), which implies that sustainability was integrated into the learning objectives and contents of each single module in a more or less explicit way. Additionally, there are modules with an explicit focus on sustainable development. In the next sections, the master courses and the ESD-approaches applied are described.

3   Master Programs "Integrated Management" and "Integrated Management Systems"

The paper deals in particular with two master programs offered at the University of Applied Sciences Zittau/Görlitz i.e. "Integrated Management" (4 semesters) and "Integrated Management Systems" (3 semesters).[1] The first students were enrolled in 2015. In the same year, the accreditation was completed successfully, without major critiques and further requirements. Both programs aim at educating and preparing its graduates for taking positions in the field of business sustainability, i.e. as quality, environmental, energy or occupational and health officers, sustainability managers or consultants. The program was developed in response to a supposed lack of adequately qualified experts for small and medium sized enterprises (SME). SME seek for employees that are able to bring the different requirements of management systems together in an integrative way.

Figure 1 shows the overall structure of the two master programs and its modules. Emphasis was paid to the link between the contents, module didactics and the supposed competencies of a sustainability manager.

The relation of the curriculum to business sustainability is given by a holistic and integrative perspective on the various management systems. Comparable study

---

[1]The study programs differ just by an advanced semester. That's why in this article the following descriptions applies to both master courses, only little distinctions are mentioned.

Table 1 Examples of competencies of a sustainability manager[a]

Generic competences
– Knowledge (theories, models, definitions, instruments, methods) – Intercultural competencies – Cognitive capabilities (analytical and reflective thinking) – Practical capabilities (transfer of theories into practice) – Social competencies (teamwork, empathy, self-esteem) – Independence – Self-directed working – Responsibility and reliability
Specific competences for sustainable development
– Knowledge of sustainability principles, issues, and problems – Systems thinking – Anticipation – Normativity, values, business ethics – Dealing with dilemmas and goal conflicts – Dealing with incomplete and complex information – Reflection of cultural models – Inter- and Transdisciplinarity
Specific competences for integrated management systems
– Familiarity with requirements, normative documents, standards and legal regulations – Stakeholder management – Knowledge of cleaner technologies – Expertise in application of methods, tools, and instruments (for sustainability accounting, reporting, process management and life cycle assessment) – Types and methods of integration – Project management – Preparing and conducting internal/external audits – Understanding the role of external audits – Definition and measuring of indicators – Preparing and communicating reports – Handling of software
Specific competences for change agents
– Strategic competence – Identification of drivers and barriers – Interpersonal competence – Participative decision making – Conflict management – Negotiating – Cooperation – Emotional intelligence, empathy, and solidarity – Self-motivation and motivation of others – Communication

[a]Own elaboration based on the accreditation self assessment report and selected literature such as Sterling and Thomas (2006), Roorda (2010), Wiek et al. (2011), de Haan (2006), Lambrechts et al. (2013). Please note that this is an overview without a claim of completeness. The categories are not disjunct

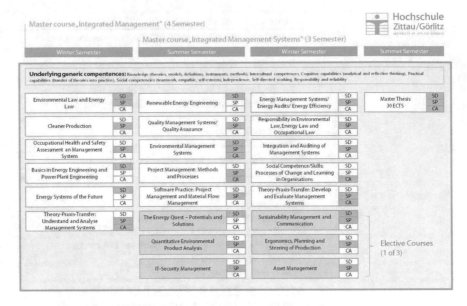

Fig. 1 Course Modules and Competences (*Abbrevations* SD = Specific Competences for Sustainable Development, SP = Specific Competences for integrated management systems, CA = Specific Competences for Change Agentry)

programs in Germany often merely refer to one or two, sometimes three special-izations (i.e. quality management and environmental management or energy man-agement and resource management). However, the master degree courses described here include all the different aspects of quality management and EHS with related issues like IT security management, project management, energy technology and sustainability management. In addition to that normative foundations of sustainable development and sustainability management are considered. Multiple perspectives from different disciplines such as engineering and economics (i.e. business admin-istration) are combined with social science. This is possible by having three departments (i.e. natural and environmental sciences, electrical engineering/ computer sciences and mechanical engineering) that provide modules for the mas-ter courses. This inter- and transdisciplinarity in both structure and content enables the students to develop a distinctive understanding of the unity of technical, social, economic and environment related aspects. Also, the group of students itself is interdisciplinary in a way, as they come from different backgrounds and sometimes with professional experience.

ESD, however, is not limited to teaching knowledge and develop competences for professional life. ESD is also about the way the education is offered, i.e. about didactics, quality assurance and evaluation as well as continual improvement. The next section will stress this.

4 Pedagogies and Didactics—Facilitating the Learning Process

During the design of the master courses, there was the trial to overcome the paradigm of transmission of knowledge, i.e. to induce a "shift from teaching to learning". Successful learning processes for sustainability are not only a question of content, infrastructure and circumstances. Instead, it is about the facilitation of a concrete learning process that provides learning settings while taking into account the diverse previous knowledge and rich experiences of students and the learning objectives.

Elements of the learning setting are:

- Lectures and seminars,
- Workshops,
- Debates and discussions,
- Project and learning groups.

Obviously the combination of different elements of creative learning spaces must not disregard the appropriation of substantial knowledge. However, it is equally important, if not more important, to provide opportunities to apply this knowledge. Since the ability of critical reflection is a key skill in the modern information society in general, it is of particular importance in the domain of sustainable development, because it is the first step of questioning habits and making change. Students should be enabled to adopt a critical attitude towards the topics as well as towards the practice, questioning both from the perspective of sustainable development. Furthermore, they should be able to reflect on themselves and their actions. In addition to the knowledge about technologies and management practices, it is the consciousness about the impact of individual actions and the possibility to act which is an important basis for more sustainable decisions. Such a reflection should not only be practiced during lectures and seminars. It also needs to be applied to the process itself, in order to be prepared for the challenges of lifelong learning, which requires the competence to identify the necessary knowledge and the suitable abilities and to appropriate both to one self. Two particular learning methods have proved to induce this kind of reflection and will be described in the next section: learning diaries and problem-based learning.

4.1 *Learning Diaries as a Tool for Reflective Learning*

An appropriate instrument to foster self-directed, autonomous and reflective learning is a learning diary (Korthagen 2001; Trif and Popescu 2013). Students are required to summarize, analyse, reflect and comment on the contents and their personal learning experiences directly after a classroom lecture. Hence the diary is a subjective view reflecting what the student has heard and learned. Only the student's individual analysis and insights are that counts. Learning diaries have several functions such as formalizing reflection, to include personal feelings, to elaborate a summary of the contents. Guiding questions may help the students to write their diaries, for example:

- What are the main points of the lecture?
- What was new during the class?
- What was of particular interest?
- Was there something that changed your (i.e. the students) views and why?
- What went against my own ideas?
- How did the class relate to your prior learning and life experience?
- What did you not understand? What was less comprehensible?
- What do you plan to do for better understanding?
- Which complementary literature do you plan to read in order to enhance your learning and understanding?
- How are you able to apply this knowledge in further studies or later career?

As long as the diaries are evaluated by the lecturer who then gives a feedback to the student, both learner and teacher gain insight into the learning process (Wetherell and Mullins 1996). However, as the learning diary supports regular self-monitoring and self-reflection, it should have a positive effect by itself (Otto and Kistner 2017). Also, the learning diary can replace or supplement other forms of final exams (such as written examinations, essays or oral presentations).

As a means of examination, the following criteria on a 1–5 scale are used for grading:

- Formal correctness (complete sentences, text beyond mere transcription of slides or description of content but reference to the individual learning process)
- Structural aspects (degree of organisation and structure, 'red line', relation between topics are recognisable)
- Elaboration (degree of deepness, relations to own expertise, examples, the degree of which own questions for further studies are derived)
- Reflection of own learning processes (monitoring of positive and negative learning episodes is described clearly, concrete measures for improvement of learning).

Learning diaries are up to now used in two modules of the study program ("Quantitative Environmental Assessment of Products/Life Cycle Assessment" and "Sustainability Management and Communication").

4.2 Problem-Based Learning

Apart from the critical position and the consciousness or reflecting process another essential aim of education for sustainable development is, that students can also put their knowledge into practice, which means, that they can recognise problems and challenges, develop solution strategies and, finally, act competently and responsibly. Hence, problem-oriented learning is as central to the design of suitable courses as a high practical relevance of the topics. In the master courses, this is practiced by using a combination of different methods.

Case studies

In addition to research-related and practical case studies in seminars, problem-based learning is realised, for example, with project works. In small projects, the students work—based on the provided new knowledge—on a task with a practical orientation (e.g. in the modules project management, energy technology or IT security management). This kind of project work is seen as reasonable especially in education for sustainable development, because in doing so the students schedule their activities to a great extent by themselves, apply their knowledge and put it into practice at (the location of) their university.

TPT-Module

The most important elements of the orientation towards practice in the curriculum are the theory practice transfer modules (TPT modules).

As seen in Fig. 2 there are two of these modules in the 4-semester-program and one in the 3-semester-program. The first module, which is located at the beginning of the study program, concentrates on the topic "Understanding and Analysis of Management Systems". Focus here is a critical assessment of business cases on the base of the transmission of knowledge. The second module, which is almost at the end of the study program, deals with the "Development and evaluation of Management Systems". In these modules groups of 2–3 students have to work on and solve an actual task in a company out of the specific range of topics of the management systems (TPT task).

The company and the lecturers settle the task in advance and reduce its complexity so that it is achievable for the students in the allowed working time. During this time the students are guided and looked after by lecturers of the specific domain.

The procedure of a TPT module always includes:

- a kick-off meeting,
- practice phases in the companies,
- consultations with the lecturers as well as
- a presentation during the project
- and a final presentation.

During the process, the students should work on the TPT task as autonomously as possible by making a project schedule of the projects aims, progress and working packages which they handle independently from then on. But the project plans as well as the interim results have to be regularly coordinated with and released by the lecturer and the company.

Therewith not only the students' way of working is oriented to the intended competencies for sustainable development but also the content of the TPT tasks has an explicit relation to sustainability, transferred into the organizations as well. In the winter semester of 2016/17, for example, the students developed sustainability approaches for a big regional sports event and a concept for a sustainability report for a regional holiday park company. The sports event is organised voluntary so that there wouldn't have been the resources for an intensive engagement with the topic, as the feedback of the organizer revealed. The students' results gave starting points for concrete measures and an impulse for the change process. For the company, in turn, the sustainability report was a totally new arena which they discovered with the work of the students. Also, in this case, the students' work has a lasting effect because now the company will realize the concept they came up with.

Guest Lectures, excursions and colloquia

For the students, the practice not only becomes concrete when they do projects but also during excursions to companies or presentations by guest lecturers. Additionally, a colloquium developed with the TÜV-Akademie Görlitz, called "IMS-breakfast", is integrated into the course of studies. The format is actually a networking event between the university and local companies, which takes places in an informal atmosphere and as a common business breakfast. At this colloquium which takes place twice a semester with about 20 persons (guests from companies, lecturers, and students), ideas for projects are developed and topics of integrated management systems are discussed. First, expert information regarding the topic of integrated management systems is imparted, then there is an experience report by a company about the implementation. This forms the basis for the exchange of experiences with the implementation, for example regarding legal, technological or management system requirements. In this way and through the cooperation of the TPT modules, a well going regional network "integrated management systems" was established, including practitioners, lecturers, students, and consultants.

This all in all high practical relevance in which the TPT module can be seen as some kind of an internship integrated into the course of studies is extraordinary for a master degree course. This was also emphasised by the reviewers during the accreditation process and results from the fact that the lecturers are bridging to practice themselves. Therefore the practical relevance of the courses not only contributes to the competency development regarding the sustainable development but also establishes the university as a place, where sustainability is explored, taught as well as practiced and from where it is even transferred into society. The students achieve a high level of professionalism, which is reflected in a very good placement rate for graduates in regional and supra-regional companies.

4.3   Evaluation and Continual Improvement

The continual improvement of the study program, the modules, contents, and didactical concepts is another important aspect in the understanding of ESD. Although course evaluation is an established element of course development

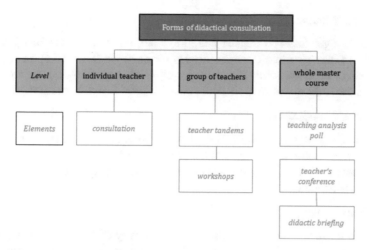

Fig. 2 Elements of didactical consultation within the master courses

policies at universities, there is some doubt and also dissatisfaction about the way and effectiveness of the utilization of evaluation information for improvement. Evaluations may serve as a "fire-alarm-function", but course development and improvement seem not to be in the foreground for several reasons (Edström 2008).

In order to deal with the professional, teaching, and didactical concept of the master courses in a credible, responsible and critical way, cooperation has been established with a didactical consultant.[2] The cooperation was carried out not only during curriculum development but also while implementation and following improvement steps. Based on the requirements of lecturers and students different consulting opportunities are provided (see Fig. 2).

- Didactics breakfast

The didactics breakfast takes place about once per semester to ask for the needs concerning the didactical support for the upcoming semester. Furthermore, it offers the possibility to swap ideas on the challenges of the ongoing semester in an informal setting and to develop solution approaches collegially.

- Workshops

Workshops, as a traditional form of the didactic training, impart knowledge in specific issues to the lecturers. Nevertheless, they also focus practical relevance by giving the opportunity to transfer immediately the impulses into the planning and practice of teaching.

---

[2]This was enabled by the project "Teaching practice in transfer (plus)" at the International Institute Zittau, a central academic unit of the TU Dresden.

- Lecturer's tandems

An even stronger focus on transfer and individual ambitions lies in the „Lecturer's tandem". In doing so two lecturers discuss for example the development of exercises, tests or tasks for seminar papers in a moderated meeting. Furthermore, there is always the offer to get an individual consulting for didactic questions and issues.

- Lecturer's conference

The lecturer's conference addresses the inter-referentiality of the lectures. Once per semester the lecturers jointly review the semester from an organisational, subject-specific and didactical perspective and plan the upcoming semester from these points of view.

- Students survey

Before the lecturer's conference takes place, a survey of the student's satisfaction with the course of studies is conducted. The survey findings are finally presented, analysed and discussed at the lecturer's conference. This combination of a critically reflecting perspective towards the master course of the lecturers and the students leads to a continuous process of further development. To make this process of further development constructive the survey uses a combination of quantitative and qualitative approaches. The qualitative questioning is a modification of the evaluation method "Teaching Analysis Poll" (TAP). Regarding the organisation, content design and the courses the students are asked for positive aspects, points of criticism and suggestions for improvement. First, they gather feedback in groups. Then there is a voting in the plenum containing each aspect so that even in this qualitative approach majorities can be identified. The quantitative survey is realised with a questionnaire. The various items ask for the current situation in the studies, the offered courses, the employment perspective and the overall impression. By this comprehensive questioning, the students experience themselves as responsible and influential participants of the master courses because their opinion is not only captured but also leads to a response of the lecturers after the lecturer's conference and therefore often to a process of change. The process of the survey and in particular it's qualitative questioning additionally has the effect that the students exchange and compare their views on the process of learning and the master course. The teachers for their part receive constructive, reasoned suggestions for improvement and can discuss the ones that were capable of securing a majority regarding their feasibility with the students. In this way, in the spirit of sustainable development, both sides practice and put into practice a critical reflection of their own actions, the competence of entering into a dialogue as well as constructively shaping a change process. The design of this survey has not only earned credit from the participants but was also appreciated by the reviewers during the accreditation. They considered it as an innovative method of evaluation and as a result encouraged to use it further.

5   Conclusion

The case study described the implementation of ESD in a Master Course study program on the level of content (ESD as the topic of the program and the modules), learning outcome (problem-based and practice-oriented learning) and didactics (expert monitoring, evaluation and improvement of teaching skills). The wide ranging topic of entrepreneurial sustainability based on the competencies of a sustainable manager is interwoven in various modules of the course and it is thematised in particular modules. In addition to the content orientation, principles of sustainability science are considered also on the didactic level, where different problem-based and learning-oriented teaching methods are applied. The goal of every study program is individual learning. Thus, students are expected not only to have replicable and applicable knowledge but also to get an understanding of the relationship of various bits of information taught in the different modules. The faculty of the course aims not at Education *for* Sustainable Development in a form of preaching. Instead, attention is paid to facilitate a process of deeper learning, where students are expected to participate intelligently in the sustainability debate by challenging the different and often incommensurable positions and arguments in the field. The education offered aims at clarifying arguments and ethical values and helping students to develop their own positions with regard to the values that are present in their potential professional workplaces. In fact, the study program enables the acquisition of professional, methodological and social skills and thereby fosters competency-based employability. Even more important is that the study program aims at providing learning settings, where students can develop their practical skills in real-life projects, including project and change management issues. Thus the approach of problem-based learning is made even more concrete and practice-oriented in forms of project-based learning. In order to communicate ESD in such an interdisciplinary manner and to make it tangible in cooperation with partners from science and business, it is not sufficient to work merely on a content level. There is a need for appropriate didactical concepts and for a critical self-assessment and a continuous improvement process. The application of new teaching and learning methods, such as learning diaries, requires the teachers to develop themselves and to try new methods in a process of long life learning. The implementation of practical and problem-oriented modules implies higher preparation and implementation than "normal" modules and thus a higher commitment of the teachers. At the same time, this form of teaching is only possible at the base of a high willingness of companies to cooperate and their confidence in the competence of the students. Walking along this thorny road requires patience as a great staying power in a continuous improvement process. Needless to say that this involves changes not only on the surface of curricula and didactics but also on the level of epistemic culture. This is also is a major constraint identified during three years of activity that the initial enthusiasm tends to decrease after some times. ESD in this context is both a challenge and satisfaction for all parties involved. It means also to share experiences and to inform and motivate other educators to also "walk the talk".

References

AtKisson A (2010) The sustainability transformation: how to accelerate positive change in challenging times. Routledge, Earthscan

Ávila LC, Leal Filho W, Brandli L, MacGregor C, Molthan-Hill P, Özuyar PG, Moreira RM (2017) Barriers to innovation and sustainability at universities around the world. J Clean Prod 164(2017):1268–1278

Bernado M, Gianni M, Gotzamani K, Simon A (2017) Is there a common pattern to integrate multiple management systems? A comparative analysis between organizations in Greece and Spain. J Clean Prod 151(2017):121–133

Davim JP, Leal Filho W (eds) (2016) Challenges in higher education for sustainability, management and industrial engineering. Springer International Publishing, Cham

de Haan G (2006) The BLK '21' programme in Germany: a 'Gestaltungskompetenz'-based model for education for sustainable development. Environ Educ Res 12(1):19–32

Disterheft A, Caeiro S, Azeiteiro MU, Leal Filho W (2013) Sustainability science and education for sustainable development in universities: a way for transition. In: Caeiro S et al (eds) Sustainability assessment tools in higher education institutions. Springer, Cham

Domingues P, Sampaio P, Arezes PM (2015) Analysis of integrated management systems from various perspectives. Total Qual Manag Bus Excellence 26(11–12):1311–1334. doi:10.1080/14783363.2014.931064

Edström K (2008) Doing course evaluation as if learning matters most. High Educ Res Dev 27 (2):95–106

EQF (2017) European Qualifications Framework. https://ec.europa.eu/ploteus/content/descriptors-page

Goshal S (2005) Bad management theories are destroying good management practices. Acad Manag Learn Educ 4(1):75–91

Hesselbarth C, Schaltegger S (2014) Educating change agents for sustainability e learnings from the first sustainability management master of business administration. J Clean Prod 62 (2014):24–36

Jickling B (2016) Losing traction and the art of slip-sliding away: or, getting over education for sustainable development. J Environ Educ 47(2):128–138

Jickling B, Wals A (2012) Debating education for sustainable development 20 Years after Rio: a conversation between Bob Jickling and Arjen Wals. In: ESD in Higher Education, the Professions and at Home, 6(1). Sage Publications, pp 49–57

Kapitulčinová D, AtKisson A, Perdue J, Will M (2017) Towards integrated sustainability in higher education—mapping the use of the Accelerator toolset in all dimensions of university practice. J Clean Prod. doi:10.1016/j.jclepro.2017.05.050

Karapetrovic S, Willborn W (1998) Integration of quality and environmental management systems. TQM Mag 10(3):204–213

Korthagen F (2001) Reflection on reflection. In: Korthagen FAJ, Kessels J, Koster B, Lagerwerf B, Wubbels T (eds) Linking practice and theory: the pedagogy of realistic teacher education. Lawrence Erlbaum Associates, Mahwah, pp 51–68

Lambrechts W, Mulà I, Ceulemans K, Molderez I, Gaeremynck V (2013) The integration of competences for sustainable development in higher education: an analysis of bachelor programs in management. J Clean Prod 48:65–73

Lambrechts W, Van Petegem P (2016) The interrelations between competences for sustainable development and research competences. Int J Sustain High Educ 17(6):776–795

Lambrechts W, de Vall IMP, van den Haute H (2011) The integration of sustainability in competence based higher education. Using competencies as a starting point to achieve sustainable higher education. In: Knowledge Collaboration & Learning for Sustainable Innovation ERSCP-EMSU conference, Delft, The Netherlands, 25–29 October 2011

Lansu A, Boon J, Sloep PB, van Dam-Mieras R (2013) Changing professional demands in sustainable regional development: a curriculum design process to meet transboundary competence. J Clean Prod 49:123–133

Laszlo C, Zhexembayeva N (2011) Embedded sustainability: the next big competitive advantage. Stanford University Press

Leal Filho W (2011) About the role of universities and their contribution to sustainable development. High Educ Policy 24:427–438

Leal Filho W, Jim Wu Y, Brandli LL, Avila LV, Azeiteiro UM, Caeiro S, Madruga (2017) Identifying and overcoming obstacles to the implementation of sustainable development at universities. J Integr Environ Sci 14(1):93–108. doi:10.1080/1943815X.2017.1362007

Lozano R, Lozano F, Mulder K, Huisingh D, Waas T (2013) Advancing higher education for sustainable development: international insights and critical reflections. J Clean Prod 48:3–9

Lozano R, Ceulemans K, Alonso-Almeida M, Huisingh D, Lozano FJ, Waas T, Lambrechts W, Lukman R, Hugé J (2015) A review of commitment and implementation of sustainable development in higher education: results from a worldwide survey. J Clean Prod 108(Part A):1–18. doi:10.1016/j.jclepro.2014.09.048

Mulder K, Segalàs J, Ferrer-Balas D (2012) How to educate engineers for/in sustainable development: ten years of discussion, remaining challenges. Int J Sustain High Educ 13 (3):211–218

Nunhes T, Barbosa L, de Olivera J (2017) Identification and analysis of the elements and functions integrable in integrated management systems. J Clean Prod 142(2017):3225–3235

Otto B, Kistner S (2017) Is there a Matthew effect in self-regulated learning and mathematical strategy application?—Assessing the effects of a training program with standardized learning diaries. Learn Individ Differ 55:75–86

Roorda N (2010) Sailing on the winds of change. The Odyssey to sustainability of the University of Applied Sciences in the Netherlands, Maastricht

Sterling S, Scott W (2008) Higher education and ESD in England: a critical commentary on recent intiatives. Environ Educ Res 14(4):386–398

Sterling S, Thomas I (2006) Education for sustainability: the role of capabilities in guiding university curricula. Int J Innov Sustain Dev 1(4):349–370

Stoof A, Martens RL, van Merriënboer JJG, Bastiaens TJ (2002) The boundary approach of competence: a constructivist aid for understanding and using the concept of competence. Human Resour Dev Rev 1(3):345–365 (Sage Publications)

Stough T, Ceulemans K, Lambrechts W, Cappuyns V (2017) Assessing sustainability in higher education curricula: a critical reflection on validity issues. J Clean Prod 2017:1–11

Sylvestre P, McNeil R, Wright T (2013) From Talloires to Turin: a critical discourse analysis of declarations for sustainability in higher education. Sustainability 5:1356–1371

Trif L, Popescu T (2013) The reflective diary, an effective professional training instrument for future teachers. Procedia - Social and Behavioral Sciences 93(2013):1070–1074

Urbanski M, Leal Filho W (2014) Measuring sustainability at universities by means of Sustainability Tracking, Assessment and Rating Systems (STARS): early findings from STARS data. Environ Dev Sustain 17(2):209–220

Van den Bergh V, Mortelmans D, Spooren P, Van Petegem P, Gijbels D, Vanthournout G (2006) New assessment modes within project-based education—the stakeholders. Stud Educ Eval 32:345–368

Waas T, Verbruggen A, Wright T (2010) University research for sustainable development: definition and characteristics explored. J Clean Prod 18:629–636

Wals A (2008) From cosmetic reform to meaningful integration: implementing education for sustainable development in higher education institutes—the state of affairs in six European countries. DHO, Amsterdam

Wals AJ (2012) Shaping the education of tomorrow: 2012 full-length report on the UN decade of education for sustainable development. Commissioned by UNESCO

Wetherell J, Mullins G (1996) The use of student journals in problem-based learning. Med Educ 30(2):105–111. doi:10.1111/j.1365-2923.1996.tb00727.x

Wiek A, Withycombe L, Redman CL (2011) Key competencies in sustainability: a reference framework for academic program development. Sustain Sci 6(2):203–219

Will M (2017) A sustainable company is possible—some case studies and a maturity model. In: Velazquez-Contreras L (ed) International sustainability stories: enhancing good practices. Universidad de Sonora

Will M, Zenker-Hoffmann A, Brauweiler J, Delakowitz B, Riedel S (2015) Every end is a new beginning—a realignment and relaunch of an environmental management system at a German University following a 15 years maintaining period. In: Leal Filho W (ed) Transformative approaches to sustainable development at universities. World sustainability series. Springer International Publishing, Switzerland. doi:10.1007/978-3-319-08837-2_21

Incorporating Sustainable Development Issues in Teaching Practice

Walter Leal Filho and Lena-Maria Dahms

Abstract This conclusions paper summarises the main development and trends related to sustainable development teaching and outlines future research needs.

Keywords Sustainabledevelopment · Highereducation-teaching · Barriers-action

1 Introduction

The emphasis to sustainable development has never been higher on national and international political agendas, especially after the launching of the Sustainable Development Goals (SDGs) in 2015. With their three mains assignments, teaching, research and community services, universities play a crucial role in the "transformation of the world" towards sustainability (UN Assembly 2015). According to Wals "Universities in particular have a responsibility in creating space for alternative thinking and emergence of new ideas, as well as in critically exploring old ones" (Wals 2010, p. 380). Therefore, integrating sustainability into college and university curricula and creating new academic programs are the required first steps.

This final paper discusses the guiding principles and characteristics of sustainability curricula in higher education, and shows some of topics and methods which may be incorporated in these curricula. The study finishes with two case studies of a Malaysian and an Australian university that have successfully implemented sustainability in their curricula. All in all one of the greatest challenge lies on "creating

W. Leal Filho (✉)
School of the Science and the Environment, Manchester Metropolitan University, Chester Street, Manchester, UK
e-mail: w.leal@mmu.ac.uk

L.-M. Dahms
Faculty of Life Sciences, Research and Transfer Centre "Sustainable Development and Climate Change Management", Hamburg University of Applied Sciences, Ulmenliet 20, 21033 Hamburg, Germany
e-mail: lena_maria@gmx.de

short-term targets to achieve long-term sustainability goals that are global in scope, but specific in result" (AASHE 2010, p. 6).

2 The International Context

In December 2002 the United Nations (UN) proclaimed the "Decade of Education for Sustainable Development"—to run from 2005 to 2014—to encourage changes in direction of a more sustainable future in terms of environmental integrity, economic viability and a just society for present and future generations (Macquarie University 2009). According of the vision of UNESCO, everyone should have "the opportunity to benefit from education and learn values, behaviour and lifestyles required for a sustainable future and for positive societal transformation" (UNESCO 2005, p. 4). UNESCO as the lead agency of the Decade, identified four main goals: Rethinking and revising education from nursery school to university to include a clear focus of current and future societies on the development of knowledge, skills, perspectives and values related to sustainability. This means that reviews of exiting curricula in terms of objectives and contents were supposed to made, in order to develop interdisciplinary approaches with a focus social, economic and environmental sustainability. Furthermore, it recommended a variety of new approaches to teaching and learning (UNESCO 2005, p. 56).

3 A Curriculum of Sustainability or a Sustainability Curriculum?

To change and reform curricula is not an easy task, as this change is not something that can be legislated or achieved through policy decisions. Instead it depends on the expertise and ability of decision-makers in faculties (AASHE 2010). Designing a curriculum of sustainability is not simply about adding new subjects, courses or topics as an add-on; much more the principles of sustainability have to be the central idea throughout the whole unit (Scott 2009). Leal Filho and Pace (2016) have documented a variety of experiences in this field.

As a syllabus the sustainable curriculum should include the goals and aims of the unit, renew the units and classes, guide teaching practice while giving practical support and above all acquiring an awareness of sustainability issues through secondary unit materials (Macquarie University 2009). The superordinate question is how to design a curriculum that leads the students to a personal and lifelong commitment to be prepared for living sustainably, professionally and personally. Throughout the curriculum the learner should deeply understand the interactions and inter-connections as well as their impact of (their) actions and decisions.

If attempts to modify a curriculum are made, there is a need to bear in mind some guiding principles.

Based on a literature review (Nichols 2007; Nichols and Adams 2011; Junyent and Geli de Ciurana 2008; Hensley 2011; Scott 2009) the process of designing a sustainable curriculum should take into account the following guiding principles:

- Interdisciplinary approach

As the concept of sustainability is interdisciplinary by nature, but since and sustainability is yet to be embedded across a wide range of disciplines, teaching sustainability must be based on a crossdisciplinary, interdisciplinary and multidisciplinary approach (Nichols 2007). The complexity of reality needs a multidisciplinary approach in which several areas of expertise, skills and perspective are put together. Collaborations and dialogues between different disciplines enable to deal with new and emerging issues (Nichols and Adams 2011). Therefore a holistic and integrated approach is appropriate.

- Systemic thinking

Based on the assumption of systemic thinking universities have to work from all perspectives (social, economic, technological, ecological etc.) to promote teaching and research programmes (Junyent and Geli de Ciurana 2008). Regardless of the specific subject sustainable thinking always means to consider all the consequences in terms of social, economic, cultural and environmental aspects of possible actions.

- Problem-based and a societal problem-solving orientation

Originally developed for medical education problem-based learning doesn't mean to find defined solutions for the discussed problem. Instead it allows learners to find different solutions and develop useful skills for the future practice.

- A place-based approach

In his article "The Future of Sustainability in Higher Education" Hensley highlights the importance of a place-based approach: "When we consider what needs to be conserved and how to be more ecologically sustainable, place comes into the forefront. Place is the platform on which relevant environmental practice can be identified and acted upon" (Hensley 2011), Place-based intergenerational knowledge and practices within both rural and urban settings are preconditions to reduce our environmental impact and to preserve the integrity of land base (Hensley 2011).

- Critical thinking and reflection

While designing a sustainable curriculum there should be a strong emphasis on developing critical thinking skills, critical thinking and reflection. Only students who learned to develop a critical distance are able to question, think critically from various viewpoints and make well-informed ethical decisions (Scott 2009).

- A consistent relationship between theory and practice

Sustainability education ought to make students to take action solving the emerging environmental problems our world is facing and helping to create a more sustainable world. For that practical action is an essential element of the teaching process. Important is that every action is the result of a preceding clear and reflective consideration. Theory and practice must go hand in hand and interact with each other at different levels: institutional, teaching and research (Junyent and Geli de Ciurana 2008).

- Cooperation and participation

It is obvious that all university sectors must work together towards improving the education for sustainable development and making the curricula more sustainable. That requires bringing together the various sustainable initiatives and stakeholders on the campus. Promoting participation within national and international networks of Higher Education and looking for partnership with local and national governments, enterprises, administration, should facilitate the pathway towards a deeper understanding of the sustainability as a priority at university institutional level (Junyent and Geli de Ciurana 2008). And so the relationship between universities and external actors, as well as between local and institutional organizations is mandatory and become a key factor in order to collaborate in finding solutions and solving problems (Junyent and Geli de Ciurana 2008).

- Methodological adaptation: new teaching and learning methodologies

A methodological adaptation is required to construct knowledge, action, responsibility, participation and commitment. At the end all teaching and learning methodologies should ensure a high level of transferability so that students are enabled to apply their knowledge to local and global realities.

- Democratic participation

Students and teachers actively participating in academic decisions and in negotiating the assessment processes is a source of learning and a rewarding experience for both. This space should lead to actions for change towards sustainability that involve all the groups of the university community. It is important to create this space at different levels: within disciplines, inter-departments, institution, etc. (Junyent and Geli de Ciurana 2008).

A number of other principles could be listed, but the above illustrate the complexity of the matter, and the need to carefully consider and plan any attempt to engage on sustainability-related curriculum changes.

A practical helpful faculty guide, that provides a general overview of sustainability at universities as well as modules related to each of the individual disciplines, was published by the Association for the Advancement of Sustainability in Higher Education (AASHE). AASHE is an US-American association of colleges and universities that collaborate to create sustainable curricula. Their efforts include

Curriculum Leadership Workshops that enable universities in the US to develop a sustainable curriculum in their home universities. The rules that AASHE provides are helpful in conducting inventories of and giving recognition to school courses that address issues of sustainability (AASHE 2010).

As far as **topics** are concerned, sustainability can be taught in all different disciplines at university there is a broad range of possible topics. A detailed list of topics and courses related to sustainability from the faculties School of Business, Fine Arts, Engineering and Computer Sciences and Arts and Sciences are e.g. presented on the website of the Concordia University of Montréal (https://www.concordia.ca/about/sustainability/study-teach/sustainable-courses.html). Issues emerge especially from ethical practice, health, climate change, planning and development, resource and land use, environmental policy and legislation (Macquarie University 2009). They all should favour the progress in transforming the relationship between society and the environment and also between people (Junyent and Geli de Ciurana 2008).

Care should also be exercised in choosing and using suitable **methods**. Among all the conceivable teaching and learning methods, some of the ones most flexible in terms of their use, are:

- Self-assessments

A self-assessment is a good tool to make students aware of what they learned throughout the unit based on the sustainable curriculum. In form of self-assessment questions e.g. they are asked to answer questions related to the subject of their course.

- Case studies

Case studies are an excellent instrument to combine theory and practice. They can exemplify both poor and good practices where students can learn from mistakes or realize how successfully sustainability can put into practice. Case studies can demonstrate how products, processes, business, policies and institutions could act in sustainable ways. Case studies may be hypothetical or real, local or global (Scott 2009).

- Dialogue forums

Dialogue forums may act as platforms where professors, students and administrative staff meet to debate about sustainable issues. On an equal basis and in a positive climate problems and their solutions can be discussed at all levels: local, national and international (Junyent and Geli de Ciurana 2008).

Future-casting

Future-casting is a powerful visionary tool within the context of sustainability for drawing the most optimistic picture of the future. How does the world look like we strive for? To cast the future means to imagine and explore the opportunities for a transformation of the world (Hensley 2017).

But no matter which methods ones chooses, they should help to provide a deeper
—subject-related understanding of scientific, engineering, economic, social, and
policy issues related to sustainability, in their own contexts.

4 Two Case Studies

A set of two case studies are, showcasing some universities which already transformed their curricula successfully.

- Universiti Sains Malaysia (USM)

At the Universiti Sains Malaysia (USM) the Center for Global Sustainability promotes and integrates sustainability subjects across the university's curriculum for
undergraduate studies (USM 2017) Since then they have integrated social, economic and environmental issues into their teaching, research, community engagement and institutional arrangements. In elective sustainability courses for
undergraduate participants learn with the use of global case studies and examples
from sustainability programmes from around the world. Furthermore, the students
get information about the latest developments in the sustainability studies agenda.
The students finish the courses with group projects, presentations and mini projects
related to the subject (UNESCO Green citizens 2017).

- Macquarie University, Sydney, Australia

In an exemplary bottom-up process a small project team consisting of academics
across the Australian Macquarie University's four faculties (Faculty of Arts,
Faculty of Human Science, Faculty of Science and Faculty of Business and
Economics) was brought together to develop a framework to clarify how sustainability can be embedded into these four undergraduate programs (Denby and
Rickards 2016). While providing clear guidelines the frameworks shows how skills
for sustainability are applicable in any program and what support is needed for
program level, interdisciplinary concept learning for those who wish to enhance
sustainability learning for themselves and/or their students (Macquarie University
2009).

There are many other well documented examples of initiatives (e.g. Leal Filho
et al. 2017), but the above serve the purpose of illustrating how simple and effective
the implementation of sustainability in the curriculum can be.

5 Conclusions

This paper aimed at presenting how sustainability can be successfully incorporated
in university curricula, complementing the various other examples on this book. It
is clear that there are plenty of different ways to pursue the implementation of

sustainability in the curriculum, but the starting point is always the same: The discussion and debate about what 'sustainability' could mean in a particular disciplinary context and how to link the topics with the concept of sustainability. One is for sure: "We cannot solve our problems with the same level of thinking that created them" (Albert Einstein).

One of the major barriers to the wide incorporation of matters related to sustainable development at higher education institutions, is the fact that sustainability is seldom systematically embedded in the curriculum. Yet, proper provisions for curricular integration of sustainability issues at part of teaching programmes across universities is an important element towards curriculum greening. Despite the central relevance of this topic, not many universities have specifically focused on identifying ways of how better teach about sustainability issues in a university-wide context. This gap, is still to be addressed and further works are needed.

References

AASHE, the Association for the Advancement of Sustainability in Higher Education (2010) Sustainability curriculum in higher education. A call to action. Denver. URL: https://oakland.edu/.../2010_Documents_A_Call_to_Action.pdf

Denby L, Rickards S (2016) An approach to embedding sustainability into undergraduate curriculum: Macquarie University Australia case study. In: Leal Filho W, Pace P (eds) Teaching education for sustainable development at university level. World sustainability series. Switzerland, pp 9–33. Available at: http://www.springer.com/de/book/9783319329260

Hensley N (2011) Curriculum studies gone wild: bioregional education and the scholarship of sustainability. Peter Lang, New York

Hensley N (2017) The future of sustainability in higher education. J Sustain Educ 13. Available at: http://www.susted.com/wordpress/content/the-future-of-sustainability-in-higher-education_2017_03/

Junyent M, Geli de Ciurana AM (2008) Education for sustainability in university studies: a model for reorienting the curriculum. Br Educ Res J 34(6):763–782. Available at: https://www.researchgate.net/publication/248994817_Education_for_sustainability_in_university_studies_A_model_for_reorienting_the_curriculum

Leal Filho W, Pace P (eds) (2016) Teaching education for sustainable development at university level. Springer, Berlin

Leal Filho W, Brandli L, Castro P, Newman J (eds) (2017) Handbook of theory and practice of sustainable development in higher education, vol 1. Springer, Berlin

Macquarie University (2009) Sustainability in the curriculum project. Sydney. Available at: https://www.mq.edu.au/lih/pdfs/039_sust_in_curric.pdf

Nichols J (2007) A hearty economy and healthy ecology can co-exist. J Inter Des 12(2):6–10. Available at: http://onlinelibrary.wiley.com/doi/10.1111/j.1939-1668.2007.tb00536.x/abstract

Nichols J, Adams E (2011) Sustainability education in the interior design curriculum. J Sustain Educ. Available at: http://www.jsedimensions.org/wordpress/content/author/jnichols/

Scott RH (2009) Sustainable curriculum, sustainable university. eCulture 2:119–129 (Edith Cowan University, Perth). Available at: http://ro.ecu.edu.au/

UN Assembly (2015) Transforming our world: the 2030 Agenda for sustainable development. New York. Available at: https://sustainabledevelopment.un.org/post2015/transformingourworld

UNESCO (2005) United Nations decade of education for sustainable development united (2005–2014). International implementation scheme. Paris. URL: http://unesdoc.unesco.org/images/0014/001486/148654E.pdf

UNESCO Green citizens (2017) Sustainability course: a powerful tool in university curriculum. URL: http://en.unesco.org/greencitizens/stories/sustainability-course-powerful-tool-university-curriculum

Universiti Sains Malaysia (USM), Centre for Global Sustainability Studies (CGSS) (2017) URL: https://sustainabledevelopment.un.org/partnership/?p=2386

Wals AEJ (2010) Mirroring, gestaltwitching and transformative social learning. Int J Sustain High Educ 11(4):380–390. Available at http://www.scirp.org/(S(351jmbntvnsjt1aadkposzje))/reference/ReferencesPapers.aspx?ReferenceID=589897

Printed in the United States
By Bookmasters